Nucleic Acids and Molecular Biology

19

Series Editor
H. J. Gross

Michael Fry Karen Usdin (Eds.)

Human Nucleotide Expansion Disorders

With 39 Figures, 1 in Color and 8 Tables

Springer

Michael Fry

Department of Biochemistry
Rappaport Faculty of Medicine
Technion – Israel Institute of Technology
P.O. Box 9649
Bat Galim
Haifa, 31096
Israel

Karen Usdin

Gene Structure and Disease Section
Laboratory of Molecular & Cellular Biology
National Institute of Diabetes, Digestive and Kidney Diseases
National Institutes of Health
8/202, 8 Center Dr MSC 0830
Bethesda MD 20892
USA

ISSN 0933-1891
ISBN-10 3-540-33335-5 Springer Berlin Heidelberg New York
ISBN-13 978-3-540-33335-7 Springer Berlin Heidelberg New York

Library of Congress Control Number: 2006927799

Springer is a part of Springer Science+Business Media

springer.com

Printed in Germany

Cover design: *Design & Production* GmbH, Heidelberg
Typesetting and Production: LE-TEX Jelonek, Schmidt & Vöckler GbR, Leipzig

Printed on acid-free paper 31/3150/YL – 5 4 3 2 1 0

Preface

The historic discoveries in 1991 that fragile X syndrome (FXS) is caused by dynamic expansion of a d(CGG) trinucleotide repeat sequence in the 5′-UTR region of the *FMR1* gene (Fu et al. 1991, Oberlé et al. 1991, Pieretti et al. 1991, Verkerk et al. 1991, Yu et al. 1991) and that Spinal and Bulbar Muscular Atrophy (SBMA) results from a d(CAG) expansion in the androgen receptor gene (LaSpada et al. 1991), launched a broad new area of human molecular genetics. It was soon appreciated that many repeats at different loci in the human genome are subject to dynamic expansion and that this novel type of mutation results in a diverse class of neurological, neuromuscular and neurodegenerative disorders known as the Nucleotide Expansion Disorders or Repeat Expansion Disorders. While many disorders can be caused by changes in the size of a nucleotide or amino acid repeat tract (Pearson, Edamura and Cleary 2005), some of these repeat tracts are meiotically stable. To date only 20 or so disorders are attributable to dynamic mutations such as those responsible for FXS and SBMA, and it is these disorders that are the subject of this book.

Efforts of many research teams world-wide have led to the identification of the genes affected by nucleotide repeat expansions. In parallel, advances have been made in elucidating the underlying molecular mechanisms of repeat expansions and the pathological consequences of these mutations. The insights gained into the molecular, cellular and organismal bases of some disorders have already generated initial ideas and experimental approaches to their therapy (Di Prospero and Fischbeck 2005).

Unlike static mutations that are stably transmitted, nucleotide repeats are dynamically expanded both upon transmission to offspring and in some instances also within different tissues of an individual. Longer repeat stretches are more prone to expansion than shorter tracts and, in most cases where the disease is not congenital, repeat length is correlated with an earlier age of onset and an increased disease severity. As a result, the expansion disorders are characterized by genetic anticipation in which each successive generation presents a more severe form of the disease.

Different disorders are characterized by differences in the sequence and length of the nucleotide repeat unit as well as by its location within the gene. The largest number of disorders is linked to expansion of trinucleotide repeat sequences. Fewer diseases are associated with expansions of four, five or 12 nucleotide repeat sequences (Table 1). The location of expanded repeats within or outside coding regions of genes is arguably their most instructive characteristic – indicative of a likely pathological mechanism of the disease. As schematically shown in Fig. 1, several disorders are linked to the expansion of different repeats in the promoter, the 5′ or 3′ untranslated regions or in introns of various genes. A different class of diseases is coupled to expansions of repeat tracts in exon sequences. The largest group within this class is associated with expansion of d(CAG) triplet repeat sequences that results in the accumulation of product proteins with abnormally long polyglutamine tracts.

Identification of affected genes, the location of the position of expanded tracts outside or within the coding regions of the genes and characterization of their protein products, has shed light in many cases on the resulting pathologies. Expansions can result in either a loss-of-gene function or a gain-of-function. Loss-of-function mutations result in reduced or abolished protein function. Gain-of-function mutations confer abnormal properties on the protein or mRNA. Most, if not all, of the expansion mutations occurring in coding regions of genes result in a gain-of-function, while many expansions in non-coding regions result in a loss-of-function. Although the underlying mechanism of a number of nucleotide expansion disorders is still unknown, those diseases that were characterized as being associated with loss- or gain-of-function, opened new vistas into the diverse pathological processes that are at the basis of repeat expansion disorders. Thus, some disorders develop as a result of gene silencing (i.e. fragile X syndrome), others are due to aberrant protein function (polyglutamine disorders such as Huntington and a large number of Spinocerebellar ataxias), whereas another set of disorders results

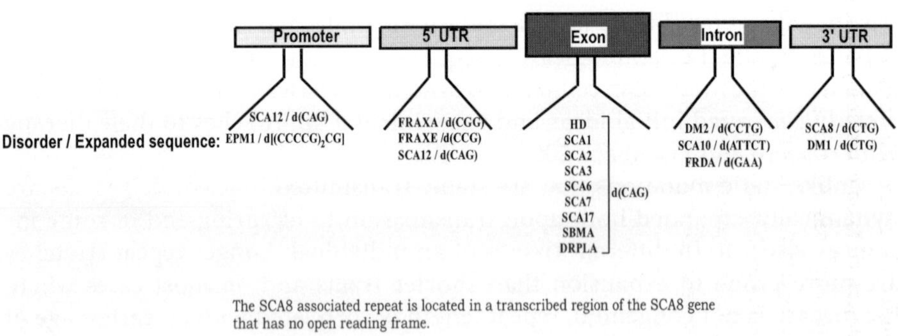

Fig. 1 Location of disorder-associated expandable nucleotide repeats. Schematically shown are locations of disease-causing nucleotide repeats and their location within coding or non-coding regions of affected genes. The repeat unit sequences are of the DNA strands that are considered to be relevant to the pathology of each disorder

Table 1 Expanded nucleotide repeat disorders. Listed are major disorders that are reviewed in this volume. A comprehensive catalogue of all the repeat expansions described to date, including those with no confirmed disorder linkage can be found elsewhere (Pearson, Edamura and Cleary, 2005). The listed repeat unit sequences are those of the DNA strands that are thought to be relevant to the pathology of the respective disorders. [a] HD- Huntington disease; SCA - Spinocerebellar ataxia; SBMA-Spinal and Bulbar Muscular Atrophy; DRPLA-Dentatorubral-pallidoluysian atrophy; HDL2 - Huntington disease-like 2; DM1 - Myotonic dystrophy type 1; FRDA - Friedreich ataxia; FXS, Fragile X Syndrome, FRAXE MR, FRAXE mental retardation; DM2 - Myotonic dystrophy type 2; EPM1 - Progressive Myoclonus Epilepsy. [c] Nucleotide repeats expanded in coding regions. [d] Nucleotide repeats expanded in non-coding regions. *the relevant strand has not been definitively determined

REPEAT SIZE	REPEAT UNIT	DISORDER[a]	AFFECTED GENE[b]	PATHOGENESIS
Trinucleotide	d(CAG)$_n$	HD	Huntingtin[c]	GOF
		SBMA	Androgen receptor	GOF
		DRPLA	*DRPLA* (or atrophin 1)	GOF
		SCA1	Ataxin-1[c]	GOF
		SCA2	Ataxin-2[c]	GOF
		SCA3	Ataxin-3[c]	GOF
		SCA6	Ca^{2+} channel a, 1A subunit[c]	?
		SCA7	Ataxin-7[c]	GOF
		SCA17	TATA box binding protein[c]	GOF
		SCA12	PPP2R2B	?
	d(CTG)$_n$	DM1	Dystrophia myotonia Protein kinase[d]	GOF (+LOF in CDM1?)
		SCA8	*SCA8*[d]	?
		HDL2*	Junctophilin-3	?
	d(GAA)	FRDA	Frataxin[d]	LOF
	d(CGG)$_n$	FXS	*FMR1*[d]	GOF (RNA)/LOF
	d(CCG)	FRAXE MR 0	*FMR2*[d]	LOF
Tetranucleotide	d(CCTG)$_n$	DM2	Zinc finger protein 9[d]	GOF (RNA)
Pentanucleotide	d(ATTCT)$_n$	SCA10	Ataxin-10[d]	?
Dodecanucleotide	d(CCCCGCCCCGCG)	EPM1	Cystatin B	LOF

from RNA toxicity (Myotonic dystrophies types 1 and 2 and possibly additional nucleotide expansion diseases).

This volume presents an updated survey of the current knowledge on a number of human nucleotide repeat expansion disorders. It is not, however, a comprehensive compilation of every known disease. Rather, the different chapters review well defined disorders whose mechanism is either already understood or is close to being elucidated. We were fortunate indeed to have leading researchers contribute chapters on the state-of-the-art of their respective areas of expertise. The different chapters cover nearly every aspect of major human nucleotide repeat expansion disorders including the molecular mechanisms of expansion, the mode of inheritance of the individual diseases and discussion of their clinical presentation, pathological mechanisms, animal models and prospective therapeutic strategies.

A volume on a group of highly divergent disorders and their different molecular and cellular mechanisms can, of course, be organized according to different criteria. We chose to dedicate one section to the general molecular mechanisms of repeat expansion and then to group different diseases in separate sections based on whether the repeat occurs in non-coding or coding regions of the affected gene. The last chapter is devoted to diseases whose location in the affected gene is as yet unresolved. Thus, the opening section of this volume consists of a comprehensive survey by R. R. Sinden and M. J. Pytlos (Texas A&M University) of the current understanding of the varied types of secondary structures of repeat DNA tracts and their roles in expansion. The second section is dedicated to disorders that result from expansion of repeat sequences in non-coding regions. Expert authors review the divergent cases of Fragile X syndrome (F. Tassone and P. J. Hagerman, University of California, Davis), FRAXE MR (D. L. Nelson, Baylor College of Medicine), Friedreich Ataxia (M. Pandolfo, Université Libre, Brussels), Progressive Myoclonus Epilepsy (M. D. Lalioti, S. E. Antonarakis and H. S. Scott, Yale and Geneva Universities and Walter and Eliza Hall Institute, Australia) Myotonic dystrophies 1 and 2 (P. Teng-umnuay and M. S. Swanson, University of Florida, Gainsville) and Spinocerebellar ataxia 10 (X. Lin and T. Ashizawa, University of Texas, Galvston). The third section is devoted to disorders that are linked to repeat expansion in protein-encoding regions of genes. Included is a review on the large body of data that is now available on the diverse group of polyglutamine expansion disorders, (M. J. Friedman, S.-H. Li and X.-J. Li, Emory University). Also in this section, M. Frontali (Institute of Neurobiology and Medicine, Rome) discusses Spinocerebellar ataxia 6 and the unresolved issue of its pathological mechanism. The fourth section deals with expansion disorders whose precise mechanisms are still under investigation This part consists of surveys of our current understanding of Spinocerebellar ataxia 8 (K. A. Dick, J. W. Day, and L. P. W. Ranum, University of Minesota) and of Spinocerebellar ataxia 12 and Huntington disease like 2 (R. L. Margolis, S. E. Holmes, E. O'Hearn, D. D. Rudnicki, J. Hwang, N. Cortez-Aperza, O. Plenikova and J. C.

Troncoso, Johns Hopkins University). In a final postscript we briefly summarize the main unanswered questions concerning the molecular mechanisms of the nucleotide repeat disorders and point to future directions of research.

Many individuals made the publication of this book possible. First and foremost, we are thankful to the authors of the different chapters for their comprehensive and lucid reviews. We are grateful to the series editor, Professor H. J. Gross for recognizing the importance of the subject matter of this book, for initiating its compilation and for his steady support. Last, but not least, we gratefully acknowledge the contribution of Ursula Gramm, Editor Springer Life Sciences whose dedicated work was vital in bringing the volume to press.

Haifa, Michael Fry
Bethesda, Karen Usdin
June 2006

References

Di Prodpero, NA, Fischbeck, KH (2005) Therapeutic development for triplet repeat expansion diseases. Nature Reviews Genetics 6:756-765

Fu YH, Kuhl DP, Pizzuti A, Pieretti M, Sutcliffe JS, Richards S, Verkerk AJ, Holden JJ, Fenwick RG, Jr., Warren ST, Oostra BA, Nelson DL, Caskey CT (1991) Variation of the CGG repeat at the fragile X site results in genetic instability: resolution of the Sherman paradox. Cell 67:1047-1058

Oberlé I, Rousseau F, Heitz D, Kretz C, Devys D, Hanauer A, Boue J, Bertheas MF, Mandel JL (1991) Instability of a 550-base pair DNA segment and abnormal methylation in fragile X syndrome. Science 252:1097-1102

Pieretti M, Zhang FP, Fu YH, Warren ST, Oostra BA, Caskey CT, Nelson DL (1991) Absence of expression of the *FMR-1* gene in fragile X syndrome. Cell 66:817-822

LaSpada AR, Wilson, EM, Lubahn, DB, Haeding, AE, Fischbeck, KH (1991) Androgen receptor gene mutations in X-linked spinal and bulbar muscular atrophy. Nature 352:77-79

Pearson CE, Edamura, KN, Cleary, JD (2005) Repeat instability: Mechanisms of dynamic mutations. Nature Reviews Genetics 6:729-742

Verkerk AJ, Pieretti M, Sutcliffe JS, Fu YH, Kuhl DP, Pizzuti A, Reiner O, Richards S, Victoria MF, Zhang FP, Eussen BE, van Ommen GJB, Blonden LAJ, Riggins GJ, Chastain JL, Kunst CB, Galjaard H, Caskey CT, Nelson DL, Oostra BA, Warren ST (1991) Identification of a gene (*FMR-1*) containing a CGG repeat coincident with a breakpoint cluster region exhibiting length variation in fragile X syndrome. Cell 65:905-914

Yu S, Pritchard M, Kremer E, Lynch M, Nancarrow J, Baker E, Holman K, Mulley JC, Warren ST, Schlessinger D, et al. (1991) Fragile X genotype characterized by an unstable region of DNA. Science 252:1179-1181

Contents

Part I Molecular Bases of Nucleotide Expansions

Mechanisms of DNA Repeat Expansion
R. R. Sinden, M. J. Pytlos, V. N. Potaman 3

1 Introduction: Repeat Expansion and Deletion Associated with Human Neurodegenerative Disease 3
2 DNA Structures Formed by Disease-Associated DNA Repeats 5
2.1 Inverted Repeats, Hairpins 5
2.2 Slipped-Strand DNA Structures and Slipped Intermediate DNA Molecules 8
2.3 Triplex DNA 9
2.4 Quadruplex DNA 10
2.5 Unwound DNA 10
2.6 Parallel-Strand DNA 11
3 Effects of Alternative DNA Conformations on Biology 12
3.1 Replication Blockage by Hairpins, Triplex, Quadruplex, and Triplet Repeats 12
3.2 Aberrant Polymerization Associated with DNA Repeats: Slippage During Primer Template Misalignment and Strand Switching 13
3.3 Helicase Activity at Hairpins, Triplex DNA, Quadruplex DNA, and Triplet Repeats 14
3.4 Recombination Associated with DNA Repeats 15
3.5 Interaction of DNA Replication, Repair, and Recombination Proteins with DNA Repeats 16
4 Pathways for Repeat Expansion 18
4.1 Primer-Template Misalignment During Replication Can Account for Repeat Length Changes Less Than Twofold in Length in Disease-Associated DNA Repeats . . . 18
4.2 Strand Switching During Synthesis of d(ATTCT) · d(AGAAT) DNA Repeats Can Result in Complex Expansion Mutations . 20

4.3 DNA Mismatch Repair May, or May Not, Participate in Small Repeat Length Changes During Replication in Dividing Cells or During Gap Repair in Nondividing Cells 23
4.4 DNA Repair of Slipped-Strand Intermediates Containing $(CTG)_n$ Hairpins or $(CAG)_n$ Loopouts 25
4.5 Recombination at Disease-Associated DNA Repeats Can Lead to Deletions and Expansions 26
4.6 Double-Strand Break Repair, Replication Restart, and Checkpoint Control Associated with Repeat Replication 27
4.7 DNA Amplification Provides a Facile Means for Repeat Expansion for SCA10 d(ATTCT) · d(AGAAT) Repeats 31
4.8 Influence of the Direction of Replication, Origin Proximity, Origin Activity, and Transcription on Repeat Instability . . . 33
5 Concluding Remarks: Mutation Mechanisms, DNA Repeats, and Human Disease— Where Have We Come in 15 Years? . . 37
References 39

Part II Disorders Associated with Non-coding Repeats

Molecular Correlates of Fragile X Syndrome and FXTAS
F. Tassone, P. J. Hagerman 57
1 Overview 57
2 Introduction and Clinical Perspective 58
2.1 Fragile X Syndrome 58
2.2 Clinical Involvement Among Carriers of Premutation Alleles of the *FMR1* Gene 59
3 Expression of the *FMR1* Gene 60
3.1 Mechanisms of CGG Repeat Expansion 60
3.2 Regulation of Expression of the *FMR1* Gene in the Normal and Premutation Ranges 63
3.3 Regulation of Transcription Start Site Selection 64
3.4 Increased Transcription in the Premutation Range 65
3.5 Mechanisms of Silencing/Reactivation of the *FMR1* Gene . . 65
3.6 Regulation of Translation of the *FMR1* mRNA 68
4 Function of the *FMR1* Protein 69
4.1 FMRP Can Function as a Negative Regulator of Translation . 70
4.2 A Possible Role for FMRP in Regulating Actin Cytoskeletal Dynamics 71
4.3 The mGluR Hypothesis as a Specific Example of How FMRP Could Regulate Synaptic Function/Plasticity . 71
5 The Molecular Basis of FXTAS 72
References 74

The Neglected Fragile X Mutations: *FRAXE* and *FRAXF*
D. L. Nelson, Y. Gu 87
1 History of the Xq27-q28 Fragile Sites 87
2 Clinical Features of *FRAXE* Disease 89
3 Repeat Dynamics in Patients and Families 90
4 The *FMR2* Locus 91
5 Mouse Model of *Fmr2* Deficiency and *Drosophila Lilli* 93
6 The *FMR3* Transcript 95
7 *FRAXF* and *FAM11A* 95
8 Future Prospects 98
References 99

Friedreich Ataxia
M. Pandolfo 103
1 Clinical Features and Pathology 103
2 Gene Structure and Expression 104
3 The d(GAA) Triplet Repeat Mutation 105
3.1 Instability of Expanded Repeats 105
3.2 Origin and Mechanisms of Expansion of the Repeat 106
4 Pathogenic Mechanisms: Triplexes and Sticky DNA 108
5 Genotype–Phenotype Correlation 110
5.1 d(GAA) Triplet Repeat Expansion 110
5.2 Point Mutations 110
6 Frataxin Structure and Function 111
7 Animal Models 112
8 Pathogenic Mechanisms in Friedreich Ataxia 113
9 Perspectives for Treatment 113
References 115

Dodecamer Repeat Expansion in Progressive Myoclonus Epilepsy 1
M. D. Lalioti, S. E. Antonarakis, H. S. Scott 121
1 Introduction and Disease Features 121
2 Linkage Analysis and Positional Cloning of the EPM1 Gene . 122
3 Cystatin B 123
4 Point Mutations and Polymorphisms of the Cystatin B Gene . 126
5 Effect of Point Mutations on Cystatin B Expression 126
6 A Dodecamer Repeat Expansion is the Most Common Mutation in EPM1 128
7 Instability of the Dodecamer Repeat 129
8 Size of the Repeat and Age of Onset of the Epilepsy 131
9 Effect of Expansion on Cystatin B Expression 131
10 Mechanism of Transcriptional Repression 132
11 Loss of Cystatin B Function and Disease Pathophysiology . . 134
References 136

Myotonic Dystrophies Types 1 and 2
P. Teng-umnuay, M. S. Swanson 143
1 The Myotonic Dystrophies: an Overview 143
2 Genetics and Clinical Presentation of the Myotonic Dystrophies: DM1 Versus DM2 Disease: Many Similarities but Significant Differences 144
3 RNA Gain-of-Function Model for Myotonic Dystrophy 146
3.1 Myotonic Dystrophy Associated Microsatellite Expansions Are Toxic at the RNA Level 146
3.2 Toxic RNAs Molecules Sequester Muscleblind-like Proteins . 148
3.3 Poly r(CUG) Toxicity Requires Expression of Specific Muscleblind-like Isoforms 150
3.4 The Splicing Connection: DM is Associated with Fetal Exon Retention in Adults 151
3.5 MBNL and CELF Proteins are Splicing Antagonists Which Regulate Fetal Exon Splicing 154
3.6 Is Myotonic Dystrophy Caused by MBNL Loss, CUGBP1 Overexpression or Both? 154
4 Congenital Myotonic Dystrophy: a Distinct Disease with a Different Molecular Etiology? . . . 155
4.1 The DMPK $(CTG)_n$ Expansion Alters the Chromatin Structure and Expression of the DM1 Locus 156
4.2 Cell Culture Models for Congenital Myotonic Dystrophy . . . 157
4.3 Mouse Transgenic and Knockout Models for Congenital Myotonic Dystrophy 158
References 159

Spinocerebellar Ataxia Type 8
K. A. Dick, J. W. Day, L. P. W. Ranum 167
1 Introduction: Repeat Expansions and SCA8 167
2 Rapid Cloning of the SCA8 Repeat Expansion 168
3 The d(CTG) Repeat Cosegregates with a Novel Form of Ataxia 168
4 Organization of the SCA8 Gene 169
5 Clinical Features of SCA8 170
6 Disease Penetrance Affected by d(CTG) Repeat Length: the MN-A Family 171
7 Reduced Penetrance of SCA8 in Other Families 172
8 SCA8 Expansions on Control Chromosomes 173
9 SCA8 Expansions Cosegregate with Ataxia in Small Families 174
10 Haplotype Analysis of SCA8 Expansion Chromosomes . . . 174
11 Factors that May Influence SCA8 Disease Penetrance 175
11.1 The d(CTA) Repeat Tract 175
11.2 Interruptions Within the d(CTG) Expansion 176

11.3 Repeat Instability During Transmission 176
11.4 En Masse d(CTG) Repeat Contractions in Sperm 178
12 Molecular Parallels with Myotonic Dystrophy 179
13 Modeling SCA8 Pathogenesis in the Fly 179
14 Conclusion 180
References 180

Recent Progress in Spinocerebellar Ataxia Type 10
X. Lin, T Ashizawa 185
1 Introduction 185
2 Clinical Presentations 186
2.1 Cerebellar Signs and Symptoms 186
2.2 Seizures 187
2.3 Other Extracerebellar Signs and Symptoms 187
3 Genetic Features 188
3.1 d(ATTCT) Pentanucleotide Repeat Expansion 188
3.2 Anticipation, Repeat Instability and Genotype–Phenotype Correlation 189
4 Molecular Studies 190
4.1 DNA Structure 190
4.2 Protein Function 191
4.3 RNA Gain of Function 192
5 Concluding Remarks 193
References 194

Part III Disorders Associated with Coding Repeats

Polyglutamine Diseases
M. J. Friedman, S. Li, X. Li 199
1 Introduction 199
2 Genetics of PolyQ Diseases 200
3 Neuropathology of PolyQ Diseases 203
4 PolyQ-Dependent Misfolding and Aggregation 205
5 Pathogenesis of the PolyQ Diseases 208
5.1 PolyQ Diseases as “Transcriptionopathies” 209
5.2 Disruption of Cytoplasmic Activities in PolyQ Disease 214
5.3 Other Contributors to Pathogenesis 216
6 Potential Therapeutic Strategies 217
7 Concluding Remarks 219
References 220

The Enigma of Spinocerebellar Ataxia Type 6
M. Frontali . . . 233
1 Introduction . . . 233
2 Genetics . . . 233
3 The Cav2.1 α1A Isoforms . . . 234
4 Epidemiology . . . 235
5 Clinical Features . . . 236
6 Similar and Discrepant Clinical Features of SCA6 and EA2 . 239
7 Neuropathology . . . 240
8 Toxic Gain of Function Versus Abnormal Channel Activity . 241
9 Conclusions . . . 243
References . . . 243

Part IV Disorders Associated with Repeats in an Undetermined Location

Spinocerebellar Ataxia Type 12 and Huntington's Disease-Like 2: Clues to Pathogenesis
R. L. Margolis, S. E. Holmes, E. O'Hearn, D. D. Rudnicki, J. Hwang, N. Cortez-Apreza, O. Pletnikova, J. C. Troncoso . . . 253
1 The Identification of Spinocerebellar Ataxia Type 12 and Huntington's Disease-Like 2 . . . 253
2 HDL2 and SCA12 Clinical Phenotype . . . 254
3 HDL2 and SCA12 Pathological Phenotype . . . 256
4 Epidemiology . . . 258
5 Phenotype–Genotype Relationship . . . 259
6 The HDL2 and the SCA12 Repeats Do Not Encode Polyglutamine . . . 261
7 HDL2 and SCA12: Clues to Pathogenesis . . . 265
7.1 HDL2 and Gain of Function . . . 265
7.2 The Function of *JPH3* and *PPP2R2B* . . . 266
7.3 Expression Level and Pathogenesis . . . 268
8 Conclusion . . . 269
References . . . 271

Part V Postscript

Current Issues and Therapeutic Prospects
K. Usdin, M. Fry . . . 279
1 Current Issues . . . 279
1.1 Mechanisms of Nucleotide Repeat Expansion . . . 279
1.2 Consequences of Repeat Expansion . . . 280
2 Therapeutic Prospects . . . 282

2.1 Gene Correction/Replacement 282
2.2 Targeting the Downstream Consequences of Expansion . . . 283
References 285

Subject Index 289

Part I
Molecular Bases of Nucleotide Expansions

Nucleic Acids and Molecular Biology, Vol. 19
M. Fry/K. Usdin (Eds.)
Human Nucleotide Expansion Disorders

Mechanisms of DNA Repeat Expansion

Richard R. Sinden (✉) · Malgorzata J. Pytlos · Vladimir N. Potaman

Laboratory of DNA Structure and Mutagenesis, Center for Genome Research,
Institute of Biosciences and Technology,
Texas A&M University System Health Science Center, 2121 West Holcombe Blvd.,
Houston, TX 77030-3303, USA
rsinden@ibt.tamhsc.edu

1 Introduction: Repeat Expansion and Deletion Associated with Human Neurodegenerative Disease

Since 1991, many genetic neurodegenerative diseases and six fragile sites have been associated with the expansion of trinucleotide $d(CTG)_n \cdot d(CAG)_n$, $d(CGG)_n \cdot d(CCG)_n$, or $d(GAA)_n \cdot d(TTC)_n$ repeats, a $d(CCTG)_n$ tetranucleotide repeat, a $d(ATTCT)_n \cdot d(AGAAT)_n$ pentanucleotide repeat, or a $d(CCCCGCCCCGCG)_n \cdot d(CGCGGGGCGGGG)_n$ dodecamer repeat. Models proposed for the expansion of these repeats involve the formation of alternative DNA structures which differ from the canonical B-form DNA. Alternative structures that can form in the disease-related nucleotide repeats include hairpins in single-stranded DNA, slipped-stranded DNA, triplex DNA, quadruplex DNA, parallel-strand DNA, and unwound DNA (Table 1).

Expansions of DNA repeats are a unique hallmark of a group of neurodegenerative diseases that now number more than 30 (reviewed in Sinden et al. 2002; Parniewski and Staczek 2002; Pearson 2003; Cleary and Pearson 2003, 2005; Lenzmeier and Freudenreich 2003; Brown and Brown 2004; Mirkin 2004; Ranum and Day 2004). Both small and large changes in repeat lengths are associated with human neurodegenerative diseases. Variation in repeat number is a classic type of mutation. The molecular mechanism explaining this type of genetic mutation was first suggested by Streisinger in 1966 in which primer-template misalignment during replication of DNA repeats can lead to addition or deletion mutations (Streisinger et al. 1966). This mechanism is widely accepted and supported by a large body of experimental data. In the case of neurodegenerative diseases, small repeat length changes are frequently observed in somatic cells throughout the life of an individual (Wong et al. 1995; Martorell et al. 1998). These changes could easily occur by primer-template misalignment during DNA replication associated with cell division. For many repeats, the slipped-out strands can form alternative structures that

Table 1 DNA repeat expansion diseases and

Disease	Gene	Repeat	Alternative structure	Repeat length Normal	Disease
Fragile X syndrome	*FRAXA*	d(CGG)$_n$	Hairpins Slipped-strand DNA quadruplex	6–52	230–2000
Myotonic dystrophy type 1 (DM1)	*DMPK*	d(CTG)$_n$	Hairpins Slipped-strand DNA	5–37	80–2000
Friedreich ataxia (FRDA)	*frataxin*	d(GAA)$_n$	Hairpins Intramolecular triplex Biduplex Parallel-strand DNA	6–29	200–2000
Spinocerebellar ataxia type 10 (SCA10)	*SCA10/E46L*	d(ATTCT)$_n$	Denatured bubble	9–23	750–4500
Myotonic dystrophy type 2 (DM2)	*ZNF9*	d(CCTG)$_n$	Hairpin Slipped-strand DNA	< 26	75–11 000
Progressive myoclonus epilepsy	*cystatin b*	d(CCCCGC CCCGCG)$_n$	Hairpins Quadruplex	12–17	80

can further direct or escape repair in a sequence and strand-specific orientation with respect to the leading or lagging strand of replication. However, repeat length changes can also occur in nondividing cells where it is believed that the removal of spontaneous DNA damage by mismatch or other DNA repair systems may be responsible for small repeat length changes (Kovtun and McMurray 2001; Pearson 2003; McMurray and Kortun 2003; Gomes-Pereira et al. 2004).

Certain diseases are associated with very large intergenerational changes in repeat tract length. First identified with fragile X syndrome (Fu et al. 1991; Kremer et al. 1991) and myotonic dystrophy type 1 (DM1) (Fu et al. 1992; Brook et al. 1992; Mahadevan et al. 1992), expansion from an unstable length of typically 30 to 100 repeats to more than 1000 copies has now been identified in several other neurodegenerative diseases (Table 1), including spinocerebellar ataxia type 10 (SCA10) and myotonic dystrophy type 2, where expansion to 4500 and 11 000 copies of the repeat, respectively, can occur. These repeat length changes are believed to occur during germ cell development (reviewed in Pearson 2003; Cleary and Pearson 2003). They could also potentially occur during the first few cell divisions following fertilization. Prior to their discovery, massive intergenerational expansion had not been identified in the experimental systems, including bacteriophage T4, *Escherichia coli*, yeast and *Drosophila*, that have provided the fundamental knowledge of the types and mechanisms of mutations in DNA (Drake 1970, 1991a,b, 1999; Drake et al. 1983). Large repeat expansions are observed in some DNA repeats in mice (Bois et al. 2001); however, disease-associated repeats introduced into mice do not show the large intergenerational changes that are observed in humans (Gomes-Pereira et al. 2001; Libby et al. 2003; Gomes-Pereira and Monckton 2004b).

There remains no confirmed mechanism that explains expansion from 30 to 100 repeats to lengths of 1000, or even 4000 to 11 000 repeats. Here we review the roles alternative DNA structures may play in both short and long expansion of repeating DNA sequences. We will also discuss possible models for repeat deletion, because once the expansion has occurred in Friedreich ataxia and SCA10 repeat deletions may be the predominant event (Bidichandani et al. 1999; Matsuura et al. 2000, 2004; Sharma et al. 2002; Pollard et al. 2004).

2
DNA Structures Formed by Disease-Associated DNA Repeats

Most of the time DNA exists as a double-stranded structure called B-DNA, where nucleobases form hydrogen-bonded base pairs stacked into a right-handed helix. Rearrangements into other DNA forms, such as structures with locally unpaired strands and four-way junctions, may occur in any DNA sequence during cellular events, such as DNA replication, transcription, and recombination. Several non-B-DNA (also called unusual or alternative) structures can only form in DNA sequences that possess appropriate symmetry elements, such as direct or inverted repeats, segregation of purine (Pu) and pyrimidine (Py) bases in complementary strands, etc. An extensive knowledge base of the formation of non-B-DNA structures, which include hairpins, cruciforms, slipped-strand DNA, triplex DNA, quadruplex DNA, and unwound DNA, has been used for the development of various polynucleotide expansion models. These are summarized in the following and are shown in Fig. 1.

2.1
Inverted Repeats, Hairpins

Duplex DNA in which the sequence reads the same from the 5′ to 3′ direction in both strands is called an inverted repeat. In a single-stranded DNA with inverted repeat symmetry, nucleobases of the two halves of the inverted repeat may form canonical AT and CG base pairs, thus forming a hairpin with

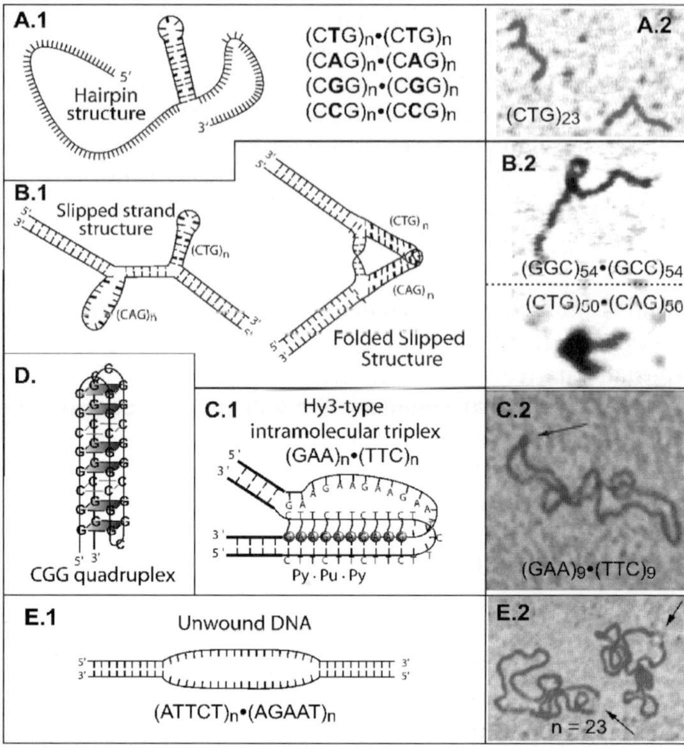

Fig. 1 Alternative DNA structures associated with unstable DNA repeats. *A.1* Hairpins can form in single-stranded tracts of repeats with the sequence motif d(CXG)$_n$, where n = T, A, G, or C, forming AA, TT, GG, or CC mismatches every third base pair. *A.2.* Atomic force microscope (*AFM*) image of a d(CTG)$_{23}$ hairpin in a heteroduplex molecule with duplex-flanking DNA (Sinden et al. 2002). *B.1* Models of slipped-strand DNA (*S-DNA*) with a d(CTG)$_n$ hairpin and a d(CAG)$_n$ unstructured loopout (*left*) and folded S-DNA (*right*) stabilized by base pairing between the loops (Sinden et al. 2002). *B.2* AFM images of purified S-DNAs formed from a fragile X d(CCG)$_{54}$ · d(CGG)$_{54}$ repeat (*top*) and a myotonic dystrophy type 1 d(CTG)$_{50}$ · d(CAG)$_{50}$ sequence (*bottom*) (Sinden et al. 2002). *C.1* PyPu*Py* intramolecular triplex DNA can form from the Friedreich ataxia d(GAA)$_n$ · d(TTC)$_n$ repeats, where $n \geq 9$. *C.2* AFM image of plasmid containing a human d(GAA)$_9$ · d(TTC)$_9$ repeat with flanking sequence from the *frataxin* gene showing the triplex structure (*arrow*) (Potaman et al. 2004). *D* Model of quadruplex DNA that may be formed by (GGC)$_n$ repeats. *E.1* Model of unwound or unpaired DNA. *E.2* AFM images of unpaired bubbles formed from a d(ATTCT)$_{23}$ · d(AGAAT)$_{23}$ tract with a flanking human sequence from the *SCA10* gene (Potaman et al. 2003). *Pu* purine, *Py* pyrimidine

three unpaired bases at the tip. Upon hairpin formation in an imperfect inverted repeat, structural destabilization due to a deficit of hydrogen-bonded base pairs in the stems may be compensated by adjustments in base stacking (Chou et al. 2003). Hairpin formation can impede the progression of tracking enzymes such as DNA and RNA polymerases, as discussed later.

The $d(CTG)_n \cdot d(CAG)_n$ and $d(CGG)_n \cdot d(CCG)_n$ repeat tracts do not contain perfect inverted repeat symmetry; however, they possess sufficient inverted repeat symmetry to fold into hairpins that are not identical in the two complementary repeat tracts. In such hairpins, every two out of the three bases of the triplet are involved in a canonical base pair with the complementary bases of another triplet (Fig. 1, panels A.1, A.2) (reviewed in Mitas 1997; Darlow and Leach 1998a,b; Pearson and Sinden 1998a). d(CTG) and d(CAG) repeats can each form only one hairpin structure with a T · T or A · A base pair mismatch (Mitas 1997; Pearson and Sinden 1998a; Darlow and Leach 1998b). $d(CTG)_n$ hairpins are more stable than $d(CAG)_n$ hairpins because the smaller T · T mispairs are better stacked in the DNA helix than are bulky A · A mispairs (Petruska et al. 1996; Mitas 1997; Gacy and McMurray 1998). In fact, in a duplex DNA single-stranded $d(CAG)_n$ loopouts of about 20 repeats form an unstructured loop rather than an intrastrand base-paired hairpin (Pearson et al. 2002). d(CGG) and d(CCG) strands can each fold into hairpin structures in two ways involving either a G · G or a C · C mismatch. Hairpins formed by the $d(CGG)_n$ repeat sequence are more stable than those formed by $d(CCG)_n$ (Mitas 1997). In this case, the smaller C · C mismatch may be more destabilizing than the larger G · G mismatch because in a long d(CCG) tract the duplex forms an e-motif (Gao et al. 1995) in which C · C mispairs are not well stacked in the helix and cytosines become extrahelical.

If the hairpin folding patterns are not dictated by flanking sequences, loops with even numbers of unpaired bases may be somewhat more stable than those with an odd number of unpaired bases because they each have one more base pair stabilizing the hairpin bend (Darlow and Leach 1995, 1998b; Petruska et al. 1998; Hartenstine et al. 2000). The pronounced preference of the even-numbered loops results in a more frequent polymerization slippage by two triplets than by one triplet (Petruska et al. 1998).

Single-stranded d(GAA) repeats and long d(CTT) repeats may also fold into hairpins with G · A pairs and A · A mismatches (Suen et al. 1999; Heidenfelder et al. 2003). Hairpin formation was initially studied in the relatively short $d(GAA)_{15}$ and $d(TTC)_{15}$ fragments. Self-annealing to form the hairpin was only detected by chemical and enzymatic probing in $d(GAA)_{15}$ at low temperature (Suen et al. 1999). In later studies with $d(GAA)_n$ and $d(TTC)_n$ ($n = 17$, 33), electron microscopy and enzymatic probing revealed that at longer repeat lengths the propensity of hairpin formation increased so that they could be detected at physiological temperature and salt concentrations (Heidenfelder et al. 2003).

The $d(CCTG)_n \cdot d(CAGG)_n$ repeats are also prone to hairpin formation (Heidenfelder and Topal 2003; Dere et al. 2004). Thermal melting, native gel electrophoresis, as well as chemical and enzymatic probing indicate that hairpins formed by the $d(CAGG)_n$ strand are much more stable than those formed by the $d(CCTG)_n$ strand. The $d(CAGG)_n$ hairpin is stabilized by two Watson–Crick G · C and two unusual G · A pairs per tetranucleotide repeat.

2.2 Slipped-Strand DNA Structures and Slipped Intermediate DNA Molecules

Slipped-strand DNA structures can form within directly repeated DNA sequences (Sinden 1994). To form slipped-strand DNA, the DNA duplex must unwind and the complementary strands then reanneal in an out-of-register alignment within the repeat region. Such misalignment will result in the formation of loops in the complementary strands. Three-way junctions are formed at the sites where the loops connect to the rest of duplex DNA (Fig. 1, panels B.1, B.2). In the case of $d(CTG)_n \cdot d(CAG)_n$ and $d(CGG)_n \cdot d(CCG)_n$ repeats, the looped-out regions can form hairpins, as described before, although some loops can remain unstructured, as does a $d(CAG)_{20}$ loopout (Pearson et al. 2002).

Slipped-strand DNA formed by $d(CTG)_n \cdot d(CAG)_n$ or $d(CGG)_n \cdot d(CCG)_n$ repeats has been extensively characterized (Pearson and Sinden 1996, 1998b; Sinden et al. 2002; Pearson et al. 2002; Tam et al. 2003). Following denaturation and renaturation of DNA molecules containing $d(CTG)_n \cdot d(CAG)_n$ or $d(CGG)_n \cdot d(CCG)_n$ repeats in vitro, a high proportion of the DNA population adopted alternative three-way junction-containing conformations as indicated by the retarded mobility of molecules in polyacrylamide gels. The amount of slipped-strand DNA structure formed was proportional to the repeat tract length and homogeneity (Pearson et al. 1998a; Pearson and Sinden 1998b). Sequence interruptions within the repeat tract reduced the overall amount of the alternative DNA structure and the heterogeneity of the products formed. Although loopouts of different sizes can potentially form anywhere within a triplet repeat tract, typically only several major products were observed (Pearson and Sinden 1996, 1998b; Pearson et al. 1998b, 2002). Biochemical, electron microscopy, and atomic force microscopy experiments mapped the site of the unusual structures within the triplet repeat region (Pearson and Sinden 1998b; Sinden et al. 2002; Pearson et al. 2002). The slipped-strand structures were stable and little conversion into the correctly annealed duplex DNA was observed (Pearson and Sinden 1996). This stability may result from the combination of base pairing within the hairpin loop and in the duplex DNA between the loopouts, which would all have to unpair for the structure to convert back to the linear form (Pearson and Sinden 1996). In some slipped-strand DNA molecules, multiple short loopouts can occur at variable sites throughout the repeat tract (Pearson et al. 2002). The stability may additionally result from loop–loop interactions which can occur between $d(CTG)_n$ and $d(CAG)_n$ hairpins (Sinden et al. 2002). In slipped intermediate DNA molecules [$d(CTG)_{30} \cdot d(CAG)_{50}$ or $d(CAG)_{30} \cdot d(CTG)_{50}$], the excess $d(CAG)_{20}$ loopout remains unpaired, while the $d(CTG)_{20}$ loopout forms a hairpin (Pearson et al. 2002). Both the junctions and the $d(CTG)_n$ and $d(CAG)_n$ loopouts are recognized by DNA-processing proteins (see later). Finally, recent experiments have shown that the myotonic dystrophy type 2

(DM2) $d(CCTG)_n \cdot d(CAGG)_n$ repeats can also form slipped-strand DNA structures (Edwards and Sinden, unpublished results).

2.3 Triplex DNA

After the formation of the canonical A · T and G · C base pairs in B-form DNA, several hydrogen bond donor and acceptor groups in nucleobases remain unused. Each Pu base has two such groups on the major groove-exposed edges. These groups can be used to form base triads that are unit blocks of triple-stranded (triplex) DNA that consist of the B-form double helix and the third strand bound in the major groove (Soyfer and Potaman 1995; Frank-Kamenetskii and Mirkin 1995). Energetically favorable triplexes have duplex Py and Pu bases segregated in complementary strands (Py · Pu duplex). Bases of the third strand form the so-called Hoogsteen-type hydrogen bonds with Pu bases in the B-form duplex. For a snug fit in the duplex major groove, the third strands must contain either only Py bases (Py · Pu · *Py* triplex) or mostly Pu bases with a fraction of Py bases (Py · Pu · *Pu* triplex) (the third strand is shown in *italics*). In the Py · Pu · *Py* triplex, the common base triads are T · A · *T* and $C \cdot G \cdot C^+$ (cytosine is protonated, and this is favored by $pH < 5$). The Py · Pu · *Pu* triplex includes the T · A · *A* and C · G · *G*, and less frequently T · A · *T* triads. Triplex DNA may form *inter*molecularly, between a duplex target and a third oligonucleotide strand. It may also form *intra*molecularly within a Py · Pu sequence of mirror-repeat symmetry. For this, half of the mirror-repeat Py · Pu sequence must unpair and one of the unpaired strands must fold back and bind as a third strand to Pu bases in the repeat's double-stranded half. The resulting local structure contains three notable features: a triple-stranded region; an unpaired fourth strand; and a short (3–4-nt) loop of unpaired bases in the loop of the fold-back strand. The presence of single-stranded regions provides the DNA molecule with local increased flexibility akin to a hinge. The triplex/single-strand combinations are termed H (H′)-DNA for the Py · Pu · *Py* and Py · Pu · *Pu* triplexes, respectively. Energy to support stable formation of H (H′)-DNA comes from the torsional stress in a topologically closed DNA. Other factors that promote H (H′)-DNA are longer lengths of Py · Pu mirror repeats, the presence of multivalent cations, and cytosine protonation in the $C \cdot G \cdot C^+$ triads in H-DNA. Protein–DNA interactions significantly change at sites of triplex formation. In particular, activities of restriction and polymerization enzymes are inhibited, as discussed later.

Out of several disease-associated triplet repeats, only $d(GAA)_n \cdot d(TTC)_n$ has the potential to form triplexes. H-DNA forms in short $d(GAA)_n \cdot d(TTC)_n$ tracts under the influence of negative supercoiling and low pH (Hanvey et al. 1988). Characterization of alternative structures in long $d(GAA)_n \cdot d(TTC)_n$ stretches proved more difficult. Unidentified alternative structures were detected in long $d(GAA)_n \cdot d(TTC)_n$ tracts at neutral pH (Ohshima et al. 1996b; Bidichandani et al. 1998). Structures formed by very long $d(GAA)_n \cdot d(TTC)_n$ tracts ($n > 75$) in negatively supercoiled DNA were interpreted as a bi-triplex structure, formed either by the association of the two Py · Pu · *Pu* triplexes (Sakamoto et al. 1999) or a single long Py · Pu · *Pu* triplex (Vetcher et al. 2002). Length-dependent triplex structures were found in $d(GAA)_n \cdot d(TTC)_n$ tracts from $n = 9$–23, which at longer lengths, $n = 42$, formed bi-triplex structures of the Py · Pu · *Py* type (Fig. 1, panels C.1, C.2) (Potaman et al. 2004). These structures were stabilized at neutral pH by additional T · A · *T* triads formed by A and T bases flanking the repeats.

2.4 Quadruplex DNA

The Hoogsteen-type hydrogen bonding between guanine bases may result in the formation of square guanine tetrads (Fig. 1, panel D). Stacking of several such tetrads produces a four-stranded tetraplex (quadruplex) DNA which is usually stabilized by potassium ions (reviewed in Sinden 1994). Appropriate DNA sequences include repeating tracts of guanine nucleotides that may be interrupted by one or two other nucleotides as occurs in the sequences at the chromosomal ends (telomeres) or in some trinucleotide repeats. Short $d(CGG)_n$ oligonucleotides, where $n = 4$–7, associate to form intermolecular quadruplex structures (Fry and Loeb 1994). A block to DNA polymerase in the $d(CGG)_{20}$ template was also interpreted as a consequence of quadruplex formation (Usdin and Woodford 1995). Similarly, replication blocks in $d(CGG)_n$, $d(AGG)_n$, and $d(TGG)_n$ repeats, as well as pure poly(G) sequences were interpreted as resulting from quadruplex structure formation (Usdin 1998). Although longer d(CGG) repeat tracts ($n = 8$, 11, 16) may preferentially form hairpins rather than a quadruplex structure (Nadel et al. 1995; Fojtik et al. 2004), the association of two hairpins may lead to a quadruplex (Weisman-Shomer et al. 2000). Such an association may potentially occur owing to interactions of two $d(CGG)_n$ loopouts formed in one DNA strand. The $d(CGG)_n$ quadruplex may be weakened by d(AGG) triplets in spite of the latter being also capable of quadruplex formation (Usdin 1998; Weisman-Shomer et al. 2000).

2.5 Unwound DNA

A+T rich DNA sequences form a less thermodynamically stable DNA duplex compared with that formed by random sequences. This is the reason for a relatively easy strand separation (DNA unwinding) in A+T rich sequences by increasing temperature, torsional stress in supercoiled DNA, and potentially by proteins. The propensity for easy unpairing of A+T rich sequences has been identified in 30–100-bp-long DNA unwinding elements (DUE) in bacterial and some eukaryotic replication origins as well as in base-unpairing regions (BUR) in chromosomal matrix attachment regions (Sheflin and Kowalski 1985; Umek and Kowalski 1988; Kowalski et al. 1988; Bode et al. 1992). In cells, two major factors determine the structures of DUEs/BURs. Under torsional stress, unwinding of the double helix occurs first in A+T-rich sequences, whose unpaired state can be supported by negative supercoiling. However, in the presence of Mg^{2+}, DUEs/BURs tend to remain double-stranded and other regions (such as inverted repeats) unwind to partially relieve superhelical tension (Sheflin and Kowalski 1985). Thus, the ability of DUEs to form denaturation bubbles may depend on the level of unrestrained supercoiling and the local ionic environment in cells.

Several experimental approaches have provided evidence of stable supercoil-induced DNA unpairing in SCA10 d(ATTCT) · d(AGAAT) repeats (Fig. 1, panels E.1, E.2) (Potaman et al. 2003). At moderate physiological levels of negative superhelical densities, unpaired regions have well-separated strands which are visible in the atomic force microscope as denatured bubbles at lengths of 11 to 29 repeats. Below lengths of 11 repeats, other A+T-rich blocks in the plasmid melted before melting of the SCA10 repeats. Chemical probe analysis also showed reactivity expanding from the center of the d(ATTCT) · d(AGAAT) tract into the flanking human A+T-rich DNA sequence as the superhelical energy increased, consistent with unpairing of an increasingly large DNA region. Finally, two-dimensional gel analyses showed a structural transition characteristic of DNA melting at a DUE.

At high superhelical densities, long unpaired repeat tracts (29 repeats) "collapse" and probably form structures with loosely intertwined strands. Unpaired single strands in the bubbles and even in the collapsed structures are accessible for normal interactions with small CAA and larger oligonucleotide molecules. These bubbles may also be accessible to proteins involved in DNA replication, including helicase, primase, and DNA polymerase.

2.6 Parallel-Strand DNA

DNA typically exists as a right-handed helix in which the orientation of the two complementary strands is antiparallel. That is, the 5′-to-3′ polarity runs in opposite directions (Sinden 1994). In an alternative parallel orientation, the polarity of the complementary DNA strands runs in the same direction. The biological implications for this alternative DNA structure are, at present, not known. Using a variety of methods including NMR, LeProust et al. (2000) have shown that the Friedreich ataxia short oligonucleotides of d(GAA) · d(TTC) repeats adopt a parallel d(GAA) · d(TTC) duplex in equilibrium with the antiparallel d(GAA) · d(TTC) duplex.

3 Effects of Alternative DNA Conformations on Biology

The formation of alternative structures may have profound effects on cell biology by changing the ways the DNA interacts with other cellular components, most importantly with proteins, during DNA replication, transcription, recombination, and repair. Interference with the normal activity of DNA polymerases is the most notable effect of alternative structures, and it is one of the key elements in the trinucleotide expansion phenomenon.

3.1 Replication Blockage by Hairpins, Triplex, Quadruplex, and Triplet Repeats

Stable DNA structures that DNA polymerases encounter while tracking on a single-stranded template may present blocks of different potency for DNA polymerization. Phage, viral, and eukaryotic polymerases pause and can even be completely inhibited by the hairpin-forming template sequences and by premade hairpins (Kaguni and Clayton 1982; Weaver and DePamphilis 1982; Hacker and Alberts 1994; Suo and Johnson 1998). Similar to polymerase dissociation from a lagging template when it reaches the $5'$ end of the next Okazaki fragment, polymerase quickly dissociates from the hairpin blockage site (Klarmann et al. 1993; Hacker and Alberts 1994). The $3'$ end of the nascent strand may potentially dissociate from the template when polymerase encounters a hairpin or during replication of a perfect or imperfect inverted repeat that can potentially form a hairpin. This may induce mutations in a DNA sequence, as documented in many systems (Gordenin et al. 1993; Rosche et al. 1997, 1998; Viswanathan et al. 2000; Yoshiyama and Maki 2003).

Inhibition of DNA polymerization has been shown for preformed $d(CGG)_n$ hairpins (Kamath-Loeb et al. 2001). The in vivo effects of $d(CGG)_n \cdot d(CCG)_n$ and $d(CTG)_n \cdot d(CAG)_n$ repeats, cloned into plasmids, have been studied by two-dimensional electrophoretic analysis of replication intermediates in bacteria and yeast (Samadashwily et al. 1997; Pelletier et al. 2003; Krasilnikova and Mirkin 2004). Replication fork stalling, albeit often at a low level, occurs within the repeats and is dependent on repeat length, repeat orientation relative to the replication origin, and the status of protein synthesis in cells. One interpretation of these results is that the formation of unusual DNA structures (likely hairpins) by trinucleotide repeats in the lagging-strand template causes the observed replication blockage.

As mentioned already, additional folding of the long $d(CGG)_n$ hairpins or side-by-side interaction of shorter $d(CGG)_n$ hairpins may result in quadruplex formation and concomitant replication blockage. Replication of $d(CGG)_n$, $d(AGG)_n$, and $d(TGG)_n$ templates by bacterial and phage DNA polymerases is most likely blocked owing to the quadruplex formation (Usdin and Woodford 1995; Usdin 1998). Preformed quadruplex structures in the $d(CGG)_n$ template were efficient blocks for eukaryotic replicative DNA polymerases α, δ, and ϵ (Kamath-Loeb et al. 2001).

Triplex structures are also strong blocks for DNA synthesis. This was first observed in vivo from slow DNA replication at the Py · Pu sites (Rao et al. 1988; Baran et al. 1991). Replication fork blockage was consistent with the folding of unreplicated template DNA onto the nascent duplex in the replicated half of the Py · Pu repeat to form a triplex. DNA polymerase was unable to unwind an unreplicated template strand from its position as a third strand and polymerization stalled. Such a structural block that forms during DNA polymerization is very thermodynamically stable, as the template folding has been detected at temperatures up to 80 °C (Baran et al. 1991; Krasilnikov et al. 1997; Potaman and Bissler 1999). Strong blocks to DNA replication were also observed when triplex structures were formed prior to polymerization (Dayn et al. 1992). In this case, DNA polymerase tracks on the template that becomes part of the triplex. Depending on the particular H (H′)-DNA isomer, it becomes either a part of the Py · Pu duplex or a third strand. In both cases, DNA polymerase cannot unfold the structure and stalls.

A wealth of data shows that $d(GAA)_n \cdot d(TTC)_n$ repeats may form triple-stranded structures (Hanvey et al. 1988; Ohshima et al. 1996b; Bidichandani et al. 1998; Sakamoto et al. 1999; Spiro et al. 1999; Vetcher et al. 2002; Potaman et al. 2004). Suppression of DNA replication in vitro and pause sites at the $d(TTC)_n$ template (Spiro et al. 1999) were in agreement with the template fold-back as in other triplex-forming sequences (Baran et al. 1991; Krasilnikov et al. 1997; Potaman and Bissler 1999). Downregulation of *frataxin* gene expression was interpreted as a result of RNA becoming trapped in a triple-stranded structure formed between the $d(GAA)_n \cdot d(TTC)_n$ duplex and the $d(GAA)_n$ transcript (Ohshima et al. 1998; Bidichandani et al. 1998), or between the $d(GAA)_n \cdot d(TTC)_n$ duplex and the nontranscribed $d(GAA)_n$ single strand of the transcription bubble (Grabczyk and Usdin 2000).

3.2 Aberrant Polymerization Associated with DNA Repeats: Slippage During Primer Template Misalignment and Strand Switching

Many reports have indicated that aberrant replication events can occur at DNA repeats. These certainly result in part from the direct repeat and the imperfect inverted repeat (quasipalindromic) sequence organization. The sequence allows DNA directed mutations, including slippage and strand switching.

DNA slippage during replication in vitro has been reported by several groups. Primer template slippage during replication of disease-associated repeats has been observed in vitro with several enzymes, including *E. coli* polymerase I Klenow fragment and human polymerase β (Petruska et al. 1998; Hartenstine et al. 2000; Heidenfelder et al. 2003; Heidenfelder and Topal

2003; Ruggiero and Topal 2004). The propensity and length of slippage and the number of bases that adopt a preferred loop vary with the repeat sequence [d(CAG) versus d(CTG)] (Petruska et al. 1998; Hartenstine et al. 2000; Ruggiero and Topal 2004).

Differences in repeat instability as a function of the orientation of the repeat with respect to the direction of replication were first shown in *E. coli* (Kang et al. 1995a) and subsequently in yeast (Maurer et al. 1996; Freudenreich et al. 1997, 1998; Miret et al. 1997, 1998). Replication slippage has also been observed in an SV40 viral replication system in HeLa cell extracts containing the replication initiator protein large T-antigen, where repeat instability recapitulated the instability seen in humans in terms of repeat length (Panigrahi et al. 2002). Repeat expansion was observed when $d(CAG)_{79}$ comprised the lagging template strand and deletions predominated when $d(CTG)_{79}$ comprised the lagging template strand (Panigrahi et al. 2002). Using the SV40 replication origin system, Cleary et al. (2002) were the first to demonstrate an origin distance dependence on instability in mammalian cells.

Intramolecular strand switching (snap-back synthesis) at d(CTG) · d(CAG) repeats, resulting in hairpin molecules, has been observed in vitro during replication by *E. coli* polymerase I Klenow fragment on plasmid DNA following alkali denaturation and renaturation (Kang et al. 1995b; Ohshima and Wells 1997). DNA polymerase may pause and dissociate during replication of the repeat, or it may encounter an alternative DNA secondary structure formed during denaturation and renaturation. Hairpin products resulting from snap-back synthesis during replication of inverted repeats cloned at a distance from the ColE1 origin where the transition between replication by Pol I and Pol III occurs has been observed in *E. coli* cells (Backman et al. 1978).

In Sect. 4.2, complex expansion mutations are described which involve intermolecular and intramolecular strand switching during replication of SCA10 $d(ATTCT)_n \cdot d(AGAAT)_n$ repeats in *E. coli*.

3.3 Helicase Activity at Hairpins, Triplex DNA, Quadruplex DNA, and Triplet Repeats

The formation of alternative DNA structures can be deleterious to cellular events that require DNA unwinding, such as DNA replication and transcription; therefore, eukaryotic cells might have developed some protective means, e.g., certain proteins that specifically recognize and unwind specific DNA alternative structures. Certain helicases, whose normal function is to unwind double-stranded DNA, can also unfold alternative DNA structures. The bacteriophage T4 DNA polymerase holoenzyme, which includes DNA helicase (T4 gene 41 protein) rather than polymerase alone, alleviates pausing on a hairpin-containing template (Bedinger et al. 1989). DNA helicases T4 dda protein and SV40 large T-antigen can unwind DNA triplexes (Maine and Kodadek 1994; Kopel et al. 1996), and SV40 large T-antigen, the human Werner and Bloom syndrome proteins and *Saccharomyces cerevisiae* Sgs1p DNA helicases can unwind quadruplexes (Baran et al. 1997; Fry and Loeb 1999; Huber et al. 2002).

Knowledge of helicase effects on alternative structures formed in trinucleotide repeated sequences is mostly limited to quadruplex unwinding. Helicases that may unwind the $(CGG)_n$ quadruplex include *S. cerevisiae* Sgs1p, human Werner syndrome helicase/exonuclease (WRN), the CArG-box binding protein A (CBF-A), and others (Khateb et al. 2004). It should be noted, however, that only the natural helicase/polymerase partnerships may prove efficient. In the early experiments on DNA replication by the bacteriophage T4 DNA polymerase holoenzyme, it was recognized that the presence of DNA helicase (T4 gene 41 protein) eliminated detectable polymerase pausing. However, another protein with helicase activity (T4 dda protein) was ineffective (Bedinger et al. 1989). While quadruplex structure can be unwound by a number of helicases, in combination with DNA polymerases, Werner helicase could only alleviate a polymerization block for DNA polymerase δ, but not α and ϵ (Kamath-Loeb et al. 2001). This likely suggests a requirement for a proper helicase–polymerase interaction, so that a concerted action of DNA unwinding by the helicase and DNA synthesis by polymerase result in an efficient progression of the replication complex.

3.4 Recombination Associated with DNA Repeats

Increased levels of genetic recombination at unstable $d(CTG) \cdot d(CAG)$ and $d(GAA) \cdot d(TTC)$ DNA repeats has been extensively studied in *E. coli* by Wells and colleagues (Jakupciak and Wells 1999, 2000a; Pluciennik et al. 2002; Napierala et al. 2002, 2004). Both intermolecular and intramolecular recombination between $d(CTG) \cdot d(CAG)$ repeats that occurs in *E. coli* was elevated 1–2 orders of magnitude relative to the case for nonrepeated control sequences (Pluciennik et al. 2002; Napierala et al. 2002). Longer repeats recombined more frequently than short ones (Napierala et al. 2002). The recombination frequency was higher when the $d(CTG)_n$ repeats comprised the lagging strand template as shown by both intermolecular and intramolecular assays (Pluciennik et al. 2002; Napierala et al. 2002). Although recombination was also stimulated by the $d(GAA)_n \cdot d(TTC)_n$ repeats, the frequency diminished at longer lengths, possibly owing to the sticky DNA formation (Napierala et al. 2004). Assuming that recombination may be initiated at single-strand and double-strand DNA breaks when the replication fork stalls at secondary structures formed in the repeat tracts, Hebert et al. (2004) developed an assay to study the influence of double-strand breaks on repeat stability. The recombinational repair of double-strand breaks within the repeats stimulated deletions in both $d(CTG)_n \cdot d(CAG)_n$ and $d(CGG)_n \cdot d(CCG)_n$ repeats. Double-strand break repair has also been studied in mammalian cells where slipped DNA structures generally underwent deletion upon repair (Marcadier and Pearson 2003). As discussed later, recombination can result in increases and decreases in repeat lengths. Recombination at DNA repeats has also been reported in yeast during meiosis, where it can be associated with the introduction of double-strand breaks (Jankowski et al. 2000; Jankowski and Nag 2002; Nag et al. 2004).

3.5
Interaction of DNA Replication, Repair, and Recombination Proteins with DNA Repeats

Many proteins, including UvrA, hMSH2, RecG, PriA, cruciform binding proteins, anti-Z-DNA antibodies, and anti-DNA antibodies exhibit altered, unexpected, and often differential binding to structures formed by $d(CTG)_n \cdot d(CAG)_n$ repeats. These interactions can provide insight into the biology that might be responsible for pathways to repeat instability in human cells.

DNA excision repair proteins. The bacterial damage-recognition protein UvrA binds to heteroduplex substrates containing $(CAG)_n$ repeat loops, where $n = 1$, 2, or 17, with a K_d of about 10–20 nM, about 2 orders of magnitude higher than that for binding to duplex $d(CTG)_n \cdot d(CAG)_n$ (Oussatcheva et al. 2001). Moreover, when plasmid containing a $d(CTG)_{23}$ or a $d(CAG)_{23}$ heteroduplex loop was introduced into *E. coli* cells, the loops were effectively excised in cells containing functional UvrA. Loops were less effectively excised in cells deficient in UvrA (Oussatcheva et al. 2001). These results imply a more comprehensive role for UvrA, in addition to the recognition of DNA damage, in maintaining the integrity of the genome. These results demonstrate that excision repair proteins can bind and mediate deletion of looped-out triplet repeats in cells.

DNA mismatch repair proteins. Because hairpins formed by d(CXG) repeats contain an $X \cdot X$ mismatch every 2 bp (Fig. 1, panels A.1, A.2), DNA mismatch repair proteins might be expected to recognize features of these mismatched hairpins. The human mismatch repair system involves a wide variety of proteins that may be specialized for different mismatches; proteins that can recognize mismatches and mismatched loops as long as 5–15 nt (Kunkel and Erie 2005). Purified hMSH2 protein binds differentially to heteroduplexes containing $d(CTG)_n$ and $d(CAG)_n$ loops in slipped-strand DNA (Pearson et al. 1997, 2002). The $d(CAG)_n$ loopout was preferentially bound by the human mismatch repair protein MSH2 and bacterial single-strand binding protein compared with the $d(CTG)_n$ loopout (Pearson et al. 1997, 2002). In addition, both loopouts were hypersensitive to cleavage by the junction-specific T7 endonuclease I (Pearson et al. 2002). Recently, Owen et al. (2005) have shown that the mismatch repair Msh2–Msh3 heterodimer binds to a $d(CAG)_n$ loopout containing an A · A mismatch that is critical for binding, leading to the suggestion that mismatch repair mediated stabilization of the hairpin could promote expansion. The binding alters the catalytic properties of the enzyme complex, resulting in inhibition of the repair activity, which might then stabilize the alternative DNA structure and prevent repair (Owen et al. 2005; Mirkin 2005). This differential binding may result in differential repair efficiencies of DNA intermediates in a process of mutagenesis. In mice, short expansions are dependent on functional Msh2, Msh3, or Pms2, as discussed later (Manley et al. 1999; Kovtun and McMurray 2001; van Den Broek et al. 2002; Watase et al. 2003; Savouret et al. 2003, 2004; Wheeler et al. 2003; Gomes-Pereira et al. 2004).

Replication restart–recombination proteins: RecG and PriA. The bacterial proteins PriA, which is required for restarting replication on the lagging strand following collapse, and RecG, which can drive fork reversal and branch migration forming a four-stranded chicken-foot structure, bind to various model DNA structures containing $d(CTG)_n \cdot d(CAG)_n$ repeats (Kim et al. 2006). PriA binds to D-loops, duplex DNA molecules with an unpaired single strand at one end, and forked DNA molecules containing nascent leading or lagging strands (McGlynn et al. 1997). RecG binds structures recognized by PriA but also binds to Holliday junctions and forked DNAs containing a nascent lagging strand. PriA and RecG, however, both bind poorly to single-stranded and duplex DNA molecules (McGlynn et al. 1997; Kim et al. 2006). In band-shift assays, PriA and RecG bound strongly to pure $d(CTG)_n$ or $d(CAG)_n$ hairpins, as well as to single-stranded and duplex DNA molecules containing $d(CTG)_n$ and/or $d(CAG)_n$ loopouts.

RecG and PriA also showed unexpected binding properties to forked DNA structures that represent potential intermediates formed during replication pausing and restart (Kim et al. 2006). Both PriA and RecG bound to a forked DNA structure, but with a surprising leading/lagging strand asymmetry when a $d(CTG)_7$ or a $d(CAG)_7$ loopout was present upstream of the fork. PriA and RecG bound when a $d(CTG)_7$ or a $d(CAG)_7$ loopout occurred in the lagging template strand, but neither protein bound when the loopout occurred in the leading strand. Binding of both proteins to a forked DNA structure containing an upstream slipped-strand structure was very strong. Thus, secondary DNA structures including hairpins and slipped-strand structures formed within the repeats influence the binding of PriA and RecG to single-stranded, duplex, and forked DNA molecules. This result was unexpected and extends the range of structures known to be bound by these proteins, and is consistent with a role for replication restart pathways in repeat instability.

Flap endonuclease (FEN-1). In 1997 Gordenin et al. (1997) proposed a model for repeat expansion that suggested the involvement of human flap endonuclease FEN-1 (reviewed in Liu et al. 2004a), which generally digests the

RNA primer from the 5′ end of an Okazaki fragment prior to ligation to a 3′ terminus of an adjacent Okazaki fragment. The model proposed that DNA repeats at the 5′ end of an Okazaki fragment formed a hairpin that could resist digestion by FEN-1 and subsequently become ligated into an expansion product. This prompted extensive work in vitro and in vivo, in yeast and mice, to understand the potential role of this enzyme in repeat instability.

The initial Gordenin FEN-1 model has been remarkably well validated. Numerous studies have shown that 5′ hairpins composed of different triplet repeats resist digestion by FEN-1 in a length- and sequence-dependent fashion that presumably reflects $d(CXG)_n$ hairpin stability (Spiro et al. 1999; Lee and Park 2002; Henricksen et al. 2002). An in vitro replication system with human polymerase β can generate very large $d(GAA)_n$ expansions when initiation occurs within a repeat tract if FEN-1 is omitted from the reaction, while addition of FEN-1 prevents this expansion (Ruggiero and Topal 2004). Ligation of a 3′ end to the 5′ end of a hairpin flap can occur if the hairpin loop is 6 nt away from the point of ligation (Veeraraghavan et al. 2003).

In yeast, mutations in the FEN-1 homologue, Rad27, cause increased rates of expansion and higher rates of chromosome breakage (Spiro et al. 1999; Callahan et al. 2003; Liu et al. 2004b). A Huntington disease model mouse heterozygous for FEN-1 showed a small preference for expansions over deletions (Spiro and McMurray 2003), although mutations in FEN-1 are not linked to the expansions associated with Huntington disease (Otto et al. 2001).

4 Pathways for Repeat Expansion

4.1 Primer-Template Misalignment During Replication Can Account for Repeat Length Changes Less Than Twofold in Length in Disease-Associated DNA Repeats

As first proposed by Streisinger, primer-template misalignment can occur within a run of direct repeats (Streisinger et al. 1966). Mutations associated with primer-template misalignment have been established in many model systems (Kunkel and Soni 1988; Ripley 1990; Kunkel 1990; Papanicolaou and Ripley 1991; Rosche et al. 1998; Sinden et al. 1999; Bebenek and Kunkel 2000; van Noort et al. 2003). Misalignment can occur within runs of repeats (Streisinger et al. 1966; Wierdl et al. 1997; Kroutil and Kunkel 1999; Hashem et al. 2002) or between distant direct repeats (Drake et al. 1983; Ripley et al. 1986) (Fig. 2). In the case of triplet repeats, a simple slippage may result in a 3-nt loopout (Fig. 2, panel A). In cells, a 3-nt slippage, and/or repair of the loopout, can be very different for opposite orientations of the repeat with respect to the origin of replication (Hashem et al. 2002). Large slippage events

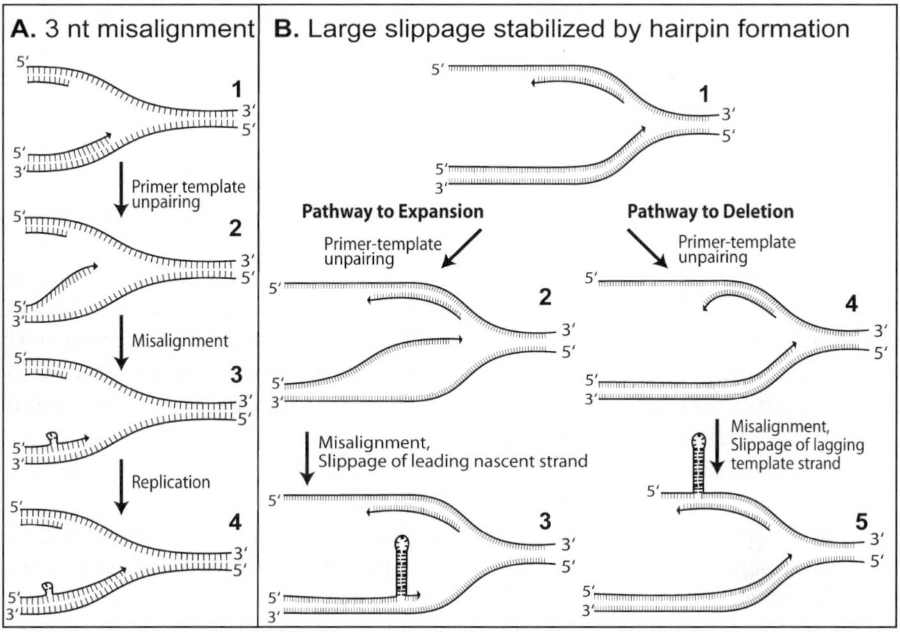

Fig. 2 Replication slippage can result in deletions and duplications (expansions). *A* A 3-bp misalignment can occur by unwinding of the primer end of the nascent strand from the template (step *2*), followed by reannealing 3 nt to the 5′ side on the template, resulting in a 3-nt loopout in the nascent strand (step *3*). Continued synthesis would result in a 3-bp expansion in the nascent strand, if not repaired by mismatch or excision repair-type activities. *B* Primer-template misalignment between short direct repeats can also occur over large distances. In the case of disease-associated repeats, misalignment can occur anywhere within the repeat tract. When DNA repeats can form stable hairpins, they can promote slippage in the nascent strand leading to expansion (steps *2*, *3*), or in the template strand leading to deletion (steps *4*, *5*)

may occur within a long repeat tract, resulting in a backwards slippage and the formation of a hairpin in the leading nascent strand (Fig. 2, panel B, pathway to expansion), and continued replication would lead to expansion by the length of the slippage. A forward slippage, perhaps directed by hairpin formation in the lagging template strand, could lead to deletion in the lagging nascent strand (Fig. 2, panel B, pathway to deletion). This type of mutation can be influenced dramatically by DNA symmetry elements, especially inverted repeats. Inverted repeats can fold into hairpins that can promote deletion between flanking direct repeats in the lagging template strand (Trinh and Sinden 1991, 1993; Rosche et al. 1995), or direct duplications when hairpins form in the leading nascent strand (Hashem and Sinden 2005).

For primer-template misalignment to occur, DNA polymerization must stop and the polymerase must presumably dissociate from the DNA. It is not known what feature of DNA, either sequence or structure, might be involved in mediating this pausing or stopping. Polymerase may pause at random, or exhibit preferred pause sites, as occurs in vitro, where pause sites are associated with misalignment mutations (Papanicolaou and Ripley 1989, 1991). In the case of disease-associated DNA repeats, pausing might be enhanced or promoted by the formation of an alternative DNA structure, of the types discussed before. Moreover, as leading and lagging strand replications are believed to be coordinated, structure formation in the lagging strand may stop leading strand synthesis, and vice versa. Once the nascent 3′ terminus dissociates from the template, it is free to anneal at any location containing complementary base pairs. Essentially nothing is known about the events and mechanics associated with polymerase dissociation. Similar frequencies of duplication and deletion between direct repeats spaced 17 bp apart lead to the suggestion that at least 20 bp become unpaired in the initial dissociation event (Trinh and Sinden 1993).

Because the size of potential expansion or deletion is limited to the length of the repeat tract minus the length of the segment used as a template, expansion by this model is necessarily less than the length of the repeat; thus, only expansion by less than a factor of 2 is possible. In previous reviews, we have discussed the possibility of reiterative DNA synthesis (Kornberg et al. 1964), perhaps caused by an alternative DNA structure block to DNA replication in the leading or lagging strand (Sinden and Wells 1992; Wells and Sinden 1993; Sinden 1999). Repeated slippage during replication has been observed in vitro with several enzymes, including human polymerase β (Petruska et al. 1998; Hartenstine et al. 2000; Kobayashi et al. 2002; Heidenfelder et al. 2003; Heidenfelder and Topal 2003; Ruggiero and Topal 2004).

Instabilities occurring throughout life in certain tissues in humans and in mice are consistent with the possibility that simple slipped misalignment occurs during replication. In humans, the expanded d(CTG) repeat is unstable and shows a bias toward continued expansion in germline and somatic tissues during life (Wong et al. 1995; Martorell et al. 1998). Mice also show repeat instabilities throughout life (Mangiarini et al. 1997; Monckton et al. 1997; Sato et al. 1999; Seznec et al. 2000; Fortune et al. 2000). Repeat heterogeneity in *E. coli*, especially in mismatch repair deficient strains, is consistent with slipped misalignment during replication of repeats in bacteria (Schumacher et al. 1998; Schmidt et al. 2000; Parniewski et al. 2000). Thus, slipped misalignment may be the simplest mechanism for repeat instability and it could be operable for all repeats.

4.2 Strand Switching During Synthesis of d(ATTCT) · d(AGAAT) DNA Repeats Can Result in Complex Expansion Mutations

Primer-template misalignment can occur forward or backward along the same template strand, resulting in duplications and deletions, respectively;

however, misalignment within a palindromic or quasi-palindromic sequence can also occur on a different template strand. This can occur in an *inter*molecular fashion within a single replication fork from the leading to the lagging strand (or from the lagging to the leading strand) or between two different chromosomes (Fig. 3). Strand switching can also occur in an *intra*molecular fashion when the nascent strand snaps back on itself, forming

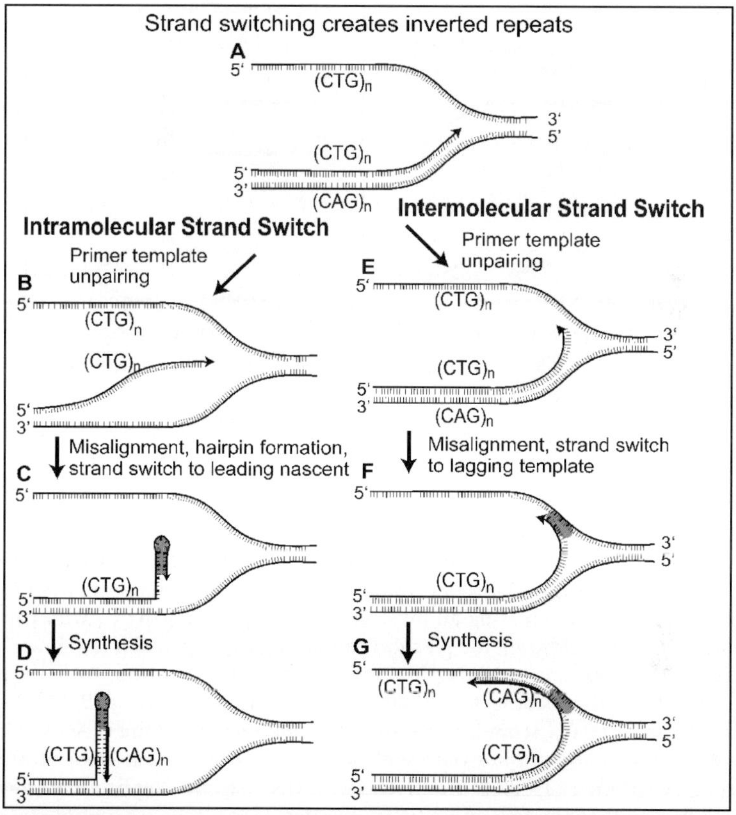

Fig. 3 Intermolecular strand switching can occur within quasi-palindromic repeats forming perfect inverted repeats. A quasi-palindromic sequence, including d(CXG)$_n$, d(CCTG)$_n \cdot$ d(CAGG)$_n$, and d(ATTCT)$_n \cdot$ d(AGAAT)$_n$ repeats can form various degrees of mispaired hairpin structures in one or both strands. The self-complementary base-pairing potential can lead to an intramolecular or an intermolecular strand switch. For the intramolecular strand switch, during leading-strand synthesis the nascent strand can unpair (step *B*) and form a mispaired hairpin region (denoted by the *shaded region* of the helix) (step *C*). Continued synthesis down the hairpin can lead to the formation of a perfect inverted repeat (denoted by the *thicker line*) (step *D*). For the intermolecular strand switch, following unwinding (step *E*), the 3′ end of the nascent strand pairs with the repeat in the lagging template strand (step *F*). Continued synthesis also leads to a perfect inverted repeat in the leading nascent strand

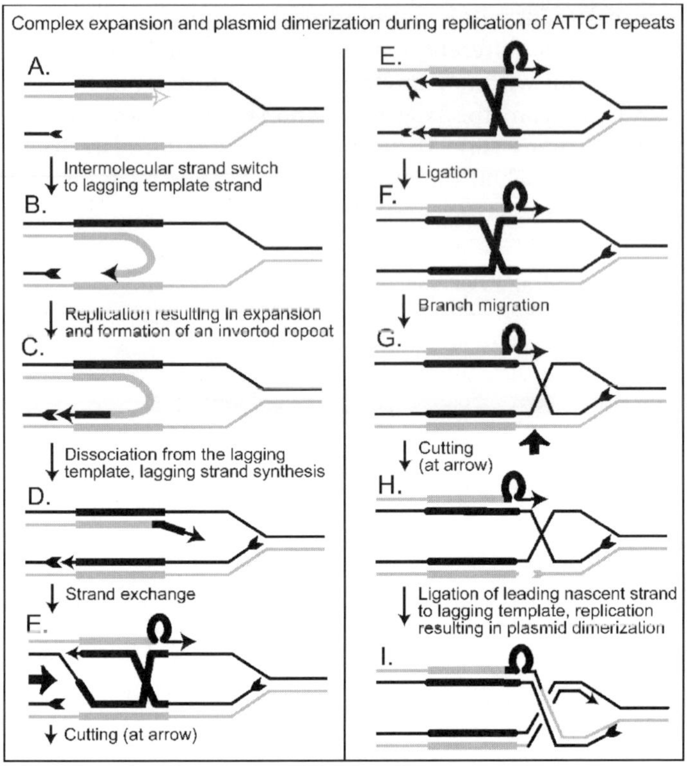

Fig. 4 Complex expansion mutation associated with an intermolecular strand switch during replication of d(ATTCT)$_n$ · d(AGAAT)$_n$ repeats. The A+T-rich spinocerebellar ataxia type 10 (*SCA10*) repeat undergoes both intermolecular and intramolecular strand switch events in *Escherichia coli*, creating an inverted repeat region associated with complex expansion mutations. In addition, the plasmid containing the expansion is a dimer. A model is presented for the expansion, inversion, and plasmid dimerization. An intermolecular strand switch from the leading to the lagging template occurs (step *B*). The *dark line* represents the d(ATTCT)$_n$ strand, while the *light line* represents the d(AGAAT)$_n$ strand. *Arrow heads at the ends of lines* represent the 3′ end of a nascent DNA strand. *Arrow tails at the ends of lines* represent the 5′ end of a DNA strand. Replication following the strand switch results in the formation of an inverted repeat region (contiguous *dark and light line*) (step *C*). Dissociation of the nascent 3′ end from the lagging template strand and reassociation with the leading template strand results in an unpaired 3′ end inherent with the expansion (step *D*). Primer-template pairing within flanking direct repeats, used for cloning the repeats, occurs concomitant with the formation of a hairpin or a loop (step *E*). A strand exchange occurs (step *E*) with the nascent lagging strand (synthesized in step *D*). Following introduction of a nick (at the *large arrow*) a Holliday junction is formed (steps *E–G*). Branch migration occurs (step *G*) and a nick is introduced into the lagging template strand (step *H*). The lagging template then becomes joined to the leading nascent strand (step *I*). Synthesis from the 3′ end of the lagging template strand restores the crossover replication fork and continued replication leads to plasmid dimerization (step *I*). This complex molecular event provides a good example of the degree to which the properties of a simple DNA repeat sequence can direct complex genetic alterations

a hairpin on continued DNA synthesis. Replication that follows the strand switch within a quasi-palindrome results in the formation of a perfect inverted repeat (Ripley 1982; Sinden et al. 1999; van Noort et al. 2003).

Results for one quasi-palindrome correction mutation in *E. coli* indicated that an intermolecular strand switch specific for the leading strand occurred (Rosche et al. 1997), while an intramolecular strand switch was implicated to explain another mutation (Viswanathan et al. 2000). Therefore, quasi-palindrome corrections occurring in either the leading or the lagging strands have been identified in different mutational systems (Rosche et al. 1997, 1998; Viswanathan et al. 2000; Yoshiyama et al. 2001; Yoshiyama and Maki 2003). During replication of plasmid in *E. coli* containing the SCA10 repeat tract, which contains weak quasi-palindromic repeat symmetry, similar complex expansion mutations with the general sequence, $d(TATTC)_{5-11} \cdot d(GAATA)_{9-35}$, were observed regardless of the initial orientation of the repeat tract (either $d(AGAAT)_{23} \cdot d(ATTCT)_{23}$ or $d(ATTCT)_{24} \cdot d(AGAAT)_{24}$). This mutation was also coupled with plasmid dimerization (Hashem VI, Edwards SF, Klysik EA, Pytlos MJ, Sinden RR, unpublished). Insight into an explanation for this result stems from the fact that only a strand switch of the nascent Py-rich strand can produce the inverted repeat found in the complex expansion mutations. For the two different orientations of the repeat tract, the nascent Py-rich strand comprises the leading nascent strand for the $d(AGAAT)_{23} \cdot d(ATTCT)_{23}$ orientation, while it comprises the lagging nascent strand for the $d(ATTCT)_{24} \cdot d(AGAAT)_{24}$ orientation. Thus, to form the inverted repeat in the two different repeat orientations, the strand switch must occur in the leading strand for one orientation and in the lagging strand for the other orientation (Hashem VI, Edwards SF, Klysik EA, Pytlos MJ, Sinden RR, unpublished). In the $d(ATTCT)_{24} \cdot d(AGAAT)_{24}$ orientation, a simple slippage to lengthen the repeat tract must occur prior to the strand switch to generate the observed product (Fig. 4). This is the first example of a DNA sequence that can support both an intermolecular and an intramolecular strand switch during leading or lagging strand synthesis. The instability associated with $d(ATTCT) \cdot d(AGAAT)$ repeats, and even disease-associated triplet or tetranucleotide repeats, in human cells may be linked to aberrant replication. In the following a model is presented in which aberrant replication initiation leading to amplification may result in repeat expansion (as well as repeat deletion).

4.3
DNA Mismatch Repair May, or May Not, Participate in Small Repeat Length Changes During Replication in Dividing Cells or During Gap Repair in Nondividing Cells

As discussed already, the ability to form hairpins of various degrees of stability, coupled with the potential for replication slippage within direct repeats, certainly contributes, in a major way, to the potential for variation in the length of triplet and possibly other types of repeats. $d(CTG)_n$, $d(CAG)_n$, $d(CGG)_n$, and $d(CCG)_n$ repeats form hairpins with a GC · GC dinucleotide interspersed with a T · T, A · A, G · G, or C · C mismatch, respectively (Fig. 1, panels A.1, A.2). The involvement of mismatch repair in repeat instability in *E. coli* has recently been reviewed (Parniewski and Staczek 2002). The stability of these hairpins decreases in the order listed. As discussed already, human mismatch repair proteins bind to $d(CAG)_n$ hairpins better than to $d(CTG)_n$ hairpins (Pearson et al. 1997; Owen et al. 2005). DNA mismatch repair systems are designed to recognize noncanonical base pairs or mismatches following errors in DNA replication and repair them in the nascent strand when not corrected by polymerase proofreading activities. Traditionally, mismatch repair activities are thought to operate in dividing cells that are actively undergoing DNA replication. Interestingly, mice deficient in mismatch repair proteins show reduced rates of repeat instability (Manley et al.

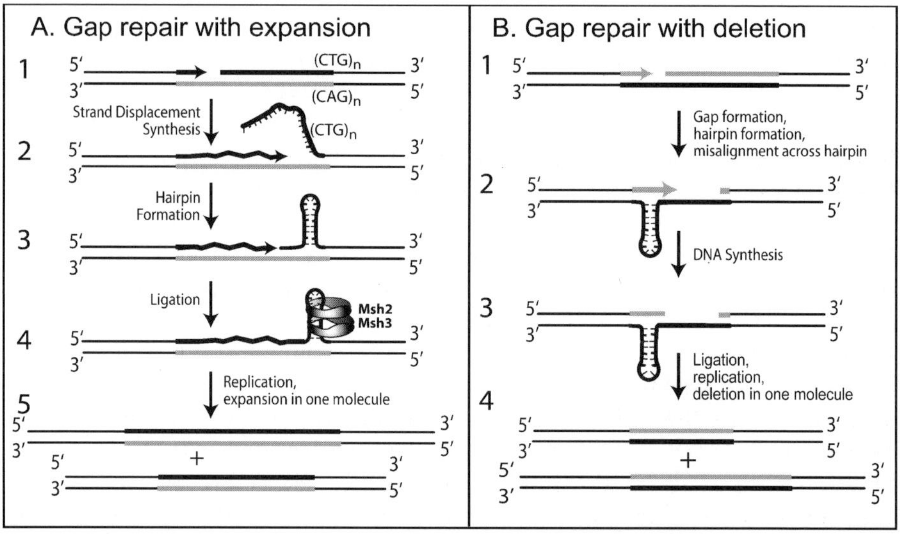

Fig. 5 Gap repair associated with expansion, deletion, or DNA repair. *A* Gap repair with expansion. Strand displacement synthesis at a nick leads to a flap that is normally digested by the flap nuclease FEN-1 (step *2*). However, the formation of a hairpin or other DNA secondary structure can interfere with digestion (step *3*). The hairpin may be stabilized by Msh2–Msh3 and subsequent ligation would lead to expansion in the nascent strand (step *4*). If the hairpin escapes subsequent mismatch repair, nucleotide excision-type repair, or other repair events, it will lead to expansion in one DNA molecule upon the next round of replication (step *5*). *B* Gap repair with deletion. If a hairpin-forming sequence is present within a gap that may be formed during repair of spontaneous or extraneous DNA damage, it may fold into a hairpin (step *2*). Slipped misalignment across a hairpin during DNA replication would lead to the loss of repeats in the nascent strand (step *3*), which would lead to deletion in one DNA molecule following the next round of replication (step *4*)

1999; Kovtun and McMurray 2001; van Den Broek et al. 2002; Watase et al. 2003; Savouret et al. 2003, 2004; Wheeler et al. 2003; Gomes-Pereira et al. 2004). These proteins have also been suggested to participate in repair of spontaneous or endogenous DNA damage in quiescent cells, even in sperm (Kovtun and McMurray 2001; McMurray and Kortun 2003; Kovtun et al. 2004).

In nondividing cells, mismatch repair proteins have been implicated in small repeat length changes, where it has been proposed they participate in the stabilization of slipped, hairpin-containing structures formed during gap repair (Fig. 5) that may be associated with spontaneous DNA damage (Kovtun and McMurray 2001; McMurray and Kortun 2003; Kovtun et al. 2004). A recent demonstration of inhibition of enzymatic activity of the Msh2–Msh3 heterodimer when bound to d(CAG)$_n$ hairpins is consistent with recognition, binding, and stabilization of loopouts that then could be ligated into an expansion event (Owen et al. 2005). Panigrahi et al. (2005), however, recently demonstrated clearly that d(CAG)$_n$ and d(CTG)$_n$ loopouts can be repaired in an orientation- and sequence-dependent fashion although mismatch repair proteins may not be involved.

4.4 DNA Repair of Slipped-Strand Intermediates Containing (CTG)$_n$ Hairpins or (CAG)$_n$ Loopouts

Evidence suggests that repair of loopouts in *E. coli* may involve excision repair proteins UvrA, UvrB, and SbcC (Parniewski et al. 1999; Oussatcheva et al. 2001). d(CTG)$_n$ and d(CAG)$_n$ loopouts in plasmids were repaired (removed) when introduced into *E. coli*, and repair was less effective, but not prevented, in cells lacking certain excision repair proteins (UvrA, SbcC) (Oussatcheva et al. 2001). The binding of UvrA to d(CAG)$_n$ loopouts in vitro supports the hypothesis that loopout structures can be repaired in *E. coli* by excision repair functions. Panigrahi et al. (2005) have carefully characterized the ability of plasmid DNA containing a slipped intermediate DNA with a d(CAG)$_n$ loopout or a d(CTG)$_n$ hairpin in a continuous template or nicked nascent strand to be repaired in mammalian cell extracts. These templates mimic products of replication slippage or strand exchange during replication restart or during double-strand break repair. The stability of the repeats was analyzed in situations where the nicks were 3′ or 5′ to the loopouts. Different substrates were repaired, or not repaired, with remarkably different efficiencies (Panigrahi et al. 2005). First, repair required a nick. Second, a substrate containing d(CAG)$_{50}$ on the continuous strand opposite d(CTG)$_{30}$ on the nicked strand was repaired to d(CAG)$_{50}$ · d(CTG)$_{50}$. Third, when d(CTG)$_{50}$ was opposite d(CAG)$_{30}$ in the nicked strand, no repair occurred in cell extracts. Fourth, when the excess d(CAG) or d(CTG) slipped-out repeats were present on the nicked strand, variable-sized products corresponding to all possible lengths from 30 to 50 repeats were observed. These events did not require mismatch or excision repair proteins or DNA polymerase β. This work clearly shows that different DNA repeat structures at various positions relative to the direction, and strand, of replication can have very different consequences.

4.5 Recombination at Disease-Associated DNA Repeats Can Lead to Deletions and Expansions

Genetic recombination between two repeat tracts occurring either as crossing over, with exchange of flanking markers, or as gene conversion can generate

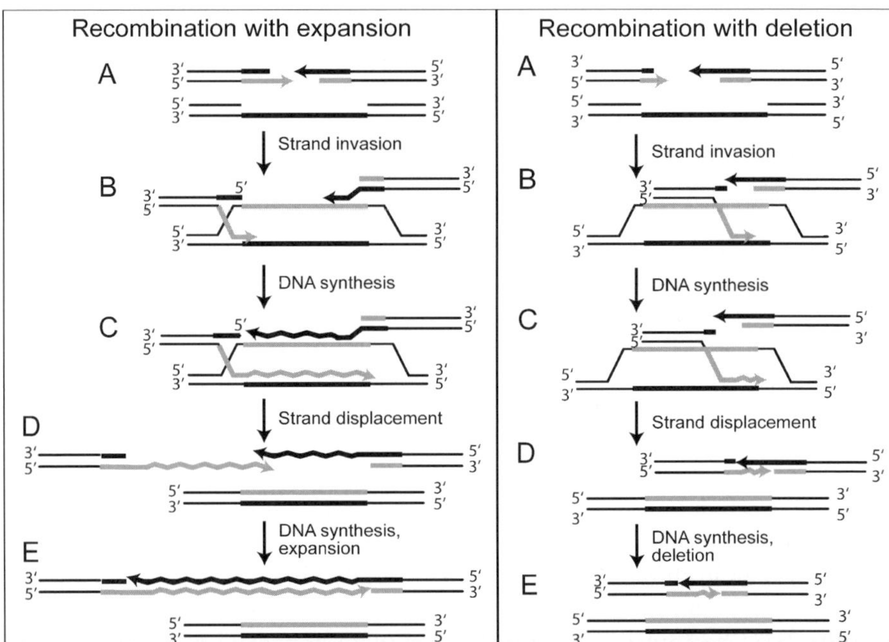

Fig. 6 Recombination associated with expansion or deletion. Recombination (gene conversion) within repeats can lead to expansion (*left*) or deletion (*right*) depending on the position within the repeat tract the strand exchange occurs. Expansion (*left*). Following a break within a DNA repeat (step *A*), strand invasion of the 3′ ends at the 3′ side of the complementary strands of a second chromosome (step *B*) would lead to expansion following DNA synthesis (step *C*). Reannealing of the complementary strands of the second duplex will displace the nascent strands, whereupon the 3′ ends of the newly synthesized strands could anneal (step *D*). Additional synthesis could lead to more than a doubling of the length of the repeat (step *E*). Deletion (*right*). If a break occurs to one side of the repeat and the strand invasion occurs near the opposite end of the repeat in the second chromosome (step *B*), following synthesis, strand displacement, and ligation, a deletion will occur in one chromosome (steps *C*–*E*)

variation in the lengths of repeats depending on where the invading strand of one DNA molecule pairs along the length of the second duplex (Fig. 6). As with simple replication slippage during synthesis of a repeat tract, recombination would be expected to generate expansions of less than a factor of 2 or 3.

A large body of evidence has been presented to show that recombination occurs at high rates at disease-associated repeats cloned into bacterial plasmids and that, as expected, this will generate variation in repeat lengths following recombination (Jakupciak and Wells 1999, 2000a,b; Pluciennik et al. 2002; Napierala et al. 2002, 2004; Vetcher and Wells 2004). The DM1 d(CAG) · d(CTG) repeats and the Friedreich ataxia d(GAA) · d(TTC) repeats stimulate either intramolecular recombination between two repeat tracts in the same plasmid or intermolecular recombination between two repeat tracts in different plasmids in the same cell. In contrast, some reports indicate that recombination does not appear to be stimulated by d(CTG) · d(CAG) repeats in yeast (Miret et al. 1997), while other reports suggest that recombination-related instability can occur in yeast (Jankowski et al. 2000; Jankowski and Nag 2002; Nag et al. 2004). d(CAG) · d(CTG) repeats have been shown to promote deletions and rearrangements when cloned into the *APRT* gene in Chinese hamster ovary cells (Meservy et al. 2003). Genetic recombination in the classic sense of gene conversion or crossing over, however, does not seem to be a major source of repeat expansion in humans, although rare instances associated with recombination have been reported (Brunner et al. 1993; van den Ouweland et al. 1994; Krahe et al. 1995; Brown et al. 1996; Losekoot et al. 1997). Recombination is an integral part of restart of paused replication forks and will be discussed in the next section in this context with respect to repeat instability.

4.6 Double-Strand Break Repair, Replication Restart, and Checkpoint Control Associated with Repeat Replication

A widely accepted model for repeat instability suggests that deletions result from primer-template misalignment, as discussed already. Large deletions have been suggested to occur by replication slippage across $d(CTG)_n$ hairpins in the lagging template strand when it is single-stranded (as shown in Fig. 2) (Kang et al. 1995a; Freudenreich et al. 1997; Schweitzer and Livingston 1998, 1999; Miret et al. 1998; Sinden 1999; Ireland et al. 2000; Rolfsmeier et al. 2001; Hashem et al. 2002; Panigrahi et al. 2002; Lee and Park 2002; Marcadier and Pearson 2003; Bhattacharyya and Lahue 2004; Liu et al. 2004b). However, functional RecA and RecB are required for the high rates of repeat instability in *E. coli* (Hashem et al. 2004b), and a simple model of replication slippage across a hairpin in the lagging template strand cannot account for the involvement of RecA and RecB. Rather, repeat deletions may result from errors occurring during replication restart following the collapse of the replication fork during synthesis of the repeats (Hashem et al. 2002, 2004b; Kim et al. 2006). At present, the molecular events responsible for replication pausing are uncertain; however, hairpins, slipped mispaired DNA, or other secondary structures may play a role in blocking or pausing replication fork progression in $d(CTG)_n \cdot d(CAG)_n$ or $d(CGG)_n \cdot d(CCG)_n$ repeats (Usdin and Woodford 1995; Kang et al. 1995b; Sinden 1999; Hartenstine et al. 2000; Kamath-Loeb et al. 2001; Heidenfelder et al. 2003), while triplex DNA formation could participate in replication pausing in $d(GAA)_n \cdot d(TTC)_n$ repeats (Gacy et al. 1998; Grabczyk and Usdin 2000; Potaman et al. 2004; Krasilnikova and Mirkin 2004). Double-strand breaks are often, but not always, associated with recombination and they can result in repeat deletion when they occur within repeats, as shown in both bacteria and mammalian cells (Marcadier and Pearson 2003; Hebert et al. 2004).

Several pathways are available for restarting a collapsed or paused fork. Here, these are described with respect to repeat deletion as understood for *E. coli* (Kim et al. 2006), but they also act in eukaryotic cells and may be responsible for spontaneous, as well as drug-induced deletion in $d(CTG)_n \cdot d(CAG)_n$ deletion in DM1 lymphoblasts (Hashem et al. 2004a). Replication restart of stalled replication forks requires DNA replication, recombination, and repair proteins (Cox et al. 2000, 2001; Marians 2000; McGlynn and Lloyd 2002). A pathway for the orientation when d(CAG) comprises the leading template strand is shown in Fig. 7. Leading-strand synthesis may be spontaneously paused during synthesis of the repeats, stalled by a short (3-bp) misalignment, or may be blocked by a stable DNA secondary structure in the leading template strand (the pause site is denoted by the asterisk in Fig. 7, step A). Following leading-strand blockage, lagging-strand replication continues (Fig. 7, step B). After fork collapse, the unwinding of stalled forks by RecG or RuvABC in *E. coli* leads to fork reversal and formation of a Holliday junction (here called a "chicken-foot" structure) through annealing of the leading and lagging nascent strands (Fig. 7, step D). Cleavage of the Holliday junction by RuvABC resolvase generates a duplex DNA (Fig. 7, step E) in which the 5′ end can be resected by RecBCD nuclease (Fig. 7, step F). RecA can then initiate recombination and restore the fork (Fig. 7, steps G–J) (Hashem et al. 2004b; Kim et al. 2006). This may be the major pathway for repeat deletion, as mutations in *recA* and *recB* can decrease deletion rates by factors of more than 1000 (Hashem et al. 2002, 2004b). The potential for d(CTG) hairpin formation when single-stranded (Fig. 7, steps G–I), and a preference for restart via the RecA- and RecBC-dependent pathway may explain the generally observed bias for deletions in this orientation, as discussed previously (Hashem et al. 2004b; Kim et al. 2006).

The stalled fork may also be rescued by other pathways. One pathway employs an exonuclease to trim the lagging nascent strand (Fig. 7, steps C–K). Alternatively, the stalled fork (Fig. 7, step A) could simply collapse (Fig. 7,

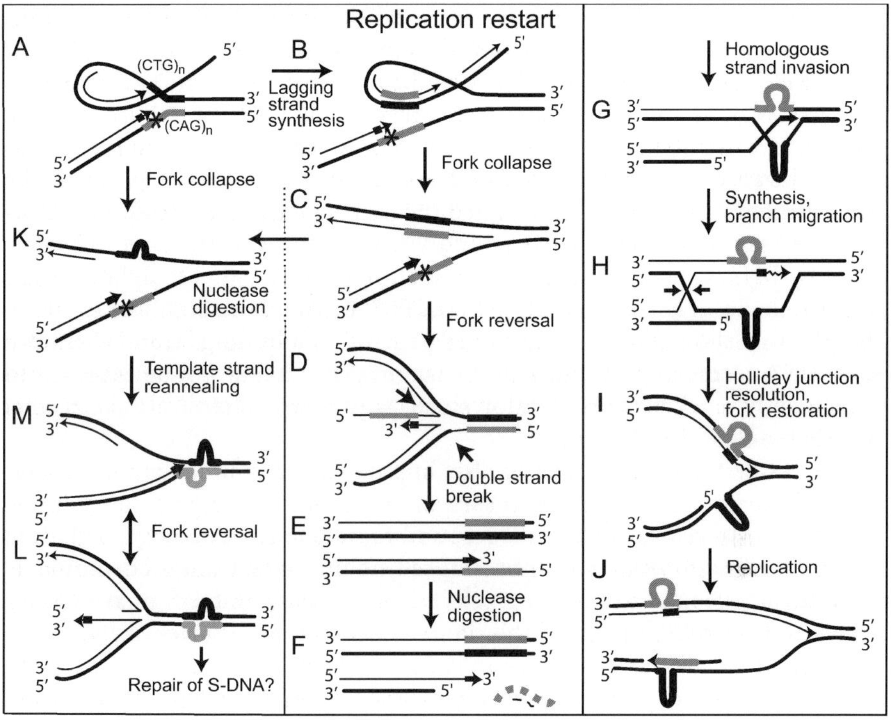

Fig. 7 Replication restart can explain orientation dependence for repeat instability. Replication restart following a block to DNA replication and fork collapse is required to complete duplication of the chromosome to ensure cell viability. Several pathways are available for this process. A major pathway involves fork reversal (step *D*) and the introduction of a double-strand break (step *E*), which is repaired by recombination functions within a cell (steps *F–I*). A high rate of repeat instability in *E. coli* is dependent on RecA and RecBC, which precludes a simple replication-based model (as shown in Fig. 2) for their participation in repeat instability. The pathways shown may account for a high rate of repeat deletion in *E. coli* and explain the orientation-dependent greater instability when the d(CTG)$_n$ tract comprises the lagging template strand. These pathways are described in detail in the text

steps A–K). During digestion, or following collapse, a hairpin may form in the d(CTG)$_n$ strand (Fig. 7, step K). Reannealing of the leading and lagging template strands would then drive the formation of slipped-strand DNA (Fig. 7, step M). Fork reversal could occur (Fig. 7, step L), which would move the slipped-strand DNA away from the Holliday junction, making it available for DNA repair, as observed in several systems (Oussatcheva et al. 2001; Panigrahi et al. 2005), and leading to changes in repeat length. Resolution of the junction shown in step L would create the double-strand break, similar to molecules shown in step F, but with slipped-strand DNA in one molecule.

Samadashwily et al. (1997) have reported the strength of replication fork pausing in *E. coli* during lagging-strand synthesis to be in the order d(CGG) > d(CCG) > d(CTG) > d(CAG). Pausing during synthesis of d(CGG) · d(CCG) tracts between 14 and 31 repeats was clearly evident; however, replication fork pausing in a $d(CAG)_{70} \cdot d(CTG)_{70}$ tract was only detected following chloramphenicol treatment to induce plasmid amplification. Moreover, pausing was only detected when d(CTG) comprised the lagging template strand. Biochemical detection of pausing in *E. coli* and yeast has been interpreted to be initiated by DNA secondary structure formation in the lagging template strand (Samadashwily et al. 1997; Pelletier et al. 2003; Krasilnikova and Mirkin 2004). The pathway shown in Fig. 7 discusses pausing as a leading strand event because DNA secondary structure in the lagging strand may not be expected to permanently block fork progression since lagging-strand replication can start on either side of the structure.

The restart of stalled forks is also important for mammalian cells, and pathways analogous to those discussed for *E. coli* may be important for instability in human cells. Human cells respond rapidly to DNA damage, including stalled replication forks, UV-light-induced photoproducts, and chemotherapeutic drug lesions, by arresting cells in the S phase (intra-S phase checkpoint) and allowing repair of the damage (Kastan and Bartek 2004). DNA damage caused by various exogenous factors leads to the activation of the DNA damage checkpoint pathways (reviewed in Melo and Toczyski 2002). These pathways are essential for preventing irreversible breakdown of replication forks stalled at the sites of DNA damage (Tercero et al. 2003). Intra-S checkpoints have also been shown to be involved in normal DNA replication (Cha and Kleckner 2002). The *S. cerevisiae* genome contains about 1500 sites where DNA replication slows, and mutations in the *MEC1* gene, a human ataxia telangectasia-related and Rad 3 related (ATR) homologue, accentuate stalling at those sites, resulting in chromosomal breakage (Cha and Kleckner 2002). Thus, intra-S checkpoints may stabilize stalled replication forks even in the absence of DNA damage. Consistent with the expectation that fork blockage during replication of repeats, or double-strand breaks generated as a consequence of replication fork restart or DNA repair events, might activate the DNA damage checkpoint response, d(CAG) · d(CTG) repeats can activate the DNA damage response in *S. cerevisiae* (Lahiri et al. 2004). Mutations in the *MEC1*, *RAD9*, or *RAD53* genes increased the rates of chromosome breakage associated with a (CAG) · (CTG) repeat tract. Deficiencies in Mec1, Ddc2, Rad17, Rad24, or Rad53 resulted in an increase in the frequency of repeat deletions. Interestingly, expansions were also increased in cells deficient in Rad24, Rad17, and Rad53. These results suggest that replication or repair events are altered when normal checkpoint controls become compromised.

4.7 DNA Amplification Provides a Facile Means for Repeat Expansion for SCA10 d(ATTCT) · d(AGAAT) Repeats

With the exception of repeated replication slippage or reiterative DNA synthesis, all the models described so far can account for small changes in repeat lengths, as observed in somatic cells, but not expansions of tenfold or greater in length. DNA amplification provides a simple, reasonable model for the large repeat expansions that occur on intergenerational transmission in some diseases. The amplification of specific DNA regions by repeated replication occurs in several systems, including the amplification of chorion genes during normal *Drosophila* development, puff II/9A in *Sciara*, and drug-resistance genes in tumor cells (Schimke 1988; Liang et al. 1993; Spradling 1999; Calvi and Spradling 1999; Tower 2004). For amplification to occur, the normal controls that limit replication to once per cell cycle must be abrogated (Spradling 1999; Calvi and Spradling 1999). Amplification has been proposed to occur by an onion-skin mechanism in which repeated initiation leads to multiple replication forks (Baran et al. 1983; Stark et al. 1989; Schimke 1992; Spradling 1999), followed by recombination, or nonhomologous end joining, to generate linear tandem arrays (Fig. 8, left panel) from the amplified DNA (Syu and Fluck 1997).

Amplification is frequently associated with replication origins, A+T-rich regions, inverted repeats, or polypurine · polypyrimidine tracts (Baran et al. 1987; Kirschner 1996; Spradling 1999). Fragile sites have also been implicated as a causative factor in oncogene amplification (Hellman et al. 2002). While commonly thought to arise by strand breakage, reinitiation at an aberrant origin could also generate abnormal DNA ends, leading to recombinational amplification (Syu and Fluck 1997).

Potaman et al. (2003) proposed that unwound DNA structures in long d(ATTCT) · d(AGAAT) repeats drive repeat amplification. The formation of an unwound DNA structure from superhelical energy in DNA may bypass the steps of pre-RC assembly that normally require the low cyclin dependent kinase (CDK) activity environment of the G1 phase, and allow polymerase α/primase to initiate replication in the high CDK environment of the S phase without the association of origin-bound checkpoint proteins. The observation that the binding affinity of the *Drosophila* replication initiator origin recognition complex (ORC) is 30-fold higher for supercoiled DNA compared with relaxed DNA (Remus et al. 2004) suggests that a topological equivalence between superhelical and unwound states could allow DUEs to act as replication switches. In addition, an increasing body of evidence suggests that transcription is a critical component of a replication origin (Ghosh et al. 2004; Kouzine et al. 2004; Jenke et al. 2004; MacAlpine et al. 2004; Danis et al. 2004; Casper et al. 2005; Nieduszynski et al. 2005). Perhaps transcription supplies the superhelical energy required to unwind

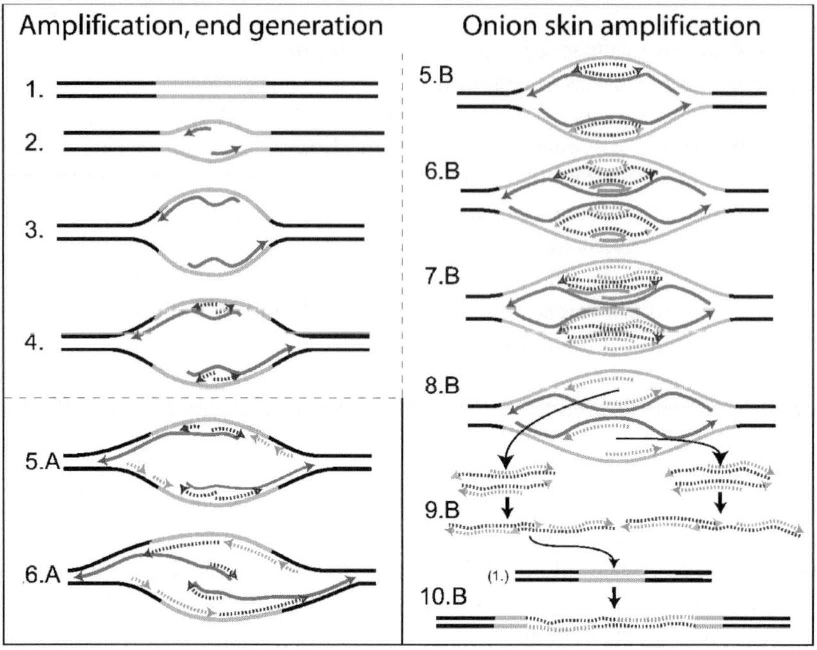

Fig. 8 Amplification models for repeat expansion. Models for repeat instability based on the utilization of an unstable DNA repeat as an aberrant replication origin initially described for the A+T-rich SCA10 d(ATTCT)$_n$ · d(AGAAT)$_n$ repeats by Potaman et al. (2003) are shown. Amplification with the generation of DNA ends (*left*). A+T-rich repeats can unwind and replication may start within the unwound bubble (step *2*). The DNA repeats are denoted by the *lighter shaded line*. Nascent strands are shown in *intermediate shading*. Following synthesis (step *3*), DNA unwinding again occurs within the repeats (step *3*) and replication again starts at the unpaired regions (step *4*). The second nascent strands are shown as the *darker dashed lines*. In step *5.A*, lagging-strand replication from the first replication event is shown as the *lighter dashed lines*. When the nascent strand from the second origin firing reaches the 3′ end of the first nascent strand, the strand will become displaced. This results in the formation of a branched molecule with free ends (step *6.A*). The DNA ends may participate in recombination leading to expansion (Cromie et al. 2001). Onion-skin amplification (*right*). The DNA molecule shown in step *5.B* follows from step *4*. Onion-skin replication can occur by repeated initiation within the A+T-rich repeat. An eightfold amplification is shown in step *7.B*. When fork movement ceases or slows at the first and second forks, a requirement for amplification, continued replication from the third fork will lead to a displacement of four DNA molecules consisting of pure repeats, (if synthesis is limited to the repeat tracts) (step *8.B*). Pairs of these molecules have complementary single-strand ends that can drive hybridization into longer repeat tracts. These can then be joined by homologous recombination into even longer repeat tracts (step *9.B*). These molecules can become integrated into the repeat tracts in the original chromosome, leading to massive expansion (step *10.B*). The length of the repeat expansion would be dependent on the number of cycles of amplification. This model alone can easily explain very large repeat expansion using a well documented biological phenomenon. Although this model was described for the A+T-rich SCA10 repeat, other DNA repeats may possibly act in a similar fashion

the DNA, allowing for the assembly of replication proteins (Gilbert 2004). The fact that unstable disease-associated repeats are associated with transcriptionally active genes suggests that transcription may reflect a significant *cis*-acting factor for repeat expansion driving aberrant replication initiation events. Alternatively, supercoiling-induced structures may be recognized as distortions by proteins involved in DNA repair (e.g., RPA, XPA, XPC, MSH2) (Pearson et al. 1997; Patrick and Turchi 1999; Wakasugi and Sancar 1999; Volker et al. 2001; Panigrahi et al. 2005; Owen et al. 2005), and may generate a 3′– OH primer by strand breakage or enzymatic nicking. Repetitive rounds of slipped mispairing during replication could then lead to repeat amplification and recombination (Cromie et al. 2001). These events may occur even more frequently during early embryogenesis or gametogenesis, where chromatin structure and replication differ from that in somatic cells, and dynamic epigenetic modifications are occurring (Fuentes-Mascorro et al. 2000; Santos et al. 2005).

Aberrant replication initiation could also be responsible for the instability observed in somatic cells. Unrestrained superhelical tension measured at active genes in living cells is sufficient to support DNA unwinding (Ljungman and Hanawalt 1992; Kramer and Sinden 1997; Kramer et al. 1999). The easily unwound pentanucleotide repeat sequence $d(ATTCT)_n \cdot d(AGAAT)_n$ is located at the transcribed SCA10 locus, and plasmids containing $d(ATTCT)_n \cdot d(AGAAT)_n$ repeats supported initiation and replication in HeLa cell extracts without the addition of a specific initiation protein (Potaman et al. 2003). In cells, unwinding of $d(ATTCT)_n \cdot d(AGAAT)_n$ repeats may support repetitive initiation of DNA replication and amplification of the repeat tract. If $d(ATTCT)_n \cdot d(AGAAT)_n$ acts as a replication origin, fractious DNA replication and amplification could lead to repeat expansion as shown in Fig. 8.

4.8 Influence of the Direction of Replication, Origin Proximity, Origin Activity, and Transcription on Repeat Instability

A minimal repeat length (usually more than 30 repeats) of defined purity is a critical factor for repeat expansion. Mutations in many genes can influence repeat expansion or contractions; however, deficiencies in replication, repair, or recombination functions are not required, a priori, for repeat instability. In one instance, a d(CTG) · d(CAG) repeat integrated at a specific site in one mouse showed many different rates of instability in different cell types, unrelated to the state of cell proliferation. In addition, when cells from different tissues from this mouse were cultured the propensity for instability persisted (Gomes-Pereira et al. 2001). In contrast, cells from other mice with repeats integrated at different locations did not show this variation (Fortune et al. 2000). Complex *cis*- and *trans*-acting factors effecting these differences are only beginning to be revealed.

Given the clear role for replication in repeat instability, as evidenced by differences in stability as a function of orientation with respect to the direction of replication (Kang et al. 1995a; Maurer et al. 1996; Freudenreich et al. 1997; Miret et al. 1997; Hashem et al. 2002; Cleary et al. 2002; Panigrahi et al. 2002), and a role for transcription (Bowater et al. 1997; Mochmann and Wells 2004), the distance and orientation of repeats with respect to replication origins might be a critical *cis*-acting factor in repeat instability. In mammalian cells and yeast genetically defined replication control regions, or replicators, overlap with biochemically defined replication initiation zones, or origins (DePamphilis 2003; Schwob 2004; Gilbert 2004; Aladjem and Fanning 2004). Though generally more expansive and less well-defined in terms of structural and functional modules than those of their yeast counterparts, several mammalian replicators have been identified (Dijkwel et al. 1991; Little et al. 1993; Aladjem et al. 1995; Kobayashi et al. 1998; Liu et al. 2003; Aladjem 2004; Paixao et al. 2004). The initiation zone neighboring the hamster *DHFR* gene encompasses more than 55 kb of DNA comprising multiple start sites firing with different efficiency in a cell population. Replication also initiates at multiple sites within the human endogenous ribosomal RNA, β-globin, and c-myc origins (Little et al. 1993; Malott and Leffak 1999; Liu et al. 2003; Aladjem 2004), and a zone of initiation accompanies translocation of the DHFR, β-globin and c-myc replicator elements to ectopic sites (Malott and Leffak 1999; Altman and Fanning 2001; Liu et al. 2003; Aladjem 2004). Origin specification results from the poorly understood interplay of sequence-directed DNA structures, histone and non-histone protein binding, and epigenetic modification of chromatin (Anglana et al. 2003; Debatisse et al. 2004; Schwob 2004; Gilbert 2004; Danis et al. 2004). Differential origin specification in murine cells provides one explanation for the variable repeat instability observed in mice and cultured cells (Fortune et al. 2000; Gomes-Pereira et al. 2001).

Numerous recent reviews have discussed the *cis* effects of origin proximity and the direction of replication of the repeat tracts on repeat instability (Mirkin and Smirnova 2002; Mirkin 2004, 2005; Cleary and Pearson 2005). Because the location and utilization of initiation sites within human origins can vary, the strand of the repeat tract that constitutes the leading or lagging strand can also vary. This has been termed the "ori switch" model (Mirkin and Smirnova 2002). This is shown in Fig. 9, panel A, as alternative directions of fork progression, defined by the site of replication initiation within an origin. Opposite polarities of replication are analogous to reversing the orientation of repeats in bacteria or yeast, with respect to their defined origins. DNA repeats that exhibit differential DNA secondary structure stabilities [e.g., $d(CAG)_n \cdot d(CTG)_n$ or $d(CCTG)_n \cdot d(CAGG)_n$], may behave in this way because structures may form in the lagging strand, given initiation from one side of the repeat (Fig. 9, panel A, 1), but not in the leading strand, given replication from the other direction (Fig. 9, panel A, 3). As discussed already, repeat instability varies depending on the direction of replication in virtually all experimental systems examined. In addition, the distance is variable between a DNA repeat and alternate potential initiation sites within an origin zone (Fig. 9, panel A, 2). This has been termed the "ori shift" model (Mirkin and Smirnova 2002). Using the SV40 viral replication origin system, Cleary et al. (2002) showed that expansions were favored when replication initiated 103 bp 3′ of a $d(CTG)_{79}$ tract, but that deletions predominated when initiation occurred 230 or 536 bp away. To complicate matters, both deletions and duplications were observed when initiation took place 667 bp away.

Several factors may be important for the ori shift (Mirkin and Smirnova 2002), also called the "fork shift" (Cleary and Pearson 2005), model. First, the neighborhood of the viral SV40 origin could be unusual in that the DNA structure, torsional stress, chromatin organization, and amount of single-stranded DNA may be unusual or unique directly adjacent to the site of replication initiation (termed replication initiation site in Fig. 9, panel A). Whether the SV40 viral replication origin is representative of the more complex human origins is not known. Unlike human replication forks, the SV40 replication fork contains T-antigen, the initiator protein and a potent replicative helicase (Borowiec et al. 1990). It is not known whether polymerases working in conjunction with endogenous human helicases will act differently during unwinding of DNA repeats. In vitro evidence indicates that the SV40 replication fork does not require ORC-dependent prereplicative complex formation, minichromosome maintenance, Cdc45, ATR proteins, or other factors that may assist in replicating alternative DNA structures (Waga and Stillman 1998). Moreover, the exact positioning of the 3′ or the 5′ end of the Okazaki fragment within the repeat tract and the length of the repeat tract may have significant consequences for repeat instability, as described by Richards and Sutherland (1994). As evident from Fig. 5, a hairpin flap would have a good opportunity to form if the 5′ end of the Okazaki fragment began within the repeat tract (Fig. 9, panel B, model 1, fragment set 2; model 2, fragment set 2,3). This is because the opportunity for DNA secondary structure formation at the 5′ end is greater than at the 3′ end, which is bound by the polymerase. Another factor is the length of the repeat tract with respect to the size of the Okazaki fragment. As the length of the repeat tract becomes longer than the length of the Okazaki fragment (approximately 140 nt), the number of nicks to be ligated increases, and this may increase the probability of structure formation and repeat instability (Richards and Sutherland 1994) (Fig. 9, panel B). The sequences of repeats may have important consequences for Okazaki fragment initiation given preferred sites for RNA synthesis (Cleary and Pearson 2005). In many DNA repeats, only one strand might easily support generation of primers by RNA primase. Thus, depending on the direction of replication, the forks could become unbalanced with the generation of an unusually long tract of single-stranded DNA in the lagging template strand. In summary, experimental evidence suggests that in addition to repeat sequence and length, the spatial relationship between

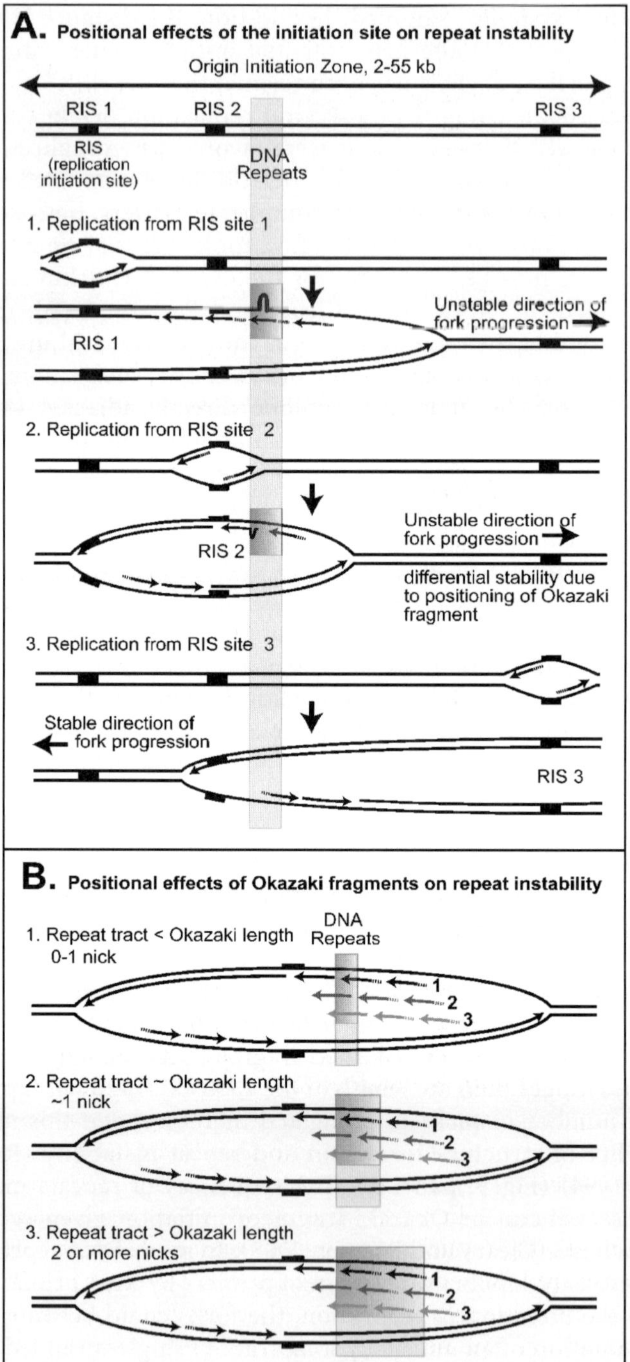

A. Positional effects of the initiation site on repeat instability
Origin Initiation Zone, 2-55 kb
RIS 1
RIS 2
RIS 3
RIS (replication initiation site)
DNA Repeats
1. Replication from RIS site 1
Unstable direction of fork progression
RIS 1
2. Replication from RIS site 2
Unstable direction of fork progression
RIS 2
differential stability due to positioning of Okazaki fragment
3. Replication from RIS site 3
Stable direction of fork progression
RIS 3
B. Positional effects of Okazaki fragments on repeat instability
DNA Repeats
1. Repeat tract < Okazaki length 0-1 nick
1
2
3
2. Repeat tract ~ Okazaki length ~1 nick
1
2
3
3. Repeat tract > Okazaki length 2 or more nicks
1
2
3

◀ **Fig. 9** *Cis* effects of replication on repeat instability: location and proximity of the origin and positioning of Okazaki fragments. In metazoan cells replication origins encompass regions ranging from about 2 kb to as much as 55 kb DNA. Within this region the initiation DNA synthesis requires unwinding of the DNA and binding of helicases and polymerase α/primase to lay down the RNA primer for extension by DNA polymerases. This occurs at multiple sites within the origin where the selection of specific sites for initiation may be a stochastic process. Site utilization may be different in different cell types. This variation can influence repeat instability. *A* Positional effects of the replication initiation site (*RIS*) on repeat instability. Replication is shown starting from *RIS 1*, *RIS 2*, or *RIS 3*, in parts *1*, *2*, and *3*, respectively. The DNA repeat tract is denoted by the *shaded section*; an unstable situation is denoted by the *gradient of shading* over the unstable strand. See text for details. *B* Positional effects of Okazaki fragments on repeat instability. The relative localization of an Okazaki fragment can vary with respect to a DNA repeat tract, as shown. Model *1* shows three different positions of Okazaki fragments across a repeat tract, where the length of the tract is shorter than the Okazaki fragment. In set *1*, the 3′ end of the middle fragment is positioned within the repeat. In set *2*, the 5′ end of the leftmost fragment is positioned within the repeat, and in set *3*, the Okazaki fragment straddles the repeat. It is not known if in a cell population only one set or multiple sets of positions will occur. Nevertheless, the number of nicks that need to be ligated within a repeat tract would range from 0 to 1. As the repeat tract lengthens the probability of nicks falling within the repeat tract increases as shown in models *2* and *3*

a DNA repeat and its origin of DNA replication may be critically important in determining repeat instability. Understanding all the factors that govern instability will require additional investigation.

5
Concluding Remarks:
Mutation Mechanisms, DNA Repeats, and Human Disease— Where Have We Come in 15 Years?

An appreciation for mutations associated with DNA repeats and the impact on human health (Cooper and Krawczak 1993) predates the excitement over the massive expansion associated with many neurodegenerative diseases. The mutations associated with many diseases caused by small changes in repeat length can be easily explained by replication slippage or, in the cases of polyalanine diseases, recombination. Both are classic, long-known mutation mechanisms (Drake 1970). Not unexpected are additional repeat destabilizing effects of mutations in genes involved in DNA replication, repair, and recombination. Though much is known, a remarkable and exciting question remains unanswered: how does a repeat expand to lengths of 1000 to 11 000 copies from an initial length of 100 copies or less during a single intergenerational transmission? The field may be only slightly closer to understanding this question now, compared with 15 years ago. Early reviews discussed many of the same models for repeat instability presented here, before there was supportive experimental data. While expandable human repeats are unstable in mice, their behavior does not accurately recapitulate the patterns of intergenerational transmission seen in humans. Human gametogenesis and early embryogenesis, where expansive instability may occur, are simply not tractable experimental systems. In the absence of a model experimental system that recapitulates intergenerational massive expansion, progress in understanding mechanisms for massive expansion may be slow.

While mysteries remain, much has been learned. One important take-home lesson is that the standard model systems, bacteriophage, bacteria, yeast, mice, and human cells, exhibit different and variable responses to long repeats. Human cells can maintain thousands of d(CGG) · d(CCG) and d(CTG) · d(CAG) repeats in quite stable fashion, while showing greater instability with d(ATTCT) · d(AGAAT) and d(GAA) · d(TTC) repeats. Bacterial cells, on the other hand, have great difficulty maintaining several hundred repeats. Different experimental systems exhibit different patterns of repeat instability, and conclusions learned from one system may not always apply to another. One must also keep in mind that during intergenerational transmission in humans a unique "window of opportunity" must exist for expansion from about 100 to thousands of repeats, and that once that window has closed, the repeats become complacent, so to speak. Either the window is missing in other systems, or it cannot be pried open. Therefore, we must continue to utilize model systems, keeping in mind the limitations and implications of each, with the ultimate goal of understanding processes that explain expansion.

A second take-home lesson is that all disease-expanding repeats are unique with their own personalities and peculiarities in terms of alternative DNA conformations (Table 1). Moreover, there is not a simple feature that correlates with expansion. In addition, different repeats can behave differently in a model system. For example, d(CTG) · d(CAG) repeats associated with DM1 or $d(CGG)_n \cdot d(CCG)_n$ repeats associated with fragile X syndrome can form slipped mispaired structures (Pearson and Sinden 1996) that may block replication, and they undergo rapid deletion in *E. coli* (Kang et al. 1995a; Bowater et al. 1996; Ohshima et al. 1996a; Hashem et al. 2002). Conversely, SCA10 d(ATTCT) · d(AGAAT) repeats do not form a structure that can block replication, but rather may support replication in human cells, and they are quite stable in *E. coli* at lengths at which d(CTG) or d(CGG) repeats are very unstable. The point to be made here is that a single pathway for massive expansion may not exist, although it cannot be presently excluded. Likewise, multiple pathways exist for the small changes in repeat length observed in somatic cells throughout life. Alternative DNA conformations associated with certain repeats are probably very important for repeat instability in some pathways; however, they may be less important for other repeats.

In summary, repeat instability remains a major problem for human health and no simple mechanism or biochemical pathway may direct massive expansion for all repeats. Moreover, given the interdependence of replication, repair, and recombination, under the global regulation and coordination of checkpoint control, many players and pathways will be expected to have an influence on repeat instability. Maybe in the last 15 years we have learned enough to know where to begin to address in new ways this complex biological phenomenon.

An additional goal is to learn how to manipulate repeat length in a therapeutic fashion to delay or prevent disease-causing expansion, or to reverse the expansion process, preventing or alleviating the genetic source of the problem. Initial investigations related to this question have recently been described (Gorbunova et al. 2003; Yang et al. 2003; Pineiro et al. 2003; Gomes-Pereira and Monckton 2004a; Hashem et al. 2004a).

Acknowledgements We thank Albino Bacolla, John J. Bissler, Sharon F. Edwards, and Michael Leffak for critically reading the manuscript. We thank Alan Maness for assistance in preparation of the figures. Work in the authors' laboratories is supported by NIH grants ES05508 to R.R.S. and CA 74175 to V.N.P.

References

Aladjem MI (2004) The mammalian beta globin origin of DNA replication. Front Biosci 9:2540–2547

Aladjem MI, Fanning E (2004) The replicon revisited: an old model learns new tricks in metazoan chromosomes. EMBO Rep 5:686–691

Aladjem MI, Groudine M, Brody LL, Dieken ES, Fournier RE, Wahl GM, Epner EM (1995) Participation of the human beta-globin locus control region in initiation of DNA replication. Science 270:815–819

Altman AL, Fanning E (2001) The Chinese hamster dihydrofolate reductase replication origin beta is active at multiple ectopic chromosomal locations and requires specific DNA sequence elements for activity. Mol Cell Biol 21:1098–1110

Anglana M, Apiou F, Bensimon A, Debatisse M (2003) Dynamics of DNA replication in mammalian somatic cells: nucleotide pool modulates origin choice and interorigin spacing. Cell 114:385–394

Backman K, Betlach M, Boyer HW, Yanofsky S (1978) Genetic and physical studies on the replication of ColE1-type plasmids. Cold Spring Harbor Symp Quant Biol 43:69–76

Baran N, Neer A, Manor H (1983) Onion skin replication of integrated polyoma virus DNA and flanking sequences in polyoma-transformed rat cells: termination within a specific cellular DNA segment. Proc Natl Acad Sci USA 80:105–109

Baran N, Lapidot A, Manor H (1987) Unusual sequence element found at the end of an amplicon. Mol Cell Biol 7:2636–2640

Baran N, Lapidot A, Manor H (1991) Formation of DNA triplexes accounts for arrests of DNA synthesis at $d(TC)_n$ and $d(GA)_n$ tracts. Proc Natl Acad Sci USA 88:507–511

Baran N, Pucshansky L, Marco Y, Benjamin S, Manor H (1997) The SV40 large T-antigen helicase can unwind four stranded DNA structures linked by G-quartets. Nucleic Acids Res 25:297–303

Bebenek K, Kunkel TA (2000) Streisinger revisited: DNA synthesis errors mediated by substrate misalignments. Cold Spring Harbor Symp Quant Biol 65:81–91

Bedinger P, Munn M, Alberts BM (1989) Sequence-specific pausing during in vitro DNA replication on double-stranded DNA templates. J Biol Chem 264:16880–16886

Bhattacharyya S, Lahue RS (2004) *Saccharomyces cerevisiae* Srs2 DNA helicase selectively blocks expansions of trinucleotide repeats. Mol Cell Biol 24:7324–7330

Bidichandani SI, Ashizawa T, Patel PI (1998) The GAA triplet-repeat expansion in Friedreich ataxia interferes with transcription and may be associated with an unusual DNA structure. Am J Hum Genet 62:111–121

Bidichandani SI, Purandare SM, Taylor EE, Gumin G, Machkhas H, Harati Y, Gibbs RA, Ashizawa T, Patel PI (1999) Somatic sequence variation at the Friedreich ataxia locus includes complete contraction of the expanded GAA triplet repeat, significant length variation in serially passaged lymphoblasts and enhanced mutagenesis in the flanking sequence. Hum Mol Genet 8:2425–2436

Bode J, Kohwi Y, Dickinson L, Joh T, Klehr D, Mielke C, Kohwi-Shigematsu T (1992) Biological significance of unwinding capability of nuclear matrix associating DNAs. Science 255:195–197

Bois PR, Southgate L, Jeffreys AJ (2001) Length of uninterrupted repeats determines instability at the unstable mouse expanded simple tandem repeat family MMS10 derived from independent SINE B1 elements. Mamm Genome 12:104–111

Borowiec JA, Dean FB, Bullock PA, Hurwitz J (1990) Binding and unwinding-How T antigen engages the SV40 origin of DNA replication. Cell 60:181–184

Bowater RP, Jaworski A, Larson JE, Parniewski P, Wells RD (1997) Transcription increases the deletion frequency of long CTG · CAG triplet repeats from plasmids in *Escherichia coli*. Nucleic Acids Res 25:2861–2868

Bowater RP, Rosche WA, Jaworski A, Sinden RR, Wells RD (1996) Relationship between *Escherichia coli* growth and deletions of CTG · CAG triplet repeats in plasmids. J Mol Biol 264:82–96

Brook JD, McCurrach ME, Harley HG, Buckler AJ, Church D, Aburatani H, Hunter K, Stanton VP, Thirion JP, Hudson T, Sohn R, Zemelman B, Snell RG, Rundle SA, Crow S, Davies J, Shelbourne P, Buxton J, Jones C, Juvonen V, Johnson K, Harper PS, Shaw DJ, Housman DE (1992) Molecular basis of myotonic dystrophy: expansion of a trinucleotide (CTG) repeat at the 3′ end of a transcript encoding a protein kinase family member. Cell 68:799–808

Brown LY, Brown SA (2004) Alanine tracts: the expanding story of human illness and trinucleotide repeats. Trends Genet 20:51–58

Brown WT, Houck GE Jr, Ding X, Zhong N, Nolin S, Glicksman A, Dobkin C, Jenkins EC (1996) Reverse mutations in the fragile X syndrome. Am J Med Genet 64:287–292

Brunner HG, Jansen G, Nillesen W, Nelen MR, de Die CE, Howeler CJ, van Oost BA, Wieringa B, Ropers HH, Smeets HJ (1993) Brief report: reverse mutation in myotonic dystrophy. N Engl J Med 328:476–480

Callahan JL, Andrews KJ, Zakian VA, Freudenreich CH (2003) Mutations in yeast replication proteins that increase CAG/CTG expansions also increase repeat fragility. Mol Cell Biol 23:7849–7860

Calvi BR, Spradling AC (1999) Chorion gene amplification in Drosophila: A model for metazoan origins of DNA replication and S-phase control. Methods 18:407–417

Casper JM, Kemp MG, Ghosh M, Randall GM, Vaillant A, Leffak M (2005) The c-myc DNA unwinding element binding protein modulates the assembly of DNA replication complexes in vitro. J Biol Chem 280:13071–13083

Cha RS, Kleckner N (2002) ATR homolog Mec1 promotes fork progression, thus averting breaks in replication slow zones. Science 297:602–606

Chou SH, Chin KH, Wang AH (2003) Unusual DNA duplex and hairpin motifs. Nucleic Acids Res 31:2461–2474
Cleary JD, Nichol K, Wang YH, Pearson CE (2002) Evidence of cis-acting factors in replication-mediated trinucleotide repeat instability in primate cells. Nat Genet 31:37–46
Cleary JD, Pearson CE (2003) The contribution of CIS-elements to disease-associated repeat instability: clinical and experimental evidence. Cytogenet Genome Res 100:25–55
Cleary JD, Pearson CE (2005) Replication fork dynamics and dynamic mutations: the fork-shift model of repeat instability. Trends Genet 21:272–280
Cooper DN, Krawczak M (1993) Human Gene Mutation. Bios Scientific Publishers, Oxford, UK
Cox MM (2001) Recombinational DNA repair of damaged replication forks in Escherichia coli: questions. Annu Rev Genet 35:53–82
Cox MM, Goodman MF, Kreuzer KN, Sherratt DJ, Sandler SJ, Marians KJ (2000) The importance of repairing stalled replication forks. Nature 404:37–41
Cromie GA, Connelly JC, Leach DR (2001) Recombination at double-strand breaks and DNA ends: conserved mechanisms from phage to humans. Mol Cell 8:1163–1174
Danis E, Brodolin K, Menut S, Maiorano D, Girard-Reydet C, Mechali M (2004) Specification of a DNA replication origin by a transcription complex. Nat Cell Biol 6:721–730
Darlow JM, Leach DRF (1995) The effects of trinucleotide repeats found in human inherited disorders on palindrome inviability in *Escherichia coli* suggest hairpin folding preferences in vivo. Genetics 141:825–832
Darlow JM, Leach DRF (1998a) Secondary structures in d(CGG) and d(CCG) repeat tracts. J Mol Biol 275:3–16
Darlow JM, Leach DF (1998b) Evidence for two preferred hairpin folding patterns in d(CGG) · d(CCG) repeat tracts in vivo. J Mol Biol 275:17–23
Dayn A, Samadashwily GM, Mirkin SM (1992) Intramolecular DNA triplexes: Unusual sequence requirements and influence on DNA polymerization. Proc Natl Acad Sci USA 89:11406–11410
Debatisse M, Toledo F, Anglana M (2004) Replication initiation in mammalian cells: changing preferences. Cell Cycle 3:19–21
DePamphilis ML (2003) Eukaryotic DNA replication origins: reconciling disparate data. Cell 114:274–275
Dere R, Napierala M, Ranum LP, Wells RD (2004) Hairpin structure-forming propensity of the (CCTG · CAGG) tetranucleotide repeats contributes to the genetic instability associated with myotonic dystrophy type 2. J Biol Chem 279:41715–41726
Dijkwel PA, Vaughn JP, Hamlin JL (1991) Mapping of replication initiation sites in mammalian genomes by two-dimensional gel analysis: stabilization and enrichment of replication intermediates by isolation on the nuclear matrix. Mol Cell Biol 11:3850–3859
Drake JW (1970) The molecular basis of mutation. Holden-Day, San Francisco
Drake JW (1999) The distribution of rates of spontaneous mutation over viruses, prokaryotes, and eukaryotes. Ann N Y Acad Sci 870:100–107
Drake JW (1991a) Spontaneous mutation. Annu Rev Genet 25:125–146
Drake JW (1991b) A constant rate of spontaneous mutation in DNA-based microbes. Proc Natl Acad Sci USA 88:7160–7164
Drake JW, Glickman BW, Ripley LS (1983) Updating the theory of mutation. Am Sci 71:621–630
Fojtik P, Kejnovska I, Vorlickova M (2004) The guanine-rich fragile X chromosome repeats are reluctant to form tetraplexes. Nucleic Acids Res 32:298–306

Fortune MT, Vassilopoulos C, Coolbaugh MI, Siciliano MJ, Monckton DG (2000) Dramatic, expansion-biased, age-dependent, tissue-specific somatic mosaicism in a transgenic mouse model of triplet repeat instability. Hum Mol Genet 9:439–445

Frank-Kamenetskii MD, Mirkin SM (1995) Triplex DNA structures. Annu Rev Biochem 64:65–95

Freudenreich CH, Kantrow SM, Zakian VA (1998) Expansion and length-dependent fragility of CTG repeats in yeast. Science 279:853–856

Freudenreich CH, Stavenhagen JB, Zakian VA (1997) Stability of a CTG/CAG trinucleotide repeat in yeast is dependent on its orientation in the genome. Mol Cell Biol 17:2090–2098

Fry M, Loeb LA (1994) The fragile X syndrome $d(CGG)_n$ nucleotide repeats form a stable tetrahelical structure. Proc Natl Acad Sci USA 91:4950–4954

Fry M, Loeb LA (1999) Human Werner syndrome DNA helicase unwinds tetrahelical structures of the fragile X syndrome repeat sequence $d(CGG)_n$. J Biol Chem 274:12797–12803

Fu YH, Pizzuti A, Fenwick RG Jr, King J, Rajnarayan S, Dunne PW, Dubel J, Nasser GA, Ashizawa T, DeJong P, Wieringa B, Korneluk R, Perryman MB, Epstein HF, Caskey CT (1992) An unstable triplet repeat in a gene related to myotonic muscular dystrophy. Science 255:1256–1258

Fu Y-H, Kuhl DPA, Pizzuti A, Pieretti M, Sutcliffe JS, Richards S, Verkerk AJMH, Holden JJA, Fenwick JRG, Warren ST, Oostra BA, Nelson DL, Caskey CT (1991) Variation of the CGG repeat at the Fragile X site results in genetic instability: resolution of the Sherman paradox. Cell 67:1047–1058

Fuentes-Mascorro G, Serrano H, Rosado A (2000) Sperm chromatin. Arch Androl 45:215–225

Gacy AM, McMurray CT (1998) Influence of hairpins on template reannealing at trinucleotide repeat duplexes: a model for slipped DNA. Biochemistry 37:9426–9434

Gacy AM, Goellner GM, Spiro C, Chen X, Gupta G, Bradbury EM, Dyer RB, Mikesell MJ, Yao JZ, Johnson AJ, Richter A, Melancon SB, McMurray CT (1998) GAA instability in Friedreich's Ataxia shares a common, DNA-directed and intraallelic mechanism with other trinucleotide diseases. Mol Cell 1:583–593

Gao X, Huang X, Smith GK, Zheng M, Liu H (1995) A new antiparallel duplex motif of DNA CCG repeats that is stabilized by extrahelical bases symetrically localized in the minor groove. J Am Chem Soc 117:8883–8884

Ghosh M, Liu G, Randall G, Bevington J, Leffak M (2004) Transcription factor binding and induced transcription alter chromosomal c-myc replicator activity. Mol Cell Biol 24:10193–10207

Gilbert DM (2004) In search of the holy replicator. Nat Rev Mol Cell Biol 5:848–855

Gomes-Pereira M, Monckton DG (2004a) Chemically induced increases and decreases in the rate of expansion of a CAG · CTG triplet repeat. Nucleic Acids Res 32:2865–2872

Gomes-Pereira M, Monckton DG (2004b) Mouse tissue culture models of unstable triplet repeats. Methods Mol Biol 277:215–228

Gomes-Pereira M, Fortune MT, Monckton DG (2001) Mouse tissue culture models of unstable triplet repeats: in vitro selection for larger alleles, mutational expansion bias and tissue specificity, but no association with cell division rates. Hum Mol Genet 10:845–854

Gomes-Pereira M, Fortune MT, Ingram L, McAbney JP, Monckton DG (2004) Pms2 is a genetic enhancer of trinucleotide CAG · CTG repeat somatic mosaicism: implications for the mechanism of triplet repeat expansion. Hum Mol Genet 13:1815–1825

Gorbunova V, Seluanov A, Dion V, Sandor Z, Meservy JL, Wilson JH (2003) Selectable system for monitoring the instability of CTG/CAG triplet repeats in mammalian cells. Mol Cell Biol 23:4485–4493

Gordenin DA, Lobachev KS, Degtyareva NP, Malkova AL, Perkins E, Resnick MA (1993) Inverted DNA repeats: A source of eukaryotic genomic instability. Mol Cell Biol 13:5315–5322

Gordenin DA, Kunkel TA, Resnick MA (1997) Repeat expansion-all in a flap? Nat Genet 16:116–118

Grabczyk E, Usdin K (2000) The GAA*TTC triplet repeat expanded in Friedreich's ataxia impedes transcription elongation by T7 RNA polymerase in a length and supercoil dependent manner. Nucleic Acids Res 28:2815–2822

Hacker KJ, Alberts BM (1994) The rapid dissociation of the T4 DNA polymerase holoenzyme when stopped by a DNA hairpin helix. A model for polymerase release following the termination of each Okazaki fragment. J Biol Chem 269:24221–24228

Hanvey JC, Shimizu M, Wells RD (1988) Intramolecular DNA triplexes in supercoiled plasmids. Proc Natl Acad Sci USA 85:6292–6296

Hartenstine MJ, Goodman MF, Petruska J (2000) Base stacking and even/odd behavior of hairpin loops in DNA triplet repeat slippage and expansion with DNA polymerase. J Biol Chem 275:18382–18390

Hashem VI, Sinden RR (2005) Duplications between direct repeats stabilized by DNA secondary structure occur preferentially in the leading strand during DNA replication. Mutat Res 570:215–226

Hashem VI, Rosche WA, Sinden RR (2002) Genetic assays for measuring rates of (CAG) · (CTG) repeat instability in *Escherichia coli*. Mutat Res 502:25–37

Hashem VI, Pytlos MJ, Klysik EA, Tsuji K, Khajavi M, Ashizawa T, Sinden RR (2004a) Chemotherapeutic deletion of CTG repeats in lymphoblast cells from DM1 patients. Nucleic Acids Res 32:6334–6346

Hashem VI, Rosche WA, Sinden RR (2004b) Genetic recombination destabilizes $(CTG)_n \cdot (CAG)_n$ repeats in *E. coli*. Mutat Res 554:95–109

Hebert ML, Spitz LA, Wells RD (2004) DNA double-strand breaks induce deletion of CTG · CAG repeats in an orientation-dependent manner in *Escherichia coli*. J Mol Biol 336:655–672

Heidenfelder BL, Topal MD (2003) Effects of sequence on repeat expansion during DNA replication. Nucleic Acids Res 31:7159–7164

Heidenfelder BL, Makhov AM, Topal MD (2003) Hairpin formation in Friedreich's ataxia triplet repeat expansion. J Biol Chem 278:2425–2431

Hellman A, Zlotorynski E, Scherer S, Cheung J, Vincent J, Smith DI, Trakhtenbrot L, Kerem B (2002) A role for common fragile site induction in amplification of human oncogenes. Cancer Cell 1:89–97

Henricksen LA, Veeraraghavan J, Chafin DR, Bambara RA (2002) DNA ligase I competes with FEN1 to expand repetitive DNA sequences in vitro. J Biol Chem 277:22361–22369

Huber MD, Lee DC, Maizels N (2002) G4 DNA unwinding by BLM and Sgs1p: substrate specificity and substrate-specific inhibition. Nucleic Acids Res 30:3954–3961

Ireland MJ, Reinke SS, Livingston DM (2000) The impact of lagging strand replication mutations on the stability of CAG repeat tracts in yeast. Genetics 155:1657–1665

Jakupciak JP, Wells RD (1999) Genetic instabilities in (CTG) · (CAG) repeats occur by recombination. J Biol Chem 274:23468–23479

Jakupciak JP, Wells RD (2000a) Gene conversion (recombination) mediates expansions of CTG · CAG repeats. J Biol Chem 275:40003–40013

Jakupciak JP, Wells RD (2000b) Genetic instabilities of triplet repeat sequences by recombination. IUBMB Life 50:355–359

Jankowski C, Nag DK (2002) Most meiotic CAG repeat tract-length alterations in yeast are SPO11 dependent. Mol Genet Genomics 267:64–70

Jankowski C, Nasar F, Nag DK (2000) Meiotic instability of CAG repeat tracts occurs by double-strand break repair in yeast. Proc Natl Acad Sci USA 97:2134–2139

Jenke AC, Stehle IM, Herrmann F, Eisenberger T, Baiker A, Bode J, Fackelmayer FO, Lipps HJ (2004) Nuclear scaffold/matrix attached region modules linked to a transcription unit are sufficient for replication and maintenance of a mammalian episome. Proc Natl Acad Sci USA 101:11322–11327

Kaguni LS, Clayton DA (1982) Template-directed pausing in in vitro DNA synthesis by DNA polymerase a from *Drosophila melanogaster* embryos. Proc Natl Acad Sci USA 79:983–987

Kamath-Loeb AS, Loeb LA, Johansson E, Burgers PM, Fry M (2001) Interactions between the Werner syndrome helicase and DNA polymerase delta specifically facilitate copying of tetraplex and hairpin structures of the d(CGG)n trinucleotide repeat sequence. J Biol Chem 276:16439–16446

Kang S, Jaworski A, Ohshima K, Wells RD (1995a) Expansion and deletion of CTG triplet repeats from human disease genes are determined by the direction of replication in *E. coli*. Nat Genet 10:213–218

Kang S, Ohshima K, Shimizu M, Amirhaeri S, Wells RD (1995b) Pausing of DNA synthesis in vitro at specific loci in CTG and CGG triplet repeats from human hereditary disease genes. J Biol Chem 270:27014–27021

Kastan MB, Bartek J (2004) Cell-cycle checkpoints and cancer. Nature 432:316–323

Khateb S, Weisman-Shomer P, Hershco I, Loeb LA, Fry M (2004) Destabilization of tetraplex structures of the fragile X repeat sequence (CGG)n is mediated by homolog-conserved domains in three members of the hnRNP family. Nucleic Acids Res 32:4145–4154

Kim S-H, Pytlos MJ, Sinden RR (2006) Replication restart: A major pathway for (CTG) · (CAG) repeat deletion in *Escherichia coli*. Mut Res 595:5–22

Kirschner LS (1996) De novo generation of simple sequence during gene amplification. Nucleic Acids Res 24:2829–2834

Klarmann GJ, Schauber CA, Preston BD (1993) Template-directed pausing of DNA synthesis by HIV-1 reverse transcriptase during polymerization of HIV-1 sequences in vitro. J Biol Chem 268:9793–9802

Kobayashi S, Valentine MR, Pham P, O'Donnell M, Goodman MF (2002) Fidelity of *Escherichia coli* DNA polymerase IV. Preferential generation of small deletion mutations by dNTP-stabilized misalignment. J Biol Chem 277:34198–34207

Kobayashi T, Rein T, DePamphilis ML (1998) Identification of primary initiation sites for DNA replication in the hamster dihydrofolate reductase gene initiation zone. Mol Cell Biol 18:3266–3277

Kopel V, Pozner A, Baran N, Manor H (1996) Unwinding of the third strand of a DNA triple helix, a novel activity of the SV40 large T-antigen helicase. Nucleic Acids Res 24:330–335

Kornberg A, Bertsch LL, Jackson JF, Khorana HG (1964) Enzymatic synthesis of deoxyribonucleic acid, XVI. Oligonucleotides as templates and the mechanism of their replication. Proc Natl Acad Sci USA 51:315–323

Kouzine F, Liu J, Sanford S, Chung HJ, Levens D (2004) The dynamic response of upstream DNA to transcription-generated torsional stress. Nat Struct Mol Biol 11:1092–1100

Kovtun IV, McMurray CT (2001) Trinucleotide expansion in haploid germ cells by gap repair. Nat Genet 27:407–411

Kovtun IV, Thornhill AR, McMurray CT (2004) Somatic deletion events occur during early embryonic development and modify the extent of CAG expansion in subsequent generations. Hum Mol Genet 13:3057–3068

Kowalski D, Natale DA, Eddy MJ (1988) Stable DNA unwinding, not breathing, accounts for single-strand-specific nuclease hypersensitivity of specific A+T-rich sequences. Proc Natl Acad Sci USA 85:9464–9468

Krahe R, Eckhart M, Ogunniyi AO, Osuntokun BO, Siciliano MJ, Ashizawa T (1995) De novo myotonic dystrophy mutation in a Nigerian kindred. Am J Hum Genet 56:1067–1074

Kramer PR, Sinden RR (1997) Measurement of unrestrained negative supercoiling and topological domain size in living human cells. Biochemistry 36:3151–3158

Kramer PR, Fragoso G, Pennie W, Htun H, Hager GL, Sinden RR (1999) Transcriptional state of the mouse mammary tumor virus promoter can affect topological domain size in vivo. J Biol Chem 274:28590–28597

Krasilnikov AS, Panyutin IG, Samadashwily GM, Cox R, Lazurkin YS, Mirkin SM (1997) Mechanisms of triplex-caused polymerization arrest. Nucleic Acids Res 25:1339–1346

Krasilnikova MM, Mirkin SM (2004) Replication stalling at Friedreich's ataxia $(GAA)_n$ repeats in vivo. Mol Cell Biol 24:2286–2295

Kremer EJ, Pritchard M, Lynch M, Yu S, Holman K, Baker E, Warren ST, Schlessinger D, Sutherland GR, Richards RI (1991) Mapping of DNA instability at the fragile X to a trinucleotide repeat sequence $p(CCG)_n$. Science 252:1711–1714

Kroutil LC, Kunkel TA (1999) Deletion errors generated during replication of CAG repeats. Nucleic Acids Res 27:3481–3486

Kunkel TA (1990) Misalignment-mediated DNA synthesis errors. Biochemistry 29:8003–8011

Kunkel TA, Erie DA (2005) DNA Mismatch Repair. Annu Rev Biochem 74:681–710

Kunkel TA, Soni A (1988) Mutagenesis by transient misalignment. J Biol Chem 263:14784–14789

Lahiri M, Gustafson TL, Majors ER, Freudenreich CH (2004) Expanded CAG repeats activate the DNA damage checkpoint pathway. Mol Cell 15:287–293

Lee S, Park MS (2002) Human FEN-1 can process the 5′-flap DNA of CTG/CAG triplet repeat derived from human genetic diseases by length and sequence dependent manner. Exp Mol Med 34:313–317

Lenzmeier BA, Freudenreich CH (2003) Trinucleotide repeat instability: a hairpin curve at the crossroads of replication, recombination, and repair. Cytogenet Genome Res 100:7–24

LeProust EM, Pearson CE, Sinden RR, Gao X (2000) Unexpected formation of parallel duplex in GAA and TTC trinucleotide repeats of Friedreich's ataxia. J Mol Biol 302:1063–1080

Liang C, Spitzer JD, Smith HS, Gerbi SA (1993) Replication initiates at a confined region during DNA amplification in Sciara DNA puff II/9A. Genes Dev 7:1072–1084

Libby RT, Monckton DG, Fu YH, Martinez RA, McAbney JP, Lau R, Einum DD, Nichol K, Ware CB, Ptacek LJ, Pearson CE, La Spada AR (2003) Genomic context drives SCA7 CAG repeat instability, while expressed SCA7 cDNAs are intergenerationally and somatically stable in transgenic mice. Hum Mol Genet 12:41–50

Little RD, Platt TH, Schildkraut CL (1993) Initiation and termination of DNA replication in human rRNA genes. Mol Cell Biol 13:6600–6613

Liu G, Malott M, Leffak M (2003) Multiple functional elements comprise a mammalian chromosomal replicator. Mol Cell Biol 23:1832–1842

Liu Y, Kao HI, Bambara RA (2004a) Flap endonuclease 1: a central component of DNA metabolism. Annu Rev Biochem 73:589–615

Liu Y, Zhang H, Veeraraghavan J, Bambara RA, Freudenreich CH (2004b) *Saccharomyces cerevisiae* flap endonuclease 1 uses flap equilibration to maintain triplet repeat stability. Mol Cell Biol 24:4049–4064

Ljungman M, Hanawalt PC (1992) Localized torsional tension in the DNA of human cells. Proc Natl Acad Sci USA 89:6055–6059

Losekoot M, Hoogendoorn E, Olmer R, Jansen CC, Oosterwijk JC, van den Ouweland AM, Halley DJ, Warren ST, Willemsen R, Oostra BA, Bakker E (1997) Prenatal diagnosis of the fragile X syndrome: loss of mutation owing to a double recombinant or gene conversion event at the FMR1 locus. J Med Genet 34:924–926

MacAlpine DM, Rodriguez HK, Bell SP (2004) Coordination of replication and transcription along a Drosophila chromosome. Genes Dev 18:3094–3105

Mahadevan M, Tsilfidis C, Sabourin L, Shutler G, Amemiya C, Jansen G, Neville C, Narang M, Barcelo J, O'Hoy K, Leblond S, Earle-McDonald J, de Jons PJ, Wieringa B, Korneluk RG (1992) Myotonic dystrophy mutation: An unstable CTG repeat in the 3′ untranslated region of the gene. Science 255:1253–1255

Maine IP, Kodadek T (1994) Efficient unwinding of triplex DNA by a DNA helicase. Biochem Biophys Res Commun 204:1119–1124

Malott M, Leffak M (1999) Activity of the c-myc replicator at an ectopic chromosomal location. Mol Cell Biol 19:5685–5695

Mangiarini L, Sathasivam K, Mahal A, Mott R, Seller M, Bates GP (1997) Instability of highly expanded CAG repeats in mice transgenic for the Huntington's disease mutation. Nat Genet 15:197–200

Manley K, Shirley TL, Flaherty L, Messer A (1999) Msh2 deficiency prevents in vivo somatic instability of the CAG repeat in Huntington disease transgenic mice. Nat Genet 23:471–473

Marcadier JL, Pearson CE (2003) Fidelity of primate cell repair of a double-strand break within a (CTG) · (CAG) tract. Effect of slipped DNA structures. J Biol Chem 278:33848–33856

Marians KJ (2000) PriA-directed replication fork restart in *Escherichia coli*. Trends Biochem Sci 25:185–189

Martorell L, Monckton DG, Gamez J, Johnson KJ, Gich I, de Munain AL, Baiget M (1998) Progression of somatic CTG repeat length heterogeneity in the blood cells of myotonic dystrophy patients. Hum Mol Genet 7:307–312

Matsuura T, Yamagata T, Burgess DL, Rasmussen A, Grewal RP, Watase K, Khajavi M, McCall AE, Davis CF, Zu L, Achari M, Pulst SM, Alonso E, Noebels JL, Nelson DL, Zoghbi HY, Ashizawa T (2000) Large expansion of the ATTCT pentanucleotide repeat in spinocerebellar ataxia type 10. Nat Genet 26:191–194

Matsuura T, Fang P, Lin X, Khajavi M, Tsuji K, Rasmussen A, Grewal RP, Achari M, Alonso ME, Pulst SM, Zoghbi HY, Nelson DL, Roa BB, Ashizawa T (2004) Somatic and germline instability of the ATTCT repeat in spinocerebellar ataxia type 10. Am J Hum Genet 74:1216–1224

Maurer DJ, O'Callaghan BL, Livingston DM (1996) Orientation dependence of trinucleotide CAG repeat instability in *Saccharomyces cerevisiae*. Mol Cell Biol 16:6617–6622

McGlynn P, Lloyd RG (2002) Recombinational repair and restart of damaged replication forks. Nat Rev Mol Cell Biol 3:859–870

McGlynn P, Al-Deib AA, Liu J, Marians KJ, Lloyd RG (1997) The DNA replication protein PriA and the recombination protein RecG bind D-loops. J Mol Biol 270:212–221

McMurray CT, Kortun IV (2003) Repair in haploid male germ cells occurs late in differentiation as chromatin is condensing. Chromosoma 111:505–508

Melo J, Toczyski D (2002) A unified view of the DNA-damage checkpoint. Curr Opin Cell Biol 14:237–245

Meservy JL, Sargent RG, Iyer RR, Chan F, McKenzie GJ, Wells RD, Wilson JH (2003) Long CTG tracts from the myotonic dystrophy gene induce deletions and rearrangements during recombination at the APRT locus in CHO cells. Mol Cell Biol 23:3152–3162

Miret JJ, Pessoabrandao L, Lahue RS (1997) Instability of CAG and CTG trinucleotide repeats in *Saccharomyces cerevisiae.* Mol Cell Biol 17:3382–3387

Miret JJ, Pessoa-Brandao L, Lahue RS (1998) Orientation-dependent and sequence-specific expansions of CTG/CAG trinucleotide repeats in Saccharomyces cerevisiae. Proc Natl Acad Sci USA 95:12438–12443

Mirkin SM (2004) Molecular models for repeat expansion. Chemtracts Biochem Mol Biol 17:639–662

Mirkin SM (2005) Toward a unified theory for repeat expansions. Nat Struct Mol Biol 12:635–637

Mirkin SM, Smirnova EV (2002) Positioned to expand. Nat Genet 31:5–6

Mitas M (1997) Trinucleotide repeats associated with human disease. Nucleic Acids Res 25:2245–2253

Mochmann LH, Wells RD (2004) Transcription influences the types of deletion and expansion products in an orientation-dependent manner from GAC · GTC repeats. Nucleic Acids Res 32:4469–4479

Monckton DG, Coolbaugh MI, Ashizawa KT, Siciliano MJ, Caskey CT (1997) Hypermutable myotonic dystrophy CTG repeats in transgenic mice. Nat Genet 15:193–196

Nadel Y, Weisman-Shomer P, Fry M (1995) The fragile X syndrome single strand $d(CGG)_n$ nucleotide repeats readily fold back to form unimolecular hairpin structures. J Biol Chem 270:28970–28977

Nag DK, Suri M, Stenson EK (2004) Both CAG repeats and inverted DNA repeats stimulate spontaneous unequal sister-chromatid exchange in Saccharomyces cerevisiae. Nucleic Acids Res 32:5677–5684

Napierala M, Parniewski P, Pluciennik A, Wells RD (2002) Long CTG · CAG repeat sequences markedly stimulate intramolecular recombination. J Biol Chem 277:34087–34100

Napierala M, Dere R, Vetcher A, Wells RD (2004) Structure-dependent recombination hot spot activity of GAA · TTC sequences from intron 1 of the Friedreich's ataxia gene. J Biol Chem 279:6444–6454

Nieduszynski CA, Blow JJ, Donaldson AD (2005) The requirement of yeast replication origins for pre-replication complex proteins is modulated by transcription. Nucleic Acids Res 33:2410–2420

Ohshima K, Wells RD (1997) Hairpin formation during DNA synthesis primer realignment in vitro in triplet repeat sequences from human hereditary disease genes. J Biol Chem 272:16798–16806

Ohshima K, Kang S, Larson JE, Wells RD (1996a) Cloning, characterization, and properties of seven triplet repeat DNA sequences. J Biol Chem 271:16773–16783

Ohshima K, Kang S, Larson JE, Wells RD (1996b) TTA · TAA triplet repeats in plasmids form a non-H bonded structure. J Biol Chem 271:16784–16791
Ohshima K, Montermini L, Wells RD, Pandolfo M (1998) Inhibitory effects of expanded GAA*TTC triplet repeats from intron I of the Friedreich ataxia gene on transcription and replication in vivo. J Biol Chem 273:14588–14595
Otto CJ, Almqvist E, Hayden MR, Andrew SE (2001) The flap endonuclease gene FEN1 is excluded as a candidate gene implicated in the CAG repeat expansion underlying Huntington disease. Clin Genet 59:122–127
Oussatcheva EA, Hashem VI, Zou Y, Sinden RR, Potaman VN (2001) Involvement of the nucleotide excision repair protein UvrA in instability of (CAG) · (CTG) repeat sequences in *E. coli*. J Biol Chem 276:30878–30884
Owen BA, Yang Z, Lai M, Gajek M, Badger JD, Hayes JJ, Edelmann W, Kucherlapati R, Wilson TM, McMurray CT (2005) (CAG)(n)-hairpin DNA binds to Msh2-Msh3 and changes properties of mismatch recognition. Nat Struct Mol Biol 12:663–670
Paixao S, Colaluca IN, Cubells M, Peverali FA, Destro A, Giadrossi S, Giacca M, Falaschi A, Riva S, Biamonti G (2004) Modular structure of the human lamin B2 replicator. Mol Cell Biol 24:2958–2967
Panigrahi GB, Cleary JD, Pearson CE (2002) In vitro (CTG) · (CAG) expansions and deletions by human cell extracts. J Biol Chem 277:13926–13934
Panigrahi GB, Lau R, Montgomery SE, Leonard MR, Pearson CE (2005) Slipped (CTG) · (CAG) repeats can be correctly repaired, escape repair or undergo error-prone repair. Nat Struct Mol Biol 12:654–662
Papanicolaou C, Ripley LS (1989) Polymerase-specific differences in the DNA intermediates of frameshift mutagenesis. In vitro synthesis errors of Escherichia coli DNA polymerase I and its large fragment derivative. J Mol Biol 207:335–353
Papanicolaou C, Ripley LS (1991) An in vitro approach to identifying specificity determinants of mutagenesis mediated by DNA misalignments. J Mol Biol 221:805–821
Parniewski P, Staczek P (2002) Molecular mechanisms of TRS instability. Adv Exp Med Biol 516:1–25
Parniewski P, Bacolla A, Jaworski A, Wells RD (1999) Nucleotide excision repair affects the stability of long transcribed (CTG) · (CAG) tracts in an orientation-dependent manner in *Escherichia coli*. Nucleic Acids Res 27:616–623
Parniewski P, Jaworski A, Wells RD, Bowater RP (2000) Length of CTG · CAG repeats determines the influence of mismatch repair on genetic instability. J Mol Biol 299:865–874
Patrick SM, Turchi JJ (1999) Replication protein A (RPA) binding to duplex cisplatin-damaged DNA is mediated through the generation of single-stranded DNA. J Biol Chem 274:14972–14978
Pearson CE, Sinden RR (1996) Alternative structures in duplex DNA formed within the trinucleotide repeats of the myotonic dystrophy and fragile X loci. Biochemistry 35:5041–5053
Pearson CE, Sinden RR (1998b) Slipped strand DNA, dynamic mutations, and human disease. In: Wells RD, Warren ST (eds) Genetic instabilities and hereditary neurological disorders. Academic, San Diego, pp 585–621
Pearson CE, Sinden RR (1998a) Trinucleotide repeat DNA structures: Dynamic mutations from dynamic DNA. Curr Opin Struct Biol 8:321–330
Pearson CE (2003) Slipping while sleeping? Trinucleotide repeat expansions in germ cells. Trends Mol Med 9:490–495
Pearson CE, Ewel A, Acharya S, Fishel RA, Sinden RR (1997) Human MSH2 binds to trinucleotide repeat DNA structures associated with neurodegenerative diseases. Hum Mol Genet 6:1117–1123

Pearson CE, Eichler EE, Lorenzetti D, Kramer SF, Zoghbi HY, Nelson DL, Sinden RR (1998a) Interruptions in the triplet repeats of SCA1 and FRAXA reduce the propensity and complexity of slipped strand DNA (S-DNA) formation. Biochemistry 37:2701–2708

Pearson CE, Wang YH, Griffith JD, Sinden RR (1998b) Structural analysis of slipped-strand DNA (S-DNA) formed in $(CTG)_n \cdot (CAG)_n$ repeats from the myotonic dystrophy locus. Nucleic Acids Res 26:816–823

Pearson CE, Tam M, Wang YH, Montgomery SE, Dar AC, Cleary JD, Nichol K (2002) Slipped-strand DNAs formed by long (CAG) · (CTG) repeats: slipped-out repeats and slip-out junctions. Nucleic Acids Res 30:4534–4547

Pelletier R, Krasilnikova MM, Samadashwily GM, Lahue R, Mirkin SM (2003) Replication and expansion of trinucleotide repeats in yeast. Mol Cell Biol 23:1349–1357

Petruska J, Arnheim N, Goodman MF (1996) Stability of intrastrand hairpin structures formed by the CAG/CTG class of DNA triplet repeats associated with neurological diseases. Nucleic Acids Res 24:1992–1998

Petruska J, Hartenstine MJ, Goodman MF (1998) Analysis of strand slippage in DNA polymerase expansions of CAG/CTG triplet repeats associated with neurodegenerative disease. J Biol Chem 273:5204–5210

Pineiro E, Fernandez-Lopez L, Gamez J, Marcos R, Surralles J, Velazquez A (2003) Mutagenic stress modulates the dynamics of CTG repeat instability associated with myotonic dystrophy type 1. Nucleic Acids Res 31:6733–6740

Pluciennik A, Iyer RR, Napierala M, Larson JE, Filutowicz M, Wells RD (2002) Long CTG · CAG repeats from myotonic dystrophy are preferred sites for intermolecular recombination. J Biol Chem 277:34074–34086

Pollard LM, Sharma R, Gomez M, Shah S, Delatycki MB, Pianese L, Monticelli A, Keats BJ, Bidichandani SI (2004) Replication-mediated instability of the GAA triplet repeat mutation in Friedreich ataxia. Nucleic Acids Res 32:5962–5971

Potaman VN, Bissler JJ (1999) Overcoming a barrier for DNA polymerization in triplex-forming sequences. Nucleic Acids Res 27:e5

Potaman VN, Bissler JJ, Hashem VI, Oussatcheva EA, Lu L, Shlyakhtenko LS, Lyubchenko YL, Matsuura T, Ashizawa T, Leffak M, Benham CJ, Sinden RR (2003) Unpaired structures in SCA10 $(ATTCT)_n \cdot (AGAAT)_n$ repeats. J Mol Biol 326:1095–1111

Potaman VN, Oussatcheva EA, Lyubchenko YL, Shlyakhtenko LS, Bidichandani SI, Ashizawa T, Sinden RR (2004) Length-dependent structure formation in Friedreich ataxia $(GAA)_n \cdot (TTC)_n$ repeats at neutral pH. Nucleic Acids Res 32:1224–1231

Ranum LP, Day JW (2004) Myotonic dystrophy: RNA pathogenesis comes into focus. Am J Hum Genet 74:793–804

Rao BS, Manor H, Martin RG (1988) Pausing in simian virus 40 DNA replication by a sequence containing $(dG\text{-}dA)_{27} \cdot (dT\text{-}dC)_{27}$. Nucleic Acids Res 16:8077–8094

Remus D, Beall EL, Botchan MR (2004) DNA topology, not DNA sequence, is a critical determinant for Drosophila ORC-DNA binding. EMBO J 23:897–907

Richards RI, Sutherland GR (1994) Simple repeated DNA is not replicated simply. Nat Genet 6:114–116

Ripley LS (1982) Model for the participation of quasi-palindromic DNA sequences in frameshift mutation. Proc Natl Acad Sci USA 79:4128–4132

Ripley LS (1990) Frameshift mutation: Determinants of specificity. Annu Rev Genet 24:189–213

Ripley LS, Clark A, deBoer JG (1986) Spectrum of spontaneous frameshift mutations. Sequences of bacteriophage T4 rII gene frameshifts. J Mol Biol 191:601–613

Rolfsmeier ML, Dixon MJ, Pessoa-Brandao L, Pelletier R, Miret JJ, Lahue RS (2001) Cis-elements governing trinucleotide repeat instability in Saccharomyces cerevisiae. Genetics 157:1569–1579

Rosche WA, Trinh TQ, Sinden RR (1995) Differential DNA secondary structure-mediated deletion mutation in the leading and lagging strands. J Bacteriol 177:4385–4391

Rosche WA, Trinh TQ, Sinden RR (1997) Leading strand specific spontaneous mutation corrects a quasipalindrome by an intermolecular strand switch mechanism. J Mol Biol 269:176–187

Rosche WA, Ripley LS, Sinden RR (1998) Primer-template misalignments during leading strand DNA synthesis account for the most frequent spontaneous mutations in a quasipalindromic region in Escherichia coli. J Mol Biol 284:633–646

Ruggiero BL, Topal MD (2004) Triplet repeat expansion generated by DNA slippage is suppressed by human flap endonuclease 1. J Biol Chem 279:23088–23097

Sakamoto N, Chastain PD, Parniewski P, Ohshima K, Pandolfo M, Griffith JD, Wells RD (1999) Sticky DNA: self-association properties of long GAA · TTC repeats in R · R · Y triplex structures from Friedreich's ataxia. Mol Cell 3:465–475

Samadashwily GM, Raca G, Mirkin SM (1997) Trinucleotide repeats affect DNA replication in vivo. Nature Genet 17:298–304

Santos F, Peters AH, Otte AP, Reik W, Dean W (2005) Dynamic chromatin modifications characterise the first cell cycle in mouse embryos. Dev Biol 280:225–236

Sato T, Oyake M, Nakamura K, Nakao K, Fukusima Y, Onodera O, Igarashi S, Takano H, Kikugawa K, Ishida Y, Shimohata T, Koide R, Ikeuchi T, Tanaka H, Futamura N, Matsumura R, Takayanagi T, Tanaka F, Sobue G, Komure O, Takahashi M, Sano A, Ichikawa Y, Goto J, Kanazawa I (1999) Transgenic mice harboring a full-length human mutant DRPLA gene exhibit age-dependent intergenerational and somatic instabilities of CAG repeats comparable with those in DRPLA patients. Hum Mol Genet 8:99–106

Savouret C, Brisson E, Essers J, Kanaar R, Pastink A, te Riele H, Junien C, Gourdon G (2003) CTG repeat instability and size variation timing in DNA repair-deficient mice. EMBO J 22:2264–2273

Savouret C, Garcia-Cordier C, Megret J, te Riele H, Junien C, Gourdon G (2004) MSH2-dependent germinal CTG repeat expansions are produced continuously in spermatogonia from DM1 transgenic mice. Mol Cell Biol 24:629–637

Schimke RT (1988) Gene amplification in cultured cells. J Biol Chem 263:5989–5992

Schimke RT (1992) Gene amplification; what are we learning? Mutat Res 276:145–149

Schmidt KH, Abbott CM, Leach DR (2000) Two opposing effects of mismatch repair on CTG repeat instability in *Escherichia coli*. Mol Microbiol 35:463–471

Schumacher S, Fuchs RPP, Bichara M (1998) Expansion of CTG repeats from human disease genes is dependent upon replication mechanisms in *Escherichia coli*—the effect of long patch mismatch repair revisited. J Mol Biol 279:1101–1110

Schweitzer JK, Livingston DM (1998) Expansions of CAG repeat tracts are frequent in a yeast mutant defective in Okazaki fragment maturation. Hum Mol Genet 7:69–74

Schweitzer JK, Livingston DM (1999) The effect of DNA replication mutations on CAG tract stability in yeast. Genetics 152:953–963

Schwob E (2004) Flexibility and governance in eukaryotic DNA replication. Curr Opin Microbiol 7:680–690

Seznec H, Lia-Baldini AS, Duros C, Fouquet C, Lacroix C, Hofmann-Radvanyi H, Junien C, Gourdon G (2000) Transgenic mice carrying large human genomic sequences with expanded CTG repeat mimic closely the DM CTG repeat intergenerational and somatic instability. Hum Mol Genet 9:1185–1194

Sharma R, Bhatti S, Gomez M, Clark RM, Murray C, Ashizawa T, Bidichandani SI (2002) The GAA triplet-repeat sequence in Friedreich ataxia shows a high level of somatic instability in vivo, with a significant predilection for large contractions. Hum Mol Genet 11:2175–2187

Sheflin LG, Kowalski D (1985) Altered DNA conformations detected by mung bean nuclease occur in promoter and terminator regions of supercoiled pBR322 DNA. Nucleic Acids Res 13:6137–6155

Sinden RR (1994) DNA structure and function. Academic, San Diego

Sinden RR (1999) Biological implications of the DNA structures associated with disease-causing triplet repeats. Am J Hum Genet 64:346–353

Sinden RR, Wells RD (1992) DNA structure, mutations, and human genetic disease. Curr Opin Biotechnol 3:612–622

Sinden RR, Hashem VI, Rosche WA (1999) DNA-directed mutations: Leading and lagging strand specificity. Ann NY Acad Sci 870:173–189

Sinden RR, Potaman VN, Oussatcheva EA, Pearson CE, Lyubchenko YL, Shlyakhtenko LS (2002) Triplet repeat DNA structures and human genetic disease: dynamic mutations from dynamic DNA. J Biosci 27:53–65

Soyfer VN, Potaman VN (1995) Triple-Helical Nucleic Acids. Springer, New York

Spiro C, McMurray CT (2003) Nuclease-deficient FEN-1 blocks Rad51/BRCA1-mediated repair and causes trinucleotide repeat instability. Mol Cell Biol 23:6063–6074

Spiro C, Pelletier R, Rolfsmeier ML, Dixon MJ, Lahue RS, Gupta G, Park MS, Chen X, Mariappan SV, McMurray CT (1999) Inhibition of FEN-1 processing by DNA secondary structure at trinucleotide repeats. Mol Cell 4:1079–1085

Spradling AC (1999) ORC binding, gene amplification, and the nature of metazoan replication origins. Genes Dev 13:2619–2623

Stark GR, Debatisse M, Giulotto E, Wahl GM (1989) Recent progress in understanding mechanisms of mammalian DNA amplification. Cell 57:901–908

Streisinger G, Okada Y, Emrich J, Newton J, Tsugita A, Terzaghi E, Inouye M (1966) Frameshift mutations and the genetic code. Cold Spring Harbor Symp Quant Biol 31:77–84

Suen IS, Rhodes JN, Christy M, McEwen B, Gray DM, Mitas M (1999) Structural properties of Friedreich's ataxia d(GAA) repeats. Biochim Biophys Acta 1444:14–24

Suo Z, Johnson KA (1998) DNA secondary structure effects on DNA synthesis catalyzed by HIV-1 reverse transcriptase. J Biol Chem 273:27259–27267

Syu LJ, Fluck MM (1997) Site-specific in situ amplification of the integrated polyomavirus genome: a case for a context-specific over-replication model of gene amplification. J Mol Biol 271:76–99

Tam M, Erin MS, Kekis M, David SB, Price GB, Pearson CE (2003) Slipped (CTG) · (CAG) repeats of the myotonic dystrophy locus: surface probing with anti-DNA antibodies. J Mol Biol 332:585–600

Tercero JA, Longhese MP, Diffley JF (2003) A central role for DNA replication forks in checkpoint activation and response. Mol Cell 11:1323–1336

Tower J (2004) Developmental gene amplification and origin regulation. Annu Rev Genet 38:273–304

Trinh TQ, Sinden RR (1991) Preferential DNA secondary structure mutagenesis in the lagging strand of replication in *E. coli*. Nature 352:544–547

Trinh TQ, Sinden RR (1993) The influence of primary and secondary DNA structure in deletion and duplication between direct repeats in *Escherichia coli*. Genetics 134:409–422

Umek RM, Kowalski D (1988) The ease of DNA unwinding as a determinant of initiation at yeast replication origins. Cell 52:559–567

Usdin K (1998) NGG-triplet repeats form similar intrastrand structures: implications for the triplet expansion diseases. Nucleic Acids Res 26:4078–4085

Usdin K, Woodford KJ (1995) CGG repeats associated with DNA instability and chromosome fragility form structures that block DNA synthesis in vitro. Nucleic Acids Res 23:4202–4209

van Den Broek WJ, Nelen MR, Wansink DG, Coerwinkel MM, te Riele H, Groenen PJ, Wieringa B (2002) Somatic expansion behaviour of the (CTG)(n) repeat in myotonic dystrophy knock-in mice is differentially affected by Msh3 and Msh6 mismatch-repair proteins. Hum Mol Genet 11:191–198

van den Ouweland AM, Deelen WH, Kunst CB, Uzielli ML, Nelson DL, Warren ST, Oostra BA, Halley DJ (1994) Loss of mutation at the FMR1 locus through multiple exchanges between maternal X chromosomes. Hum Mol Genet 3:1823–1827

van Noort V, Worning P, Ussery DW, Rosche WA, Sinden RR (2003) Strand misalignments lead to quasipalindrome correction. Trends Genet 19:365–369

Veeraraghavan J, Rossi ML, Bambara RA (2003) Analysis of DNA replication intermediates suggests mechanisms of repeat sequence expansion. J Biol Chem 278:42854–42866

Vetcher AA, Napierala M, Iyer RR, Chastain PD, Griffith JD, Wells RD (2002) Sticky DNA, a long GAA · GAA · TTC triplex that is formed intramolecularly, in the sequence of intron 1 of the frataxin gene. J Biol Chem 277:39217–39227

Vetcher AA, Wells RD (2004) Sticky DNA formation in vivo alters the plasmid dimer/monomer ratio. J Biol Chem 279:6434–6443

Viswanathan M, Lacirignola JJ, Hurley RL, Lovett ST (2000) A novel mutational hotspot in a natural quasipalindrome in *Escherichia coli*. J Mol Biol 302:553–564

Volker M, Mone MJ, Karmakar P, Van Hoffen A, Schul W, Vermeulen W, Hoeijmakers JH, van Driel R, Van Zeeland AA, Mullenders LH (2001) Sequential assembly of the nucleotide excision repair factors in vivo. Mol Cell 8:213–224

Waga S, Stillman B (1998) The DNA replication fork in eukaryotic cells. Annu Rev Biochem 67:721–751

Wakasugi M, Sancar A (1999) Order of assembly of human DNA repair excision nuclease. J Biol Chem 274:18759–18768

Watase K, Venken KJ, Sun Y, Orr HT, Zoghbi HY (2003) Regional differences of somatic CAG repeat instability do not account for selective neuronal vulnerability in a knock-in mouse model of SCA1. Hum Mol Genet 12:2789–2795

Weaver DT, DePamphilis ML (1982) Specific sequences in native DNA that arrest synthesis by DNA polymerase alpha. J Biol Chem 257:2075–2086

Weisman-Shomer P, Cohen E, Fry M (2000) Interruption of the fragile X syndrome expanded sequence d(CGG)(n) by interspersed d(AGG) trinucleotides diminishes the formation and stability of d(CGG)$_n$ tetrahelical structures. Nucleic Acids Res 28:1535–1541

Wells RD, Sinden RR (1993) Defined ordered sequence DNA, DNA structure, and DNA-directed mutation. In: Davies K, Warren S (eds) Genome analysis, vol 7: genome rearrangement and stability. Cold Spring Harbor Laboratory, Cold Spring Harbor, pp 107–138

Wheeler VC, Lebel LA, Vrbanac V, Teed A, te Riele H, MacDonald ME (2003) Mismatch repair gene Msh2 modifies the timing of early disease in Hdh(Q111) striatum. Hum Mol Genet 12:273–281

Wierdl M, Dominska M, Petes TD (1997) Microsatellite instability in yeast—dependence on the length of the microsatellite. Genetics 146:769–779

Wong L-JC, Ashizawa T, Monckton DG, Caskey CT, Richards CS (1995) Somatic heterogeneity of the CTG repeat in myotonic dystrophy is age and size dependent. Am J Hum Genet 56:114–122

Yang Z, Lau R, Marcadier JL, Chitayat D, Pearson CE (2003) Replication inhibitors modulate instability of an expanded trinucleotide repeat at the myotonic dystrophy type 1 disease locus in human cells. Am J Hum Genet 73:1092–1105

Yoshiyama K, Maki H (2003) Spontaneous hotspot mutations resistant to mismatch correction in Escherichia coli: transcription-dependent mutagenesis involving template-switching mechanisms. J Mol Biol 327:7–18

Yoshiyama K, Higuchi K, Matsumura H, Maki H (2001) Directionality of DNA replication fork movement strongly affects the generation of spontaneous mutations in Escherichia coli. J Mol Biol 307:1195–1206

Part II
Disorders Associated with Non-coding Repeats

Nucleic Acids and Molecular Biology, Vol. 19
M. Fry/K. Usdin (Eds.)
Human Nucleotide Expansion Disorders

Molecular Correlates of Fragile X Syndrome and FXTAS

Flora Tassone · Paul J. Hagerman (✉)

Department of Biochemistry and Molecular Medicine, School of Medicine,
University of California, Davis, One Shields Avenue, Davis, CA 95616-8635, USA
ftassone@ucdavis.edu, pjhagerman@ucdavis.edu

1 Overview

The fragile X mental retardation 1 (*FMR1*) gene, responsible for fragile X syndrome (FXS), is a paradigm for trinucleotide repeat expansion disorders. A particularly intriguing and important aspect of the *FMR1* gene is that it gives rise to diverse clinical syndromes, affecting different groups of people, depending on the size of the d(CGG) repeat expansion in the 5′ untranslated region (5′-UTR) of the gene. On the basis of the repeat expansion, individuals are classified as having normal alleles [5–44 d(CGG) repeats], intermediate or gray zone alleles [45–54 d(CGG) repeats], premutation alleles [55–200 d(CGG) repeats], or full-mutation alleles [more than 200 d(CGG) repeats]. Whereas full-mutation expansions generally result in FXS, and often in autism, smaller repeat expansions in the premutation range give rise to at least three separate forms of clinical involvement: (1) behavioral, physical, emotional, and cognitive problems in some children who are premutation carriers; (2) premature ovarian failure (POF) in approximately one fifth of all carrier women; and (3) fragile X-associated tremor/ataxia syndrome (FXTAS) in some older adults (predominantly, although not exclusively men). These forms of clinical involvement will be addressed in a brief clinical perspective in the first section of this chapter; the pathogenic mechanisms underlying these forms of clinical involvement will be presented in subsequent sections.

In particular, we will discuss a novel mechanism for the adult-onset FXTAS among adult carriers. It now appears that FXTAS, and perhaps POF as well, may be due to a toxic "gain of function" of the expanded r(CGG) messenger RNA (mRNA) itself. Evidence for this RNA-based mechanism, as well as associated aspects of *FMR1* expression from premutation alleles, will be presented.

The dynamic d(CGG) repeat instability associated with the *FMR1* gene is emblematic of the instability that underlies the genetic anticipation associated with many of the other trinucleotide repeat disorders. Moreover, the epigenetic mechanisms that lead to transcriptional silencing, for d(CGG) repeats exceeding approximately 200 trinucleotide units, give rise to the leading inherited form of mental retardation (FXS), and the leading known single-gene form of autism. Nevertheless, many aspects of both the expansion process per se and the resultant gene silencing remain obscure. These issues will be discussed in the middle section of this chapter.

Although the emphasis of this chapter is on the expression of the gene itself, we will also discuss several aspects of *FMR1* protein (FMRP) function, including a unifying model for the role of FMRP in synaptic plasticity, the "metabotropic glutamate receptor (mGluR) hypothesis".

2 Introduction and Clinical Perspective

2.1 Fragile X Syndrome

FXS is the leading heritable form of mental retardation (Hagerman and Hagerman 2002a), with a prevalence for cognitive impairment of approximately 1 : 4000 men and 1 : 6000 women (Turner et al. 1992; Sherman 2002). FXS is almost always due to expansion of a trinucleotide d(CGG) repeat in the 5′-UTR of the *FMR1* gene (Verkerk et al. 1991; Pieretti et al. 1991; Oberlé et al. 1991; Yu et al. 1991; Fu et al. 1991). *FMR1* alleles with more than 200 d(CGG) repeats generally become hypermethylated in the promoter region, with consequent transcriptional silencing and loss of the FMRP (Pieretti et al. 1991). Although FXS is typically described as a mental retardation syndrome (with mild physical features generally thought to reflect connective/elastic tissue laxity; Hagerman 2002), it is actually a spectrum disorder, with associated disorders of mood, behavior, and socialization. Patients with the most severe clinical involvement tend to be autistic and nonverbal (Hagerman 2002). However, not all individuals with FXS have mental retardation; approximately 15% percent of men, and 70% of women, have IQs greater than 70. Among this latter group, many women, and some men, may suffer from anxiety disorder, selective mutism, or Asperger syndrome (Hagerman 2002; Freund et al. 1993; Hagerman et al. 1999). A significant portion of fragile X children are autistic, with early estimates of approximately 15–25% (Brown et al. 1982; Hagerman et al. 1986; Reiss and Freund 1990; Bailey et al. 1993), increasing to approximately 30–35% in more recent studies (Rogers et al. 2001; Kau et al. 2004; Philofsky et al. 2004; Kaufmann et al. 2004) as better diagnostic tools have become available.

Mild clinical involvement (IQ > 70) generally occurs with FMRP levels that are only moderately reduced, as is often the case for individuals with alleles in the high premutation range (Hagerman and Hagerman 2004), or for individuals with alleles in the low full-mutation range that remain transcriptionally active (Tassone et al. 2000c). Thus, since the spectrum of involvement of FXS is quite broad, including individuals who do not meet the formal criterion for mental retardation (IQ < 70), the prevalence of FXS is likely to be much greater than 1 : 4000. Carrier frequencies in the general population range from 1 : 260 to 1 : 110 for women (Rousseau et al. 1995; Pesso et al. 2000; Toledano-Alhadef et al. 2001), and are approximately 1 : 800 for men (Dombrowski et al. 2002).

On the basis of the emerging phenotype and the spectrum of involvement for FXS, it is perhaps most correct to think of the disorder as a primary protein (FMRP) deficiency disorder, with other genetic modifiers contributing to the more variable features of the disorder, such as autism. In this view, FXS can occur for alleles that are either in the premutation or in the full-mutation ranges whether or not they are fully methylated, provided that FMRP levels are reduced.

2.2 Clinical Involvement Among Carriers of Premutation Alleles of the *FMR1* Gene

It is now clear that some carriers of premutation alleles do experience various forms of clinical involvement on the fragile X spectrum; such involvement can include mild physical features (prominent ears, hyperflexible finger joints) (Riddle et al. 1998; Hagerman and Hagerman 2002b) and/or emotional problems (Loesch et al. 1994; Franke et al. 1998; Hagerman and Hagerman 2002b, Sobesky et al. 1996). These problems, which are more likely to occur for *FMR1* alleles exceeding 100 d(CGG) repeats (Johnston et al. 2001), probably reflect the moderately reduced FMRP levels found in the upper half of the premutation range (Tassone et al. 2000a, b; Kenneson et al. 2001). The association between lowered FMRP levels and mental retardation (and/or autism) in the premutation range was first noted by Tassone et al. (2000). Not surprisingly, these forms of mental impairment appear to be more common in male carriers, likely due to the presence of a second X chromosome and random X-inactivation in female carriers (Berry-Kravis et al. 2005; Jacquemont et al. 2005; Aziz et al. 2003). Taken together, these observations suggest that the cognitive impairment and behavioral/emotional involvement are on the FXS clinical spectrum.

In contrast to FXS, which spans both the premutation and the full-mutation ranges, two disorders that are unique to the premutation range are POF, seen in approximately 20% of women who carry premutation alleles (Allingham-Hawkins et al. 1999; Marozzi et al. 2000), and FXTAS, which may affect as many as one third of older adult men with premutation alleles (Hagerman et al. 2001; Brunberg et al. 2002; Berry-Kravis et al. 2003; Jacquemont et al. 2003, 2004a, b; Leehey et al. 2003). Over the last 10–15 years, it has become clear that the women carriers of the *FMR1* premutation have an increased likelihood of having POF, defined as the complete cessation of menstruation before 40 years of age (Cronister et al. 1991; Schwartz et al. 1994; Murray et al. 1998; Allingham-Hawkins et al. 1999; Sullivan et al. 2005 – latest paper on POF). This association does not appear to be related to the parental origin of the premutation (reviewed in Sherman 2000). Despite the clear association of the *FMR1* premutation with abnormal ovarian development and function, the molecular basis for this association is still unknown.

More recently, the neurodegenerative disorder, FXTAS, has been described in older (primarily male) carriers of the fragile X premutation (Hagerman et al. 2001; Brunberg et al. 2002; Jacquemont et al. 2003). The core features of FXTAS include progressive intention tremor, gait ataxia, and parkinsonism; associated features include peripheral neuropathy as well as cognitive deficits involving loss of memory and executive function. Approximately 60% of carriers with clinical features of FXTAS display symmetric hyperintensities on T2-weighted magnetic resonance images of the middle cerebellar peduncles (Brunberg et al. 2002; Jacquemont et al. 2004b). Furthermore, Greco et al. (2002, 2006) identified ubiquitin-positive intranuclear inclusions in both neurons and astrocytes, broadly distributed throughout the brain, in postmortem brain tissue from adult male premutation carriers with FXTAS. Although female premutation carriers tend to be spared from FXTAS (Berry-Kravis et al. 2003; Jacquemont et al. 2004a), some do suffer from this disorder (Berry-Kravis et al. 2005; Jacquemont et al. 2005; Hagerman et al. 2004). Although FXTAS exists within the premutation range, where the *FMR1* gene is active, it has not been observed in the full-mutation range (Tassone et al. 2004), where the gene is generally silent. This observation has led us to propose that FXTAS is due to an RNA toxic "gain of function", by analogy to the RNA toxic gain of function proposed for myotonic dystrophy (reviewed in Ranum and Day 2004) whereby the excess, r(CGG)-expanded mRNA itself leads to cellular dysregulation (Hagerman et al. 2001; Greco et al. 2002, 2006). The molecular basis for FXTAS will be discussed in more detail in a subsequent section.

3 Expression of the *FMR1* Gene

3.1 Mechanisms of CGG Repeat Expansion

One of the most striking features of genetic anticipation associated with the *FMR1* gene is the propensity for large expansions of the d(CGG) repeat, and the occurrence of such expansions almost exclusively through maternal transmissions (Nolin et al. 2003; Rife et al. 2004); any model for repeat instability/expansion must explain this fundamental observation. Furthermore, models for d(CGG) repeat expansion must account for both the relative stability of repeat size in differentiated somatic tissues (Wohrle et al. 1993; Reyniers et al. 1999; reviewed in Pearson 2003; Cleary and Pearson 2003) and the near-certainty of transmitting the expanded allele as a full mutation if the carrier mother has a premutation allele that exceeds approximately 100 d(CGG) repeats. To date, no animal or in vitro model has successfully recapitulated the large expansions observed in humans in association with germline transmission.

A number of different animal models have been utilized to study the behavior of exogenous d(CGG) repeat elements. In a study of premutation-sized alleles [81 and 160 d(CGG) repeats] in the yeast *Saccharomyces cerevisiae*, Balakumaran et al. (2000) placed the repeat tracts onto a yeast chromosome in both orientations with respect to nearest replication origin. They observed an orientation bias for deletions, with a C-rich continuous (sense) strand (replication proceeding through the repeat element in the 5′-to-3′ direction with respect to the sense strand) more stable than the G-rich orientation. Not surprisingly, for both orientations, deletions were more common with the larger repeat; expansions were rare for either repeat, with no expansions observed for the larger repeat. In addition, they noted that both repeat tracts were recombinogenic, although in no cases were large expansions observed. More recently, Peier and Nelson (2002) examined the effects of sequences flanking the d(CGG) repeat by constructing yeast artificial chromosomes (YACs) with a 400-kb region surrounding the entire human *FMR1* locus [92 d(CGG) repeats including d(AGG) interruptions]. Length-dependent instability was observed in YAC transgenic mice as small expansions and contractions in both male and female transmissions over five generations; however, no large expansions were detected.

Several mouse models have been generated to investigate repeat tract instability in a mammalian host. Bontekoe et al. (2001) created a knockin transgenic animal in which a 98 d(CGG) repeat element was placed in the context of the mouse *Fmr1* gene, replacing the native $d(CGG)_8$ tract. Over several generations, moderate repeat instability was observed for both maternal and paternal transmissions. Such instability included small deletions and expansions; however, no large expansions have been observed to date. To examine the possible roles of *trans*-acting factors in modifying repeat stability, and the host-specificity of such factors, Fleming et al. (2003) examined the influence on d(CGG) repeat stability of two human *trans*-acting factors, the Werner's syndrome helicase and p53, in the mouse context. The frequencies of small/large deletions were independent of the presence of either the helicase or the p53 gene product. Thus, the lack of large d(CGG) repeat expansions in the mouse does not appear to be due to more efficient helicase- or p53-mediated error-correction mechanisms. Those investigators noted contractions occurring in the absence of expansions, which suggest that different mechanisms are operating for the two types of event. Finally, a single report by Baskaran et al. (2002) did describe transgenic mouse lines with dramatic increases in the size of the repeat element (from 26 to more than 300 repeats) in three generations. These authors speculated that the presence of an SV40 origin in the transgene might facilitate repeat expansion, perhaps through exclusion of nucleosomes from the region of the origin. If confirmed, the reported observations would be an exciting development. However, it will be important to directly demonstrate that the expanded sequence element is, indeed, a d(CGG) sequence. The surprising and unusual aspect of the findings of Baskaran et al. (2002) is that, following the massive expansion in two generations, the expanded alleles apparently become quite stable.

Several models have been proposed for the physical basis for repeat instability, including slippage/mispairing due to the formation of higher-order structure within the d(CGG) repeat (Weisman-Shomer et al. 2000, 2002; Uliel et al. 2000; Bowater and Wells 2001). Although there is currently debate as to the participation of specific structural motifs (Fojtik et al. 2004), it is plausible that the intrinsic propensity for d(CGG) repeat tracts to form secondary structures, such as hairpins, renders sequence elements more prone to minor degrees of genetic instability, like those observed within the gray zone and low premutation ranges. However, such models cannot readily account for the large expansions that occur during transmissions from *FMR1* premutation alleles, since the models cannot account for the relative stability of large alleles in somatic tissues.

In a recent study of d(CGG) repeat instability in primate cells, with d(CGG) replication templates under the control of the SV40 promoter, Edamura et al. (2005) made several important observations that bear on the issue of *cis* factors that control repeat instability, at least for somatic tissues. They noted that instability was nearly always in the direction of deletions from a 53 d(CGG) starting repeat element; rare expansions were always relatively small, and were not larger than equivalent expansion/instability directed by a bacterial origin. They found that the number of deletions depended both on the orientation of the d(CGG) repeat with respect to the SV40 origin [greater number of deletions with d(CGG) as the lagging strand template] and on the distance separating the origin from the repeat element, with more deletions generated for a 74-nucleotide (nt) separation versus 497-nt separation. Interestingly, the promoter-proximal repeat element, with the largest number of deletions, was the most efficiently replicated. Premethylation of the replication template reduced both the number of deletions and the efficiency of replication of the template, supporting the general observation that hypermethylated alleles in fragile X patients tend to be more stable than unmethylated alleles (Wohrle et al. 2001). One of the most significant observations of this study is that there is no block in the replication fork as it encounters the d(CGG) repeat, a finding that is consistent with earlier observations in cells from patients with FXS (Hansen et al. 1993). This result stands in contrast to the d(CGG)-mediated replication blocks observed in both yeast and bacterial systems (Samadashwily et al. 1997; Pelletier et al. 2003), and also the in vitro studies of Kamath-Loeb et al. (2001), suggesting that the model systems may not possess all of the *trans* factors required to properly replicate through a d(CGG) repeat element.

In aggregate, studies of repeat instability using model systems suggest that either recombination/repair or replication pathways (or both) are capable of generating the moderate germline and somatic instability that is observed in mammalian systems. However, the absence of any model system in which large expansions have been clearly established suggests that (1) such expansions are likely to occur exclusively during meiotic recombination, (2) the large expansions may occur only in primates, which normally harbor long d(CGG) repeat tracts (Garcia Arocena et al. 2003) and may possess the appropriate *trans*-acting factors, and (3) the propensity for expansion may depend on genomic context. Finally, a model for expansion that involves meiotic recombination needs to account for two related observations: the rarity of cases where large deletions are transmitted, and the generally large magnitude of the expansions [e.g., from approximately 100 to 1000 d(CGG) repeats] in a single transmission.

3.2 Regulation of Expression of the *FMR1* Gene in the Normal and Premutation Ranges

The *FMR1* gene (L29074) spans approximately 38 kb of genomic DNA, and contains 17 exons and an unusually large (9.9-kb) first intron; such introns have been implicated in both transcriptional and splicing regulation (Liu et al. 2000; Morishita et al. 2001). The gene is widely expressed in both neural and nonneural tissues, although at different levels in different tissues. High expression of a 4.4-kb transcript is observed by Northern blot analysis in brain, placenta, testis, lung, and kidney (Hinds et al. 1993). Lower expression is observed in liver, skeletal muscle, and pancreas. Multiple truncated transcripts of 1.4 kb have been observed in human heart (Hinds et al. 1993). In fetal human brain, *FMR1* expression has been observed early in the development in proliferating and migrating cells of the nervous system, while in older brain tissues higher expression levels were detected in cholinergic and pyramidal neurons (Abitbol et al. 1993).

Extensive alternative splicing of the *FMR1* gene, demonstrated by reverse transcription PCR (RT-PCR) analysis, can give rise to as many as 20 possible protein isoforms, which differ in various internal segments. Several of these isoforms have been observed on Western blots of both human and mouse tissues, including fetal brain neurons (Ashley et al. 1993; Verheij et al. 1993; Verkerk et al. 1993; Sittler et al. 1996; Huang et al. 1996). Alternative splicing has not been found in the amino-terminal half of the *FMR1* gene, and the splice isoforms do not appear to be tissue-specific; similar ratios of transcripts were found in several fetal tissues, including brain and testis (Verkerk et al. 1993).

The *FMR1* promoter, encompassing a GC-rich island, possesses consensus binding sequences for multiple transcription factors, including several Sp1 sites, AP2, αPAL/Nrf-1, Myc, and H4TF1/Sp1-like, which are generally characteristic of housekeeping genes (Drouin et al. 1997). In vitro and functional experiments have indicated that four of these sites, including the αPAL/Nrf-1 site, two GC boxes (Sp1 and Sp3), and an E box (USF1/2), may serve as *cis* elements for the regulation of normal *FMR1* promoter activity (Kumari and Usdin 2001; Kumari et al. 2005). DNA binding of Nrf-1 and USF1/USF2 transcription factors is influenced by CpG methylation (Kumari and Usdin 2001), while binding of both Sp1 and Sp3 does not appear to be affected (Harrington et al. 1988; Holler et al. 1988).

The binding of these transcription factors has been observed in normal human cells in vivo (Drouin et al. 1997; Schwemmle et al. 1997); however, the binding of Sp1 and Nrf-1 to the human *FMR1* promoter in vivo appears to be disrupted in fragile X human cells, suggesting that the expansion and the methylation of the d(CGG) repeat element within the 5′-UTR can prevent the transcriptional activation of the *FMR1* gene (Smith et al. 2004). More recent studies have demonstrated that Sp1 and Sp3 proteins play an important role in the regulation of the *FMR1* promoter. Kumari et al. (2005) have suggested that these two proteins, together with USF1, USF2, and Nrf-1, can induce bending of free DNA. They propose that such helix distortions may bring the 5′ and the 3′ ends of the promoter into closer proximity, thus allowing distally located factors, important for transcription initiation, to more easily interact with other components of the transcriptional machinery. The significance of this observation, in the context of native chromatin, remains to be determined.

3.3 Regulation of Transcription Start Site Selection

The promoter region of the *FMR1* gene is very CG-rich and lacks the canonical TATA box (Kumari and Usdin 2001). A single transcription initiation site was originally identified using a primer-extension approach (Hwu et al. 1993). The initiation site was located at position – 264 (+ 1 indicates the translational start site) upstream of the d(CGG) repeat element and downstream of a TATA-like sequence d(TTACA). More recent studies, using RNA ligase-mediated rapid amplification of 5′ complementary DNA ends, demonstrated that the *FMR1* promoter region possesses multiple initiator regions (Inr) that are active start sites for transcription (Beilina et al. 2004). Three of these sites, designated sites 1, 2, and 3, are active in normal and premutation lymphoblastoid lines. The presence of multiple start sites is not surprising, since multiple start sites have been observed for many other TATA-less promoters with long, GC-rich 5′-UTRs (Kawai et al. 2003). However, what is surprising is that the choice of the initiation site for the *FMR1* gene appears to be modulated by the size of the d(CGG) repeat.

Site 1, which is close to the start of the previously reported 5′-UTR (Hwu et al. 1993), represents the major transcription start site in normal *FMR1* alleles, whereas site 3, approximately 50 nt upstream of site 1, is the major start site for large premutation alleles [e.g., 160 d(CGG) repeat units]. Thus, alleles with a high number of d(CGG) repeats preferentially express the longer *FMR1* mRNA. The nucleotide sequence of all three transcriptional initiation sites was found to be highly similar to the consensus sequence of pyrimidine-rich initiator (Inr) elements [consensus sequence YYAN(T/A)YY] (Javahery et al. 1994) that are usually located near the start site and have been implicated in transcription initiation in TATA-less genes (Chow et al. 1995). The sequence between site 1 and site 2 contains another Inr sequence, which is active in human hippocampus and cerebellum from both normal and carrier men (Carosi et al. 2004); thus, this fourth Inr may have a brain-specific regulatory function. An important implication of Inr selection, based on d(CGG) repeat size and tissue type, is that the downstream d(CGG) repeat element in the *FMR1* gene directly modulates transcription initiation and, therefore, influences not only the level of transcription, but also the type of transcript expressed.

3.4 Increased Transcription in the Premutation Range

Although *FMR1* mRNA and FMRP levels were initially reported to be normal in the premutation range (Pieretti et al. 1991; Devys et al. 1993; Feng et al. 1995a, b; Hmadcha et al. 1998), it recently demonstrated that *FMR1* transcription is elevated for premutation alleles. Specifically, levels of abnormal [expanded r(CGG) repeats] *FMR1* mRNA are elevated by as much as fivefold to tenfold in the upper premutation range in peripheral blood leukocytes (Tassone et al. 2000a, c) of both female and male carriers of the fragile X premutation, despite the presence of normal or near-normal detectable FMRP levels (Tassone et al. 2000a, b; Kenneson et al. 2001). Higher levels of mRNA are due to higher transcription rates of the *FMR1* gene (Tassone et al., unpublished results), with reduced FMRP levels being due to decreased translational efficiency (Primerano et al. 2002). While the basic mechanisms leading to increased transcription of premutation alleles are not known, the presence of such an abnormal molecular phenotype appears to be associated with POF and FXTAS (reviewed in Hagerman and Hagerman 2004), two unique pathological phenotypes observed only in individual carriers of the *FMR1* premutation.

3.5 Mechanisms of Silencing/Reactivation of the *FMR1* Gene

Approximately 50–60% of all genes, including *FMR1*, contain a CpG island in the 5′-UTR region (Antequera and Bird 1993a, b; Pieretti et al. 1991). With some exceptions, CpG dinucleotides in CG-rich islands are normally unmethylated, while most CpGs outside of the CpG islands are methylated (Bird 1992); thus, patterns of methylation appear to be important for creating zones of transcriptional activity (or inactivity) within the genome.

In the case of the *FMR1* gene, expansions of over approximately 200 d(CGG) repeats in the 5′-UTR region of the gene are generally accompanied by hypermethylation of the d(CGG) repeat element and of the upstream CpG island. This hypermethylation usually results in transcriptional silencing of the *FMR1* gene, absence of FMRP and, as a consequence, the fragile X phenotype. Methylation of full-mutation *FMR1* alleles occurs early in embryonic development, and is believed to mitotically stabilize the expansion (Devys et al. 1992). Because of the assumed (causal) association between methylation and silencing, it is generally believed that methylation of the *FMR1* gene causes transcriptional silencing at that locus. Evidence in support of this notion comes from the demonstration that treatment of fragile X cells with the DNA methylation inhibitor 5′-azadeoxycytidine (5′-aza dC) leads both to a loss of methylation at expanded *FMR1* DNA and to a partial reactivation of the gene (Chiurazzi et al. 1998, 1999; Coffee et al. 1999). However, transcription is not always repressed on hypermethylated *FMR1* alleles, as transcriptional activity has been observed in men with a full mutation (Tassone et al. 2000c, 2001). It is not clear why some methylated full-mutation alleles continue to produce mRNA, despite their resistance to cleavage by methylation-sensitive restriction enzymes, whereas others do not; nor is it clear why some full-mutation alleles remain unmethylated and, therefore, transcriptionally active (Tassone et al. 2000c, 2001).

In addition to hypermethylation of the *FMR1* promoter region, hypoacetylation of associated histones and chromatin condensation—all characteristics of transcriptionally inactive genes—are also observed; however, the basic mechanisms underlying the specificity of the *FMR1* transcriptional silencing are not known. There is abundant evidence that an interplay exists between cytosine methylation and histone modifications, although the nature of this association is still being defined. Such modifications include acetylation/deacetylation and methylation of target lysine residues of the histone tails (Kuo and Allis 1998; Kouzarides 2002).

One of the consequences of these posttranslational modifications may be to modulate the binding of various regulatory factors through their chromatin-binding domains (chromodomains) to the histone tails (Turner 2000; Strahl and Allis 2000). Recent work has underscored the importance of posttranslational modification of histone proteins as another epigenetic mechanism in the organization of chromosomal domains and gene regulation (Litt et al. 2001; Nakayama et al. 2001).

Acetylation of lysine residues within the N-terminal tails of H3 and H4 (e.g., acetylation of H3-Lys9) is associated with the normal (active) *FMR1* allele, but not with fragile X full-mutation alleles (Coffee et al. 1999). The H3-K9 residue appears to be particularly important for epigenetic regulation, as it can be both acetylated in active chromatin and methylated in inactive chromatin (Lachner et al. 2003; Grewal and Moazed 2003). Methylation of H3-Lys4 appears to be associated with active chromatin; whereas, methylation of H3-Lys9 is considered to be a marker of condensed, inactive chromatin of the form associated with the inactive X-chromosome and pericentromeric heterochromatin (Heard et al. 2001; Boggs et al. 2002; Maison et al. 2002). There is also evidence that histone methylation can direct DNA methylation, which leads to gene silencing (Tamaru and Selker 2001). In fragile X cells in culture, there is a decrease in methylation of histone H3-Lys4, with a large increase in methylation at H3-Lys9 (Coffee et al. 2002). However, partially deacetylated histones and methylated H3-K9 were reported in one case of a fragile X cell line with an unmethylated *FMR1* allele, consistent with silent chromatin, but which also had high levels of methylation at H3-Lys4, characteristic of normal (active) *FMR1* alleles (Pietrobono et al. 2005). This last observation raises an intriguing possibility; namely, that the methylation state, and the state of modifications in the chromatin associated with the *FMR1* gene, may be at least partially uncoupled.

Treatment of cultured cells from fragile X patients with 5′-aza dC only partially reactivates hypermethylated *FMR1* full-mutation alleles (Chiurazzi et al. 1998; Pietrobono et al. 2002). The level of *FMR1* expression following reactivation reaches only 15% of the normal level; thus, more complex interactions between epigenetic factors must exist in mediating the organization of chromatin structure and for the regulation of gene expression.

One of the mechanisms by which methylation leads to transcriptional silencing involves the CpG binding protein MeCP2, which can indirectly inhibit the binding of transcription factors by limiting the access to regulatory elements (Nan et al. 1998; Bird 1999). It has been shown that MeCP2 associates with methylated histones, specifically, with histone H3 at Lys9 (Fuks et al. 2003). This association provides a link between DNA methylation and histone methylation. MeCP2, is part of a corepressor complex involving the human Brahma (Brm), a component of the SWI/SNF-related chromatin remodeling family. MeCP2 and Brm are assembled on the promoter of methylated genes, including the *FMR1* gene, promoting silencing (Harikrishnan et al. 2005). The resulting corepressor complex is recruited to the inactive *FMR1* gene, but is released upon treatment with 5′-aza dC, which partially restores transcriptional activity (Harikrishnan et al. 2005). The association of this complex is markedly reduced in normal cells, and knockdown of Brm and MeCP2 gene activity relieves transcription repression (Harikrishnan et al. 2005). Thus, the recruitment of the SWI/SNF complexes appears to facilitate transcriptional repression at the *FMR1* gene. In addition, MeCP2 represses gene activity by recruiting Sin3A, which interacts with histone deacetylase 1 (HDAC1), again resulting in chromatin remodeling and silencing. Thus, MeCP2 reinforces a repressive chromatin state by acting as a bridge between two global epigenetic modifications: DNA methylation and histone methylation.

It is thus clear that several related and perhaps sequential events take place so that histone deacetylation, followed by methylation of H3-Lys9, leads to transcriptional silencing in a manner that is mediated by MeCP2 (Rea et al. 2000). MeCP2 then recruits proteins such as heterochromatin protein 1 (HP1) (Lachner et al. 2001; Bannister et al. 2001), which selectively recognizes H3-Lys9 tails through its chromodomain (Fischle et al. 2003; reviewed in Eissenberg and Elgin 2000).

3.6 Regulation of Translation of the *FMR1* mRNA

Elevation of *FMR1* mRNA levels occurs for unmethylated alleles both within the premutation range and extending into the full-mutation range (Tassone et al. 2000a–c; Salat et al. 2000; Kenneson et al. 2001). The concomitant deficit in FMRP was originally suggested to be the stimulus for increased *FMR1* mRNA production, in the absence of any increase in mRNA stability, essentially as a feedback response to lowered protein (FMRP) levels (Tassone et al. 2000a). Recently, increased levels of run-on transcription in a premutation cell line (compared with a normal control) have been observed, providing direct evidence of transcriptional activation for expanded (premutation) alleles (Tassone et al., unpublished results). Moreover, using both quantitative RT-PCR and RNA in situ hybridization experiments, Tassone et al. (unpublished results) demonstrated that higher *FMR1* mRNA levels are not due to nuclear sequestration. In particular, *FMR1* mRNA is not retained in the nucleus, but is mainly localized in the cytoplasm of lymphocytes carrying either normal or premutation alleles.

As noted already, although *FMR1* mRNA levels are increased in the premutation range, FMRP expression is decreased. The FMRP deficit is r(CGG)-dependent and is due to decreased translational efficiency (Primerano et al. 2002). Reduced translational efficiency was observed both in cell lines and in transient transfection experiments using expanded alleles spanning the entire premutation range (Primerano et al. 2002; Chen et al. 2003). Particularly for premutation alleles, a smaller fraction of *FMR1* mRNA was found to be associated with polysomes, while the majority of the expanded-repeat mRNA was associated with inactive ribonucleoprotein particles. These findings, namely, increased *FMR1* mRNA expression levels and deficit in translation efficiency in premutation alleles, have also been confirmed by in vivo translation experiments using a reporter (luciferase) mRNA with the 5′-UTR of the *FMR1* gene, the latter harboring varying numbers of r(CGG) repeats. Interestingly, the decreased translation efficiency, evident in the premutation range, was also observed for an allele near the gray zone (45–54 CGG repeats). Translation efficiency gradually decreased with an increasing r(CGG) repeat number (Chen et al. 2003).

The precise mechanism by which the expanded r(CGG) repeat impedes translation is not understood at present. What is surprising is that translation occurs at all for larger premutation alleles, since the predicted free energies of stabilization of the r(CGG) repeat element would be expected to completely block translation. In this regard, an internal ribosome entry site (IRES) was identified near the 5′ end of the 5′-UTR, upstream of the r(CGG) repeat (Chiang et al. 2001). *FMR1* IRES activity was found to be of moderate strength compared with that of other known IRESs (Chiang et al. 2001); its role in the regulation of FMRP expression is not known at present. Interestingly, cellular IRESs have been shown to increase the translational efficiency of several dendritically localized mRNAs, including the microtubule-associated protein 2 (MAP2), the α-subunit of the Ca^{2+}/calmodulin-dependent protein kinase II (α-CaMKII), cytoskeleton-associated protein, arc, dendrin, and neurogranin (RC3) (Pinkstaff et al. 2001). IRESs that can mediate cap-independent translation could be used for a rapid and local synthesis of proteins in dendrites. Although translation at dendrites occurs by both cap-dependent and cap-independent mechanisms, the translation mediated by IRES in the *RC3* gene is relatively more efficient in dendrites than in the cell body (Pinkstaff et al. 2001). The finding that five different neuronal mRNAs are translated in dendrites by an IRES-mediated mechanism suggests that IRES sequences may control translation in specific neuronal regions.

4
Function of the *FMR1* Protein

Central to our understanding of the pathogenesis of FXS is a detailed description of the role(s) played by FMRP, the protein product of the *FMR1* gene. This subject will only be touched on in this chapter, since a number of excellent reviews have detailed the properties of FMRP (Jin and Warren 2000; Bardoni et al. 2001; O'Donnell and Warren 2002; Bardoni and Mandel 2002; Bagni and Greenough 2005), including its structural organization, its interacting partners, and its putative functional role(s). The principal clinical features of FXS are caused by the absence of functional FMRP. Absence of the protein is almost always due to transcriptional silencing, although in rare instances its absence results from mutations within the coding portion of the *FMR1* gene (Gedeon et al. 1992; Wohrle et al. 1992; Tarleton et al. 1993; Gu et al. 1994; Meijer et al. 1994; Trottier et al. 1994; Hirst et al. 1995; Quan et al. 1995; Wang et al. 1997).

FMRP is known to be an RNA binding protein with at least three recognized RNA binding motifs; two hnRNP K-homology (KH) domains, and an arginine–glycine-rich domain (Siomi et al. 1993). There is also a carboxy-terminal RNA binding region that is relatively non-sequence-specific for RNA binding (Adinolfi et al. 1999). These features of FMRP, coupled with the presence of nuclear localization and nuclear export signals (Liu et al. 1996; Eberhart et al. 1996; Fridell et al. 1996; Tamanini et al. 1999), suggest that the protein is involved with nuclear export and/or transport of a subset of mRNAs (Bassell and Kelic 2004); however, a specific functional role of FMRP in nuclear export has not been demonstrated. Over the past several years, most attention has been focused on the role of FMRP as a translational modulator/repressor of specific mRNAs that are important for synapse maturation and plasticity (Rudelli et al. 1985; Weiler et al. 1997; Irwin et al. 2000; Nimchinsky et al. 2001). In particular, FMRP may modulate the function of the mGluR in synaptic function and plasticity (mGluR hypothesis) (Bear et al. 2004; see later).

4.1 FMRP Can Function as a Negative Regulator of Translation

Two studies (Laggerbauer et al. 2001; Li et al. 2001) provided evidence for at least one function of FMRP, namely, the repression of translation of a subset of mRNAs. In the first of these studies, Laggerbauer et al. (2001) demonstrated that recombinant FMRP strongly inhibited the translation of selected mRNAs in both rabbit reticulocyte lysates and *Xenopus laevis* oocytes. Although the focus placed on this study is usually the observed effect of FMRP on translation, perhaps equally important is the authors' observation that neither of the two FMRP paralogs, FXR1P and FXR2P (reviewed in Hoogeveen et al. 2002), is capable of inhibiting translation of the mRNAs inhibited by FMRP. This finding is significant in that the latter two proteins both contain the KH domains found in FMRP. Further, the archetypal KH protein, hnRNP K (Siomi et al. 1994), which is also capable of binding to the test mRNA, nevertheless does not inhibit translation. Finally, although a mutant form of FMRP with an I304N substitution, associated with severe clinical involvement (De Boulle et al. 1993), retains its RNA binding capacity, it has lost the ability to inhibit translation. This failure to repress translation appears to be due to abnormal protein–protein interactions involving the second KH domain, not loss of RNA binding.

The second study, by Li et al. (2001), also used recombinant FMRP (produced in baculovirus) to demonstrate that FMRP is capable of substantial, relatively nonspecific suppression of translation. Utilizing the rabbit reticulocyte lysate assay, Li et al. (2001) showed that recombinant FMRP was capable of up to 90% suppression of all brain- or liver-derived (rat) poly(A) mRNA for protein levels comparable to those found in peripheral blood leucocytes. FMRP did not suppress translation from specific mRNAs (e.g., globin) that do not display direct FMRP binding; an observation that militates against a role for FMRP in global translation suppression as might occur through pathways involving phosphorylation of eIF2α. Li et al. (2001) also observed that nearly all inhibition could be reversed by the addition of the 3′-UTR portion of *Fmr1* mRNA, suggesting that a major point of FMRP–mRNA interaction is the 3′-UTR. The broad suppression of translation, coupled to a direct FMRP–mRNA interaction, would appear to be at variance with the observation that FMRP appears to bind only about 4% of brain-derived poly(A) mRNAs (Brown et al. 1998, 2001). Li et al. (2001) speculate that the relatively nonspecific inhibition found in their in vitro assay may reflect more widespread, weaker interactions that also occur in vivo. However, an alternative interpretation is that the suppression observed in the in vitro studies simply reflects the absence of one or more additional *trans*-acting factors that may further modulate the effects of FMRP. Thus, it may be that, in vivo, FMRP could be either a suppressor or an inducer of translation, depending on the message as well as on the cell type and/or developmental stage. In this regard, it is interesting that evidence for both upregulation and downregulation has come from the expression-profiling study of Brown et al. (2001). More recent in vivo studies have also demonstrated that FMRP can facilitate the trapping of mRNAs into cytoplasmic granules with concomitant suppression of translation (Mazroui et al. 2002).

4.2 A Possible Role for FMRP in Regulating Actin Cytoskeletal Dynamics

Castets et al. (2005) have reported a very exciting set of observations that may provide a link between FMRP and the actin dynamics likely to determine the shape of dendritic spines. In particular, they have demonstrated that FMRP appears to bind directly to the 5′-UTR of mRNA for phosphoserine/threonine phosphatase (PP2A; Zolnierowicz 2000), thus diminishing its translation. PP2A is a mediator of Rac1-coupled actin remodeling (Hall 1998; Janssens et al. 2005), which operates by shifting the balance from the inactive (phospho) to the active (dephospho) form of cofilin, an actin depolymerizing protein (Ambach et al. 2000; Meberg and Bamburg 2000; Samstag and Nebl 2003; Paavilainen et al. 2004). The principal finding of Castets et al. (2005) was that Rac1-induced actin remodeling was enhanced in cells that either lacked FMRP or possessed mutant forms of FMRP in which either the KH1 or KH2 domains were altered. These observations provide a possible basis for the altered dendritic spine morphology found in individuals with FXS.

4.3 The mGluR Hypothesis as a Specific Example of How FMRP Could Regulate Synaptic Function/Plasticity

Another exciting discovery relates to the function of FMRP is its role in regulating long-term depression (LTD) of synaptic strength in hippocampal neurons (reviewed in Bear et al. 2004). LTD is associated with a decrease in the number of ionotropic (AMPA) glutamate receptors on the postsynaptic surface through internalization of the receptors. This form of LTD, involving stimulation of the mGluR, requires protein synthesis (Huber et al. 2002). A key finding by Huber et al. (2002) is that mGluR-coupled hippocampal LTD is enhanced in *Fmr1* knockout mice. In this model, enhanced AMPA receptor internalization in the knockout mouse is due to the absence of FMRP inhibition of the synthesis of one or more as-yet-undefined proteins that facilitate receptor internalization. Thus, FMRP, by virtue of its postulated role as a translational inhibitor, would normally be acting as a "governor" to modulate the degree of LTD. This model is quite powerful inasmuch as it links the function of FMRP to dendritic function and clinical outcome. In the simplest form of the model, FMRP is proposed to act as a translational inhibitor; however, this hypothesis is based in part on in vitro data (Laggerbauer et al. 2001; Li et al. 2001), as well as on reports of inhibition of specific mRNAs (Zhang et al. 2001; Zalfa et al. 2003; Paavilainen et al. 2004). It is entirely possible that the function of FMRP in this instance may be to stimulate translation of selected mRNAs, as may be the case with the synaptic protein PSD95 (Todd et al. 2003). Thus, what is needed at this point is a concerted effort to identify the targets of FMRP function that modulate mGluR-coupled LTD.

5
The Molecular Basis of FXTAS

Although the precise molecular mechanism of FXTAS is not known, two observations led to the proposal of an RNA toxic gain-of-function model for disease pathogenesis in FXTAS (Hagerman et al. 2001; Greco et al. 2002; Jacquemont et al. 2003; Hagerman and Hagerman 2004). First, (expanded-repeat) *FMR1* mRNA levels are elevated by up to eightfold in premutation carriers, even though FMRP levels are normal to slightly reduced in the premutation range (Tassone et al. 2000a, b; Kenneson et al. 2001). Second, FXTAS has not been reported in adults who harbor hypermethylated (silenced) full-mutation alleles, where little or no *FMR1* mRNA is produced. In support of this RNA toxicity model, Jin et al. (2003) demonstrated that an expanded approximately 90 r(CGG) repeat, when expressed in the 5′-UTR context of an unrelated reporter gene, was still capable of inducing neuropathology (including inclusion formation) in the eye of the fly (*Drosophila*). This observation provided a direct demonstration that the expanded r(CGG) repeat in mRNA is capable of inducing neuropathology. The principal neuropathologic feature of FXTAS is the presence of ubiquitin-positive intranuclear inclusions, in both neurons and astrocytes, broadly distributed throughout the brain and spinal cord (Greco et al. 2002, 2006). The inclusions are immunohistochemically negative for tau isoforms, α-synuclein, and polyglutamine peptides, and appear to reflect a new class of inclusion disorder (reviewed in Hagerman and Hagerman 2004). Much of the focus of recent research efforts have been on the composition of the inclusions, which should provide important clues as to the pathogenesis of the disease itself.

Myotonic dystrophy (DM) is another noncoding, trinucleotide d(CTG) repeat-expansion disorder that is thought to result from an RNA toxic gain of function (Finsterer 2002; Mankodi and Thornton 2002), where the expanded r(CUG) repeat, located in the 3′-UTR of the DM protein kinase (*DMPK*) mRNA, sequesters one or more protein mediators of the disease process. Thus, in DM, the RNA "toxicity" arises as a result of dysregulation of the function(s) of those proteins owing to their excessive binding to the expanded r(CUG) repeat. One such mediator is believed to be *MBNL1*, the human homolog of *Drosophila* muscleblind (Miller et al. 2000; Fardaei et al. 2001; Ranum and Day 2004). Both *DMPK* mRNA and *MBNL1* are found within intranuclear foci in DM, the sequestration of *MBNL1* protein in turn leading to dysregulation of the splicing of several other mRNAs. Consistent with this RNA toxicity model, we have recently found *FMR1* mRNA itself within the intranuclear inclusions of FXTAS patients (Tassone et al. 2004). This last observation gives added impetus to study the protein complement of the inclusions, since, by analogy with DM, one or more (potential) protein mediators should be present within the inclusions themselves.

An analysis of the protein composition of the inclusions in FXTAS, through a combination of fluorescence-based particle sorting of inclusions from postmortem tissue and mass spectroscopic and immunochemical approaches for protein identification, has revealed more than 20 protein species to date (Iwahashi et al. 2006). At least one of the identified proteins, hnRNP A2, is a well-known RNA binding protein (Dreyfuss et al. 2002) that could serve as a mediator of the expanded-repeat *FMR1* mRNA in FXTAS. MBNL1 was also identified within the inclusions; however, the roles played by these two proteins await further investigation. Iwahashi et al. (2006) did not observe a dominant protein species within the inclusions, which argues against the simple accretion of specific, abnormal proteins that is thought to occur with many other inclusion disorders (Paulson 1999; Zoghbi and Orr 2000; Tarlac and Storey 2003; Taylor et al. 2002; Ross and Poirier 2004). No FMRP was detected within the inclusions. Interestingly, the inclusions do appear to contain several intermediate filament proteins, including lamin A/C. Although the lamin A/C isoforms are not believed to interact directly with *FMR1* mRNA, they are believed to be involved with the regulation of RNA synthesis and processing (Hutchison and Worman 2004; Zastrow et al. 2004). Thus, the lamins could also be involved in mediating the effects of the expanded-repeat *FMR1* mRNA (Arocena et al. 2006).

Acknowledgements This work was supported by the National Institutes of Child Health and Development, grant HD40661 (P.J.H.), by the National Institutes of Neurological Diseases and Stroke. grant NS43532 (P.J.H.), and by a National Fragile X Foundation Research Grant (F.T.).

References

Abitbol M, Menini C, Delezoide AL, Rhyner T, Vekemans M, Mallet J (1993) Nucleus basalis magnocellularis and hippocampus are the major sites of FMR-1 expression in the human fetal brain. Nat Genet 4:147–153

Adinolfi S, Bagni C, Musco G, Gibson T, Mazzarella L, Pastore A (1999) Dissecting FMR1, the protein responsible for fragile X syndrome, in its structural and functional domains. RNA 5:1248–1258

Allingham-Hawkins DJ, Babul-Hirji R, Chitayat D, Holden JJ, Yang KT, Lee C, Hudson R, Gorwill H, Nolin SL, Glicksman A, Jenkins EC, Brown WT, Howard-Peebles PN, Becchi C, Cummings E, Fallon L, Seitz S, Black SH, Vianna-Morgante AM, Costa SS, Otto PA, Mingroni-Netto RC, Murray A, Webb J, MacSwinney F, Dennis N, Jacobs PA, Syrrou M, Georgiou I, Patsalis PC, Uzielli MLG, Guarducci S, Lapi E, Cecconi A, Ricci U (1999) Fragile X premutation is a significant risk factor for premature ovarian failure: the International Collaborative POF in fragile X study—preliminary data. Am J Med Genet 83:322–325

Ambach A, Saunus J, Konstandin M, Wesselborg S, Meuer SC, Samstag Y (2000) The serine phosphatases PP1 and PP2A associate with and activate the actin-binding protein cofilin in human T lymphocytes. Eur J Immunol 30:3422–3431

Antequera F, Bird A (1993a) Number of CpG islands and genes in human and mouse. Proc Natl Acad Sci USA 90:11995–11999

Antequera F, Bird A (1993b) CpG islands. EXS 64:169–185

Arocena DG, Iwahashi CK, Won N, Beilina A, Ludwig AL, Tassone F, Schwartz PH, Hagermann PJ (2005) Induction of inclusion formation and disruption of lamin A/C structure by premutation CGG-repeat RNA in human cultured neural cells. Hum Mol Genet 14:3661–3671. Epub 2005 Oct 3620

Ashley CT, Sutcliffe JS, Kunst CB, Leiner HA, Eichler EE, Nelson DL, Warren ST (1993) Human and murine FMR-1: alternative splicing and translational initiation downstream of the CGG-repeat. Nat Genet 4:244–251

Aziz M, Stathopulu E, Callias M, Taylor C, Turk J, Oostra B, Willemsen R, Patton M (2003) Clinical features of boys with fragile X premutations and intermediate alleles. Am J Med Genet B Neuropsychiatr Genet 121:119–127

Bagni C, Greenough WT (2005) From mRNP trafficking to spine dysmorphogenesis: the roots of fragile X syndrome. Nat Rev Neurosci 6:376–387

Bailey A, Bolton P, Butler L, Le Couteur A, Murphy M, Scott S, Webb T, Rutter M (1993) Prevalence of the fragile X anomaly amongst autistic twins and singletons. J Child Psychol Psychiatry 34:673–688

Balakumaran BS, Freudenreich CH, Zakian VA (2000) CGG/CCG repeats exhibit orientation-dependent instability and orientation-independent fragility in Saccharomyces cerevisiae. Hum Mol Genet 9:93–100

Bannister AJ, Zegerman P, Partridge JF, Miska EA, Thomas JO, Allshire RC, Kouzarides T (2001) Selective recognition of methylated lysine 9 on histone H3 by the HP1 chromo domain. Nature 410:120–124

Bardoni B, Mandel JL (2002) Advances in understanding of fragile X pathogenesis and FMRP function, and in identification of X linked mental retardation genes. Curr Opin Genet Dev 12:284–293

Bardoni B, Schenck A, Mandel JL (2001) The fragile X mental retardation protein. Brain Res Bull 56:375–382

Baskaran S, Datta S, Mandal A, Gulati N, Totey S, Anand RR, Brahmachari V (2002) Instability of CGG repeats in transgenic mice. Genomics 80:151–157

Bassell GJ, Kelic S (2004) Binding proteins for mRNA localization and local translation, and their dysfunction in genetic neurological disease. Curr Opin Neurobiol 14:574–581

Bear MF, Huber KM, Warren ST (2004) The mGluR theory of fragile X mental retardation. Trends Neurosci 27:370–377

Beilina A, Tassone F, Schwartz PH, Sahota P, Hagerman PJ (2004) Redistribution of transcription start sites within the FMR1 promoter region with expansion of the downstream CGG-repeat element. Hum Mol Genet 13:543–549. Epub 2004 Jan 2013

Berry-Kravis E, Lewin F, Wuu J, Leehey M, Hagerman R, Hagerman P, Goetz CG (2003) Tremor and ataxia in fragile X premutation carriers: blinded videotape study. Ann Neurol 53:616–623

Berry-Kravis E, Potanos K, Weinberg D, Zhou L, Goetz CG (2005) Fragile X-associated tremor/ataxia syndrome in sisters related to X-inactivation. Ann Neurol 57:144–147

Bird A (1992) The essentials of DNA methylation. Cell 70:5–8

Bird A (1999) DNA methylation de novo. Science 286:2287–2288

Boggs BA, Cheung P, Heard E, Spector DL, Chinault AC, Allis CD (2002) Differentially methylated forms of histone H3 show unique association patterns with inactive human X chromosomes. Nat Genet 30:73–76

Bontekoe CJ, Bakker CE, Nieuwenhuizen IM, van der Linde H, Lans H, de Lange D, Hirst MC, Oostra BA (2001) Instability of a $(CGG)_{98}$ repeat in the Fmr1 promoter. Hum Mol Genet 10:1693–1699

Bowater RP, Wells RD (2001) The intrinsically unstable life of DNA triplet repeats associated with human hereditary disorders. Prog Nucleic Acid Res Mol Biol 66:159–202

Brown V, Small K, Lakkis L, Feng Y, Gunter C, Wilkinson KD, Warren ST (1998) Purified recombinant FMRP exhibits selective RNA binding as an intrinsic property of the fragile X mental retardation protein. J Biol Chem 273:15521–15527

Brown V, Jin P, Ceman S, Darnell JC, O'Donnell WT, Tenenbaum SA, Jin X, Feng Y, Wilkinson KD, Keene JD, Darnell RB, Warren ST (2001) Microarray identification of FMRP-associated brain mRNAs and altered mRNA translational profiles in fragile X syndrome. Cell 107:477–487

Brown WT, Jenkins EC, Friedman E, Brooks J, Wisniewski K, Raguthu S, French J (1982) Autism is associated with the fragile-X syndrome. J Autism Dev Disord 12:303–308

Brunberg JA, Jacquemont S, Hagerman RJ, Berry-Kravis EM, Grigsby J, Leehey MA, Tassone F, Brown WT, Greco CM, Hagerman PJ (2002) Fragile X premutation carriers: characteristic MR imaging findings of adult male patients with progressive cerebellar and cognitive dysfunction. AJNR Am J Neuroradiol 23:1757–1766

Carosi C, Beilina A, Bagni C, Schwartz P, Hagerman P, Tassone F (2004) Alternative transcription start site and polyadenylation site usage in the FMR1 gene. Paper presented at the American Society of Human Genetics, Toronto, 26–30 October

Castets M, Schaeffer C, Bechara E, Schenck A, Khandjian EW, Luche S, Moine H, Rabilloud T, Mandel JL, Bardoni B (2005) FMRP interferes with the Rac1 pathway and controls actin cytoskeleton dynamics in murine fibroblasts. Hum Mol Genet 14:835–844. Epub 2005 Feb 2009

Chen LS, Tassone F, Sahota P, Hagerman PJ (2003) The $(CGG)_n$ repeat element within the 5′ untranslated region of the FMR1 message provides both positive and negative cis effects on in vivo translation of a downstream reporter. Hum Mol Genet 12:3067–3074

Chiang PW, Carpenter LE, Hagerman PJ (2001) The 5′-untranslated region of the FMR1 message facilitates translation by internal ribosome entry. J Biol Chem 276:37916–37921

Chiurazzi P, Pomponi MG, Willemsen R, Oostra BA, Neri G (1998) In vitro reactivation of the FMR1 gene involved in fragile X syndrome. Hum Mol Genet 7:109–113

Chiurazzi P, Pomponi MG, Pietrobono R, Bakker CE, Neri G, Oostra BA (1999) Synergistic effect of histone hyperacetylation and DNA demethylation in the reactivation of the FMR1 gene. Hum Mol Genet 8:2317–2323

Chow CW, Clark MP, Rinaldo JE, Chalkley R (1995) Multiple initiators and C/EBP binding sites are involved in transcription from the TATA-less rat XDH/XO basal promoter. Nucleic Acids Res 23:3132–3140

Cleary JD, Pearson CE (2003) The contribution of cis-elements to disease-associated repeat instability: clinical and experimental evidence. Cytogenet Genome Res 100:25–55

Coffee B, Zhang F, Warren ST, Reines D (1999) Acetylated histones are associated with FMR1 in normal but not fragile X-syndrome cells. Nat Genet 22:98–101

Coffee B, Zhang F, Ceman S, Warren ST, Reines D (2002) Histone modifications depict an aberrantly heterochromatinized FMR1 gene in fragile x syndrome. Am J Hum Genet 71:923–932

Cronister A, Schreiner R, Wittenberger M, Amiri K, Harris K, Hagerman RJ (1991) Heterozygous fragile X female: historical, physical, cognitive, and cytogenetic features. Am J Med Genet 38:269–274

De Boulle K, Verkerk AJ, Reyniers E, Vits L, Hendrickx J, Van Roy B, Van den Bos F, de Graaff E, Oostra BA, Willems PJ (1993) A point mutation in the FMR-1 gene associated with fragile X mental retardation. Nat Genet 3:31–35

Devys D, Biancalana V, Rousseau F, Boue J, Mandel JL, Oberlé I (1992) Analysis of full fragile X mutations in fetal tissues and monozygotic twins indicate that abnormal methylation and somatic heterogeneity are established early in development. Am J Med Genet 43:208–216

Devys D, Lutz Y, Rouyer N, Bellocq JP, Mandel JL (1993) The FMR-1 protein is cytoplasmic, most abundant in neurons and appears normal in carriers of a fragile X premutation. Nat Genet 4:335–340

Dombrowski C, Levesque S, Morel ML, Rouillard P, Morgan K, Rousseau F (2002) Premutation and intermediate-size FMR1 alleles in 10572 males from the general population: loss of an AGG interruption is a late event in the generation of fragile X syndrome alleles. Hum Mol Genet 11:371–378

Dreyfuss G, Kim VN, Kataoka N (2002) Messenger-RNA-binding proteins and the messages they carry. Nat Rev Mol Cell Biol 3:195–205

Drouin R, Angers M, Dallaire N, Rose TM, Khandjian W, Rousseau F (1997) Structural and functional characterization of the human FMR1 promoter reveals similarities with the hnRNP-A2 promoter region. Hum Mol Genet 6:2051–2060

Eberhart DE, Malter HE, Feng Y, Warren ST (1996) The fragile X mental retardation protein is a ribonucleoprotein containing both nuclear localization and nuclear export signals. Hum Mol Genet 5:1083–1091

Edamura K, Leonard M, Pearson C (2005) Role of replication and CpG methylation in fragile X syndrome CGG deletions in primate cells. Am J Hum Genet 76:302–311

Eissenberg JC, Elgin SC (2000) The HP1 protein family: getting a grip on chromatin. Curr Opin Genet Dev 10:204–210

Fardaei M, Larkin K, Brook JD, Hamshere MG (2001) In vivo co-localisation of MBNL protein with DMPK expanded-repeat transcripts. Nucleic Acids Res 29:2766–2771

Feng Y, Lakkis L, Devys D, Warren ST (1995a) Quantitative comparison of FMR1 gene expression in normal and premutation alleles. Am J Hum Genet 56:106–113

Feng Y, Zhang F, Lokey LK, Chastain JL, Lakkis L, Eberhart D, Warren ST (1995b) Translational suppression by trinucleotide repeat expansion at FMR1. Science 268:731–734

Finsterer J (2002) Myotonic dystrophy type 2. Eur J Neurol 9:441–447

Fischle W, Wang Y, Jacobs SA, Kim Y, Allis CD, Khorasanizadeh S (2003) Molecular basis for the discrimination of repressive methyl-lysine marks in histone H3 by Polycomb and HP1 chromodomains. Genes Dev 17:1870–1881

Fleming K, Riser DK, Kumari D, Usdin K (2003) Instability of the fragile X syndrome repeat in mice: the effect of age, diet and mutations in genes that affect DNA replication, recombination and repair proficiency. Cytogenet Genome Res 100:140–146

Fojtik P, Kejnovska I, Vorlickova M (2004) The guanine-rich fragile X chromosome repeats are reluctant to form tetraplexes. Nucleic Acids Res 32:298–306

Franke P, Leboyer M, Gansicke M, Weiffenbach O, Biancalana V, Cornillet-Lefebre P, Croquette MF, Froster U, Schwab SG, Poustka F, Hautzinger M, Maier W (1998) Genotype-phenotype relationship in female carriers of the premutation and full mutation of FMR-1. Psychiatry Res 80:113–127

Freund LS, Reiss AL, Abrams MT (1993) Psychiatric disorders associated with fragile X in the young female. Pediatrics 91:321–329

Fridell RA, Benson RE, Hua J, Bogerd HP, Cullen BR (1996) A nuclear role for the Fragile X mental retardation protein. EMBO J 15:5408–5414

Fu YH, Kuhl DP, Pizzuti A, Pieretti M, Sutcliffe JS, Richards S, Verkerk AJ, Holden JJ, Fenwick RG, Jr., Warren ST, Oostra BA, Nelson DL, Caskey CT (1991) Variation of the CGG repeat at the fragile X site results in genetic instability: resolution of the Sherman paradox. Cell 67:1047–1058

Fuks F, Hurd PJ, Wolf D, Nan X, Bird AP, Kouzarides T (2003) The methyl-CpG-binding protein MeCP2 links DNA methylation to histone methylation. J Biol Chem 278:4035–4040. Epub 2002 Nov 4039

Garcia Arocena D, Breece KE, Hagerman PJ (2003) Distribution of CGG repeat sizes within the fragile X mental retardation 1 (FMR1) homologue in a non-human primate population. Hum Genet 113:371–376

Gedeon AK, Baker E, Robinson H, Partington MW, Gross B, Manca A, Korn B, Poustka A, Yu S, Sutherland GR, et al. (1992) Fragile X syndrome without CCG amplification has an FMR1 deletion. Nat Genet 1:341–344

Greco C, Berman R, Martin R, Tassone F, Schwartz P, Brunberg J, Grigsby J, Hessl D, Becker E, Papazian J, Bacalman S, Leehey M, Hagerman R, Hagerman P (2006) Neuropathology of fragile X-associated tremor/ataxia syndrome (FXTAS). Brain 129:243–255

Greco CM, Hagerman RJ, Tassone F, Chudley AE, Del Bigio MR, Jacquemont S, Leehey M, Hagerman PJ (2002) Neuronal intranuclear inclusions in a new cerebellar tremor/ataxia syndrome among fragile X carriers. Brain 125:1760–1771

Grewal SI, Moazed D (2003) Heterochromatin and epigenetic control of gene expression. Science 301:798–802

Gu Y, Lugenbeel KA, Vockley JG, Grody WW, Nelson DL (1994) A de novo deletion in FMR1 in a patient with developmental delay. Hum Mol Genet 3:1705–1706

Hagerman PJ, Hagerman RJ (2004) The fragile-X premutation: a maturing perspective. Am J Hum Genet 74:805–816

Hagerman RJ (2002) Physical and behavioral phenotype. In: Hagerman RJ, Hagerman PJ (eds) Fragile X syndrome: diagnosis, treatment and research, 3rd edn. The Johns Hopkins University Press, Baltimore, pp 3–109

Hagerman RJ, Hagerman PJ (eds) (2002a) Fragile X syndrome: diagnosis, treatment and research. 3rd edn. The Johns Hopkins University Press, Baltimore

Hagerman RJ, Hagerman PJ (2002b) The fragile X premutation: into the phenotypic fold. Curr Opin Genet Dev 12:278–283

Hagerman RJ, Jackson AWD, Levitas A, Rimland B, Braden M (1986) An analysis of autism in fifty males with the fragile X syndrome. Am J Med Genet 23:359–374

Hagerman RJ, Hills J, Scharfenaker S, Lewis H (1999) Fragile X syndrome and selective mutism. Am J Med Genet 83:313–317

Hagerman RJ, Leehey M, Heinrichs W, Tassone F, Wilson R, Hills J, Grigsby J, Gage B, Hagerman PJ (2001) Intention tremor, parkinsonism, and generalized brain atrophy in male carriers of fragile X. Neurology 57:127–130

Hagerman RJ, Leavitt BR, Farzin F, Jacquemont S, Greco CM, Brunberg JA, Tassone F, Hessl D, Harris SW, Zhang L, Jardini T, Gane LW, Ferranti J, Ruiz L, Leehey MA, Grigsby J, Hagerman PJ (2004) Fragile-X-Associated Tremor/Ataxia Syndrome (FXTAS) in Females with the FMR1 Premutation. Am J Hum Genet 74:1051–1056. Epub 2004 Apr 1052

Hall A (1998) Rho GTPases and the actin cytoskeleton. Science 279:509–514

Hansen RS, Canfield TK, Lamb MM, Gartler SM, Laird CD (1993) Association of fragile X syndrome with delayed replication of the FMR1 gene. Cell 73:1403–1409

Harikrishnan KN, Chow MZ, Baker EK, Pal S, Bassal S, Brasacchio D, Wang L, Craig JM, Jones PL, Sif S, El-Osta A (2005) Brahma links the SWI/SNF chromatin-remodeling complex with MeCP2-dependent transcriptional silencing. Nat Genet 37:254–264. Epub 2005 Feb 2006

Harrington MA, Jones PA, Imagawa M, Karin M (1988) Cytosine methylation does not affect binding of transcription factor Sp1. Proc Natl Acad Sci USA 85:2066–2070

Heard E, Rougeulle C, Arnaud D, Avner P, Allis CD, Spector DL (2001) Methylation of histone H3 at Lys-9 is an early mark on the X chromosome during X inactivation. Cell 107:727–738

Hinds HL, Ashley CT, Sutcliffe JS, Nelson DL, Warren ST, Housman DE, Schalling M (1993) Tissue specific expression of FMR-1 provides evidence for a functional role in fragile X syndrome. Nat Genet 3:36–43

Hirst M, Grewal P, Flannery A, Slatter R, Maher E, Barton D, Fryns JP, Davies K (1995) Two new cases of FMR1 deletion associated with mental impairment. Am J Hum Genet 56:67–74

Hmadcha A, De Diego Y, Pintado E (1998) Assessment of FMR1 expression by reverse transcriptase-polymerase chain reaction of KH domains. J Lab Clin Med 131:170–173

Holler M, Westin G, Jiricny J, Schaffner W (1988) Sp1 transcription factor binds DNA and activates transcription even when the binding site is CpG methylated. Genes Dev 2:1127–1135

Hoogeveen AT, Willemsen R, Oostra BA (2002) Fragile X syndrome, the Fragile X related proteins, and animal models. Microsc Res Tech 57:148–155

Huang T, Li LY, Shen Y, Qin XB, Pang ZL, Wu GY (1996) Alternative splicing of the FMR1 gene in human fetal brain neurons. Am J Med Genet 64:252–255

Huber KM, Gallagher SM, Warren ST, Bear MF (2002) Altered synaptic plasticity in a mouse model of fragile X mental retardation. Proc Natl Acad Sci USA 99:7746–7750

Hutchison CJ, Worman HJ (2004) A-type lamins: guardians of the soma? Nat Cell Biol 6:1062–1067

Hwu WL, Lee YM, Lee SC, Wang TR (1993) In vitro DNA methylation inhibits FMR-1 promoter. Biochem Biophys Res Commun 193:324–329

Irwin SA, Galvez R, Greenough WT (2000) Dendritic spine structural anomalies in fragile-X mental retardation syndrome. Cereb Cortex 10:1038–1044

Iwahashi CK, Yasui DH, An H-J, Greco CM, Tassone F, Nannen K, Babineau B, Lebrilla CB, Hagerman RJ, Hagerman PJ (2006) Protein composition of the intranuclear inclusions of FXTAS. Brain 129:256–271

Jacquemont S, Hagerman RJ, Leehey M, Grigsby J, Zhang L, Brunberg JA, Greco C, Des Portes V, Jardini T, Levine R, Berry-Kravis E, Brown WT, Schaeffer S, Kissel J,

Tassone F, Hagerman PJ (2003) Fragile X premutation tremor/ataxia syndrome: molecular, clinical, and neuroimaging correlates. Am J Hum Genet 72:869–878

Jacquemont S, Hagerman RJ, Leehey MA, Hall DA, Levine RA, Brunberg JA, Zhang L, Jardini T, Gane LW, Harris SW, Herman K, Grigsby J, Greco CM, Berry-Kravis E, Tassone F, Hagerman PJ (2004a) Penetrance of the fragile X-associated tremor/ataxia syndrome in a premutation carrier population. JAMA 291:460–469

Jacquemont S, Farzin F, Hall D, Leehey M, Tassone F, Gane L, Zhang L, Grigsby J, Jardini T, Lewin F, Berry-Kravis E, Hagerman PJ, Hagerman RJ (2004b) Aging in individuals with the FMR1 mutation. Am J Ment Retard 109:154–164

Jacquemont S, Orrico A, Galli L, Sahota PK, Brunberg JA, Anichini C, Leehey M, Schaeffer S, Hagerman RJ, Hagerman PJ, Tassone F (2005) Spastic paraparesis, cerebellar ataxia, and intention tremor: a severe variant of FXTAS? J Med Genet 42:e14

Janssens V, Goris J, Van Hoof C (2005) PP2A: the expected tumor suppressor. Curr Opin Genet Dev 15:34–41

Javahery R, Khachi A, Lo K, Zenzie-Gregory B, Smale ST (1994) DNA sequence requirements for transcriptional initiator activity in mammalian cells. Mol Cell Biol 14:116–127

Jin P, Warren ST (2000) Understanding the molecular basis of fragile X syndrome. Hum Mol Genet 9:901–908

Jin P, Zarnescu DC, Zhang F, Pearson CE, Lucchesi JC, Moses K, Warren ST (2003) RNA-mediated neurodegeneration caused by the fragile X premutation rCGG repeats in Drosophila. Neuron 39:739–747

Johnston C, Eliez S, Dyer-Friedman J, Hessl D, Glaser B, Blasey C, Taylor A, Reiss A (2001) Neurobehavioral phenotype in carriers of the fragile X premutation. Am J Med Genet 103:314–319

Kamath-Loeb AS, Loeb LA, Johansson E, Burgers PM, Fry M (2001) Interactions between the Werner syndrome helicase and DNA polymerase delta specifically facilitate copying of tetraplex and hairpin structures of the d(CGG)n trinucleotide repeat sequence. J Biol Chem 276:16439–16446. Epub 12001 Feb 16438

Kau AS, Tierney E, Bukelis I, Stump MH, Kates WR, Trescher WH, Kaufmann WE (2004) Social behavior profile in young males with fragile X syndrome: characteristics and specificity. Am J Med Genet A 126:9–17

Kaufmann WE, Cortell R, Kau AS, Bukelis I, Tierney E, Gray RM, Cox C, Capone GT, Stanard P (2004) Autism spectrum disorder in fragile X syndrome: communication, social interaction, and specific behaviors. Am J Med Genet A 129:225–234

Kawai Y, Asai K, Miura Y, Inoue Y, Yamamoto M, Moriyama A, Yamamoto N, Kato T (2003) Structure and promoter activity of the human glia maturation factor-gamma gene: a TATA-less, GC-rich and bidirectional promoter. Biochim Biophys Acta 1625:246–252

Kenneson A, Zhang F, Hagedorn CH, Warren ST (2001) Reduced FMRP and increased FMR1 transcription is proportionally associated with CGG repeat number in intermediate-length and premutation carriers. Hum Mol Genet 10:1449–1454

Kouzarides T (2002) Histone methylation in transcriptional control. Curr Opin Genet Dev 12:198–209

Kumari D, Usdin K (2001) Interaction of the transcription factors USF1, USF2, and (alpha)-PAL/Nrf-1 with the FMR1 promoter. J Biol Chem 276:4357–4364

Kumari D, Gabrielian A, Wheeler D, Usdin K (2005) The roles of Sp1, Sp3, USF1/USF2 and NRF-1 in the regulation and three-dimensional structure of the Fragile X mental retardation gene promoter. Biochem J 386:297–303

Kuo MH, Allis CD (1998) Roles of histone acetyltransferases and deacetylases in gene regulation. Bioessays 20:615–626

Lachner M, O'Carroll D, Rea S, Mechtler K, Jenuwein T (2001) Methylation of histone H3 lysine 9 creates a binding site for HP1 proteins. Nature 410:116–120

Lachner M, O'Sullivan RJ, Jenuwein T (2003) An epigenetic road map for histone lysine methylation. J Cell Sci 116:2117–2124

Laggerbauer B, Ostareck D, Keidel EM, Ostareck-Lederer A, Fischer U (2001) Evidence that fragile X mental retardation protein is a negative regulator of translation. Hum Mol Genet 10:329–338

Leehey MA, Munhoz RP, Lang AE, Brunberg JA, Grigsby J, Greco C, Jacquemont S, Tassone F, Lozano AM, Hagerman PJ, Hagerman RJ (2003) The fragile X premutation presenting as essential tremor. Arch Neurol 60:117–121

Li Z, Zhang Y, Ku L, Wilkinson KD, Warren ST, Feng Y (2001) The fragile X mental retardation protein inhibits translation via interacting with mRNA. Nucleic Acids Res 29:2276–2283

Litt MD, Simpson M, Gaszner M, Allis CD, Felsenfeld G (2001) Correlation between histone lysine methylation and developmental changes at the chicken beta-globin locus. Science 293:2453–2455

Liu Q, Siomi H, Siomi MC, Fischer U, Zhang Y, Wan L, Dreyfuss G (1996) Molecular characteristics of the protein products of the fragile X syndrome gene and the survival of motor neurons gene. Cold Spring Harbor Symp Quant Biol LXI:689–697

Liu Y, Li H, Tanaka K, Tsumaki N, Yamada Y (2000) Identification of an enhancer sequence within the first intron required for cartilage-specific transcription of the alpha2(XI) collagen gene. J Biol Chem 275:12712–12718

Loesch DZ, Hay DA, Mulley J (1994) Transmitting males and carrier females in fragile X–revisited. Am J Med Genet 51:392–399

Maison C, Bailly D, Peters AH, Quivy JP, Roche D, Taddei A, Lachner M, Jenuwein T, Almouzni G (2002) Higher-order structure in pericentric heterochromatin involves a distinct pattern of histone modification and an RNA component. Nat Genet 30:329–334. Epub 2002 Feb 2019

Mankodi A, Thornton CA (2002) Myotonic syndromes. Curr Opin Neurol 15:545–552

Marozzi A, Vegetti W, Manfredini E, Tibiletti MG, Testa G, Crosignani PG, Ginelli E, Meneveri R, Dalpra L (2000) Association between idiopathic premature ovarian failure and fragile X premutation. Hum Reprod 15:197–202

Mazroui R, Huot ME, Tremblay S, Filion C, Labelle Y, Khandjian EW (2002) Trapping of messenger RNA by Fragile X Mental Retardation protein into cytoplasmic granules induces translation repression. Hum Mol Genet 11:3007–3017

Meberg PJ, Bamburg JR (2000) Increase in neurite outgrowth mediated by overexpression of actin depolymerizing factor. J Neurosci 20:2459–2469

Meijer H, de Graaff E, Merckx DM, Jongbloed RJ, de Die Smulders CE, Engelen JJ, Fryns JP, Curfs PM, Oostra BA (1994) A deletion of 1.6 kb proximal to the CGG repeat of the FMR1 gene causes the clinical phenotype of the fragile X syndrome. Hum Mol Genet 3:615–620

Miller JW, Urbinati CR, Teng-Umnuay P, Stenberg MG, Byrne BJ, Thornton CA, Swanson MS (2000) Recruitment of human muscleblind proteins to $(CUG)_n$ expansions associated with myotonic dystrophy. EMBO J 19:4439–4448

Morishita M, Kishino T, Furukawa K, Yonekura A, Miyazaki Y, Kanematsu T, Yamashita S, Tsukazaki T (2001) A 30-base-pair element in the first intron of SOX9 acts as an enhancer in ATDC5. Biochem Biophys Res Commun 288:347–355

Murray A, Webb J, Grimley S, Conway G, Jacobs P (1998) Studies of FRAXA and FRAXE in women with premature ovarian failure. J Med Genet 35:637–640

Nakayama J, Rice JC, Strahl BD, Allis CD, Grewal SI (2001) Role of histone H3 lysine 9 methylation in epigenetic control of heterochromatin assembly. Science 292:110–113

Nan X, Ng HH, Johnson CA, Laherty CD, Turner BM, Eisenman RN, Bird A (1998) Transcriptional repression by the methyl-CpG-binding protein MeCP2 involves a histone deacetylase complex. Nature 393:386–389

Nimchinsky EA, Oberlander AM, Svoboda K (2001) Abnormal development of dendritic spines in FMR1 knock-out mice. Journal of Neuroscience 21:5139–5146

Nolin SL, Brown WT, Glicksman A, Houck GE, Jr., Gargano AD, Sullivan A, Biancalana V, Brondum-Nielsen K, Hjalgrim H, Holinski-Feder E, Kooy F, Longshore J, Macpherson J, Mandel JL, Matthijs G, Rousseau F, Steinbach P, Vaisanen ML, von Koskull H, Sherman SL (2003) Expansion of the fragile X CGG repeat in females with premutation or intermediate alleles. Am J Hum Genet 72:454–464

Oberle I, Rousseau F, Heitz D, Kretz C, Devys D, Hanauer A, Boue J, Bertheas MF, Mandel JL (1991) Instability of a 550-base pair DNA segment and abnormal methylation in fragile X syndrome. Science 252:1097–1102

O'Donnell WT, Warren ST (2002) A decade of molecular studies of fragile X syndrome. Annu Rev Neurosci 25:315–338

Paavilainen V, Bertling E, Falck S, Lappalainen P (2004) Regulation of cytoskeletal dynamics by actin-monomer-binding proteins. Trends Cell Biol 14:386–394

Paulson HL (1999) Protein fate in neurodegenerative proteinopathies: polyglutamine diseases join the (mis)fold. Am J Hum Genet 64:339–345

Pearson CE (2003) Slipping while sleeping? Trinucleotide repeat expansions in germ cells. Trends Mol Med 9:490–495

Peier AM, Nelson DL (2002) Instability of a premutation-sized CGG repeat in FMR1 YAC transgenic mice. Genomics 80:423–432

Pelletier R, Krasilnikova MM, Samadashwily GM, Lahue R, Mirkin SM (2003) Replication and expansion of trinucleotide repeats in yeast. Mol Cell Biol 23:1349–1357

Pesso R, Berkenstadt M, Cuckle H, Gak E, Peleg L, Frydman M, Barkai G (2000) Screening for fragile X syndrome in women of reproductive age. Prenat Diagn 20:611–614

Philofsky A, Hepburn SL, Hayes A, Hagerman R, Rogers SJ (2004) Linguistic and cognitive functioning and autism symptoms in young children with fragile X syndrome. Am J Ment Retard 109:208–218

Pieretti M, Zhang FP, Fu YH, Warren ST, Oostra BA, Caskey CT, Nelson DL (1991) Absence of expression of the FMR-1 gene in fragile X syndrome. Cell 66:817–822

Pietrobono R, Pomponi MG, Tabolacci E, Oostra B, Chiurazzi P, Neri G (2002) Quantitative analysis of DNA demethylation and transcriptional reactivation of the FMR1 gene in fragile X cells treated with 5-azadeoxycytidine. Nucleic Acids Res 30:3278–3285

Pietrobono R, Tabolacci E, Zalfa F, Zito I, Terracciano A, Moscato U, Bagni C, Oostra B, Chiurazzi P, Neri G (2005) Molecular dissection of the events leading to inactivation of the FMR1 gene. Hum Mol Genet 14:267–277. Epub 2004 Nov 2024

Pinkstaff JK, Chappell SA, Mauro VP, Edelman GM, Krushel LA (2001) Internal initiation of translation of five dendritically localized neuronal mRNAs. Proc Natl Acad Sci USA 98:2770–2775

Primerano B, Tassone F, Hagerman RJ, Hagerman P, Amaldi F, Bagni C (2002) Reduced FMR1 mRNA translation efficiency in fragile X patients with premutations. RNA 8:1482–1488

Quan F, Zonana J, Gunter K, Peterson KL, Magenis RE, Popovich BW (1995) An atypical case of fragile X syndrome caused by a deletion that includes the FMR1 gene. Am J Hum Genet 56:1042–1051

Ranum LP, Day JW (2004) Myotonic dystrophy: RNA pathogenesis comes into focus. Am J Hum Genet 74:793–804

Rea S, Eisenhaber F, O'Carroll D, Strahl BD, Sun ZW, Schmid M, Opravil S, Mechtler K, Ponting CP, Allis CD, Jenuwein T (2000) Regulation of chromatin structure by site-specific histone H3 methyltransferases. Nature 406:593–599

Reiss AL, Freund L (1990) Fragile X syndrome, DSM-III-R, and autism. J Am Acad Child Adolesc Psychiatry 29:885–891

Reyniers E, Martin JJ, Cras P, Van Marck E, Handig I, Jorens HZ, Oostra BA, Kooy RF, Willems PJ (1999) Postmortem examination of two fragile X brothers with an FMR1 full mutation. Am J Med Genet 84:245–249

Riddle JE, Cheema A, Sobesky WE, Gardner SC, Taylor AK, Pennington BF, Hagerman RJ (1998) Phenotypic involvement in females with the FMR1 gene mutation. Am J Ment Retard 102:590–601

Rife M, Badenas C, Quinto L, Puigoriol E, Tazon B, Rodriguez-Revenga L, Jimenez L, Sanchez A, Mila M (2004) Analysis of CGG variation through 642 meioses in fragile X families. Mol Hum Reprod 10:773–776

Rogers SJ, Wehner DE, Hagerman R (2001) The behavioral phenotype in fragile X: symptoms of autism in very young children with fragile X syndrome, idiopathic autism, and other developmental disorders. J Dev Behav Pediatr 22:409–417

Ross CA, Poirier MA (2004) Protein aggregation and neurodegenerative disease. Nat Med 10(Suppl):S10–17

Rousseau F, Rouillard P, Morel ML, Khandjian EW, Morgan K (1995) Prevalence of carriers of premutation-size alleles of the FMRI gene – and implications for the population genetics of the fragile X syndrome. Am J Hum Genet 57:1006–1018

Rudelli RD, Brown WT, Wisniewski K, Jenkins EC, Laure-Kamionowska M, Connell F, Wisniewski HM (1985) Adult fragile X syndrome. Clinico-neuropathologic findings. Acta Neuropathol 67:289–295

Salat U, Bardoni B, Wohrle D, Steinbach P (2000) Increase of FMRP expression, raised levels of FMR1 mRNA, and clonal selection in proliferating cells with unmethylated fragile X repeat expansions: a clue to the sex bias in the transmission of full mutations? J Med Genet 37:842–850

Samadashwily GM, Raca G, Mirkin SM (1997) Trinucleotide repeats affect DNA replication in vivo. Nat Genet 17:298–304

Samstag Y, Nebl G (2003) Interaction of cofilin with the serine phosphatases PP1 and PP2A in normal and neoplastic human T lymphocytes. Adv Enzyme Regul 43:197–211

Schwartz CE, Dean J, Howard Peebles PN, Bugge M, Mikkelsen M, Tommerup N, Hull C, Hagerman R, Holden JJ, Stevenson RE (1994) Obstetrical and gynecological complications in fragile X carriers: a multicenter study. Am J Med Genet 51:400–402

Schwemmle S, de Graaff E, Deissler H, Glaser D, Wohrle D, Kennerknecht I, Just W, Oostra BA, Doerfler W, Vogel W, Steinbach P (1997) Characterization of FMR1 promoter elements by in vivo-footprinting analysis. Am J Hum Genet 60:1354–1362

Sherman S (2002) Epidemiology. In: Hagerman RJ, Hagerman PJ (eds) Fragile X syndrome: diagnosis, treatment and research, 3rd edn. The Johns Hopkins University Press, Baltimore

Sherman SL (2000) Premature ovarian failure in the fragile X syndrome. Am J Med Genet (Semin Med Genet) 97:189–194

Siomi H, Siomi MC, Nussbaum RL, Dreyfuss G (1993) The protein product of the fragile X gene, FMR1, has characteristics of an RNA-binding protein. Cell 74:291–298

Siomi H, Choi M, Siomi MC, Nussbaum RL, Dreyfuss G (1994) Essential role for KH domains in RNA binding: impaired RNA binding by a mutation in the KH domain of FMR1 that causes fragile X syndrome. Cell 77:33–39

Sittler A, Devys D, Weber C, Mandel JL (1996) Alternative splicing of exon 14 determines nuclear or cytoplasmic localisation of fmr1 protein isoforms. Hum Mol Genet 5:95–102

Smith KT, Coffee B, Reines D (2004) Occupancy and synergistic activation of the FMR1 promoter by Nrf-1 and Sp1 in vivo. Hum Mol Genet 13:1611–1621

Sobesky WE, Taylor AK, Pennington BF, Bennetto L, Porter D, Riddle J, Hagerman RJ (1996) Molecular/clinical correlations in females with fragile X. Am J Med Genet 64:340–345

Strahl BD, Allis CD (2000) The language of covalent histone modifications. Nature 403:41–45

Sullivan AK, Marcus M, Epstein MP, Allen EG, Anido AE, Paquin JJ, Yadav-Shah M, Sherman SL (2005) Association of FMR1 repeat size with ovarian dysfunction. Hum Reprod 20:402–412. Epub 2004 Dec 2017

Tamanini F, Bontekoe C, Bakker CE, van Unen L, Anar B, Willemsen R, Yoshida M, Galjaard H, Oostra BA, Hoogeveen AT (1999) Different targets for the fragile X-related proteins revealed by their distinct nuclear localizations. Hum Mol Genet 8:863–869

Tamaru H, Selker EU (2001) A histone H3 methyltransferase controls DNA methylation in Neurospora crassa. Nature 414:277–283

Tarlac V, Storey E (2003) Role of proteolysis in polyglutamine disorders. J Neurosci Res 74:406–416

Tarleton J, Richie R, Schwartz C, Rao K, Aylsworth AS, Lachiewicz A (1993) An extensive de novo deletion removing FMR1 in a patient with mental retardation and the fragile X syndrome phenotype. Hum Mol Genet 2:1973–1974

Tassone F, Hagerman RJ, Chamberlain WD, Hagerman PJ (2000a) Transcription of the FMR1 gene in individuals with fragile X syndrome. Am J Med Genet (Semin Med Genet) 97:195–203

Tassone F, Hagerman RJ, Taylor AK, Gane LW, Godfrey TE, Hagerman PJ (2000b) Elevated levels of FMR1 mRNA in carrier males: a new mechanism of involvement in fragile X syndrome. Am J Hum Genet 66:6–15

Tassone F, Hagerman RJ, Loesch DZ, Lachiewicz A, Taylor AK, Hagerman PJ (2000c) Fragile X males with unmethylated, full mutation trinucleotide repeat expansions have elevated levels of FMR1 messenger RNA. Am J Med Genet 94:232–236

Tassone F, Hagerman RJ, Taylor AK, Hagerman PJ (2001) A majority of fragile X males with methylated, full mutation alleles have significant levels of FMR1 messenger RNA. J Med Genet 38:453–456

Tassone F, Iwahashi C, Hagerman PJ (2004) FMR1 RNA within the intranuclear inclusions of fragile X-associated tremor/ataxia syndrome (FXTAS). RNA Biology 1:103–105

Taylor JP, Hardy J, Fischbeck KH (2002) Toxic proteins in neurodegenerative disease. Science 296:1991–1995

Todd PK, Mack KJ, Malter JS (2003) The fragile X mental retardation protein is required for type-I metabotropic glutamate receptor-dependent translation of PSD-95. Proc Natl Acad Sci USA 100:14374–14378

Toledano-Alhadef H, Basel-Vanagaite L, Magal N, Davidov B, Ehrlich S, Drasinover V, Taub E, Halpern GJ, Ginott N, Shohat M (2001) Fragile-X carrier screening and the prevalence of premutation and full-mutation carriers in Israel. Am J Hum Genet 69:351–360

Trottier Y, Imbert G, Poustka A, Fryns JP, Mandel JL (1994) Male with typical fragile X phenotype is deleted for part of the FMR1 gene and for about 100 kb of upstream region. Am J Med Genet 51:454–457

Turner BM (2000) Histone acetylation and an epigenetic code. Bioessays 22:836–845

Turner G, Robinson H, Laing S, van den Berk M, Colley A, Goddard A, Sherman S, Partington M (1992) Population screening for fragile X. Lancet 339:1210–1213

Uliel L, Weisman-Shomer P, Oren-Jazan H, Newcomb T, Loeb LA, Fry M (2000) Human Ku antigen tightly binds and stabilizes a tetrahelical form of the fragile X syndrome d(CGG)$_n$ expanded sequence. J Biol Chem 275:33134–33141

Verheij C, Bakker CE, de Graaff E, Keulemans J, Willemsen R, Verkerk AJ, Galjaard H, Reuser AJ, Hoogeveen AT, Oostra BA (1993) Characterization and localization of the FMR-1 gene product associated with fragile X syndrome. Nature 363:722–724

Verkerk AJ, Pieretti M, Sutcliffe JS, Fu YH, Kuhl DP, Pizzuti A, Reiner O, Richards S, Victoria MF, Zhang FP, Eussen BE, van Ommen GJB, Blonden LAJ, Riggins GJ, Chastain JL, Kunst CB, Galjaard H, Caskey CT, Nelson DL, Oostra BA, Warren ST (1991) Identification of a gene (FMR-1) containing a CGG repeat coincident with a breakpoint cluster region exhibiting length variation in fragile X syndrome. Cell 65:905–914

Verkerk AJ, de Graaff E, De Boulle K, Eichler EE, Konecki DS, Reyniers E, Manca A, Poustka A, Willems PJ, Nelson DL, Oostra BA (1993) Alternative splicing in the fragile X gene FMR1. Hum Mol Genet 2:1348

Wang YC, Lin ML, Lin SJ, Li YC, Li SY (1997) Novel point mutation within intron 10 of FMR-1 gene causing fragile X syndrome. Hum Mutat 10:393–399

Weiler IJ, Irwin SA, Klintsova AY, Spencer CM, Brazelton AD, Miyashiro K, Comery TA, Patel B, Eberwine J, Greenough WT (1997) Fragile X mental retardation protein is translated near synapses in response to neurotransmitter activation. Proc Natl Acad Sci USA 94:5395–5400

Weisman-Shomer P, Cohen E, Fry M (2000) Interruption of the fragile X syndrome expanded sequence d(CGG)$_n$ by interspersed d(AGG) trinucleotides diminishes the formation and stability of d(CGG)$_n$ tetrahelical structures. Nucleic Acids Res 28:1535–1541

Weisman-Shomer P, Cohen E, Fry M (2002) Distinct domains in the CArG-box binding factor A destabilize tetraplex forms of the fragile X expanded sequence d(CGG)$_n$. Nucleic Acids Res 30:3672–3681

Wohrle D, Kotzot D, Hirst MC, Manca A, Korn B, Schmidt A, Barbi G, Rott HD, Poustka P, Davis KE, Steinbach P (1992) A microdeletion of less than 250 kb, including the proximal part of the FMR-I gene and the fragile-X site, in a male with the clinical phenotype of fragile-X syndrome. Am J Hum Genet 51:299–306

Wohrle D, Hennig I, Vogel W, Steinbach P (1993) Mitotic stability of fragile X mutations in differentiated cells indicates early post-conceptional trinucleotide repeat expansion [see comments]. Nat Genet 4:140–142

Wohrle D, Salat U, Hameister H, Vogel W, Steinbach P (2001) Demethylation, reactivation, and destabilization of human fragile X full-mutation alleles in mouse embryocarcinoma cells. Am J Hum Genet 69:504–515. Epub 2001 Jul 2013

Yu S, Pritchard M, Kremer E, Lynch M, Nancarrow J, Baker E, Holman K, Mulley JC, Warren ST, Schlessinger D, Sutherland GR, Richards RI (1991) Fragile X genotype characterized by an unstable region of DNA. Science 252:1179–1181

Zalfa F, Giorgi M, Primerano B, Moro A, Di Penta A, Reis S, Oostra B, Bagni C (2003) The fragile X syndrome protein FMRP associates with BC1 RNA and regulates the translation of specific mRNAs at synapses. Cell 112:317–327

Zastrow MS, Vlcek S, Wilson KL (2004) Proteins that bind A-type lamins: integrating isolated clues. J Cell Sci 117:979–987

Zhang YQ, Bailey AM, Matthies HJ, Renden RB, Smith MA, Speese SD, Rubin GM, Broadie K (2001) Drosophila fragile X-related gene regulates the MAP1B homolog Futsch to control synaptic structure and function. Cell 107:591–603

Zoghbi HY, Orr HT (2000) Glutamine repeats and neurodegeneration. Annu Rev Neurosci 23:217–247

Zolnierowicz S (2000) Type 2A protein phosphatase, the complex regulator of numerous signaling pathways. Biochem Pharmacol 60:1225–1235

Nucleic Acids and Molecular Biology, Vol. 19
M. Fry/K. Usdin (Eds.)
Human Nucleotide Expansion Disorders

The Neglected Fragile X Mutations: *FRAXE* and *FRAXF*

David L. Nelson (✉) · Yanghong Gu

Department of Molecular and Human Genetics, Baylor College of Medicine,
Room 904E, One Baylor Plaza, Houston, TX 77030, USA
nelson@bcm.edu

Abstract The first triplet-repeat mutation to be reported was the large d(CGG) repeat expansion found in individuals with Fragile X syndrome (Verkerk et al. 1991). Significant attention has been paid to this common disease, the dynamics of its d(CGG) repeat, and the function of its associated gene, *FMR1* (Bagni and Greenough 2005). However, two other folate-sensitive fragile sites in the region have received much less attention since they were found to be distinct from the sequence affected in Fragile X syndrome. These sites, *FRAXE* and *FRAXF* are reviewed here. *FRAXE* expansion results in a mild learning deficit, can lead to more severe mental retardation, and has been associated with other mental disorders. *FRAXF* is apparently benign when expanded despite affecting at least one nearby gene. Interestingly, each of the fragile site loci contains a d(CGG) or d(CCG) repeat that is expressed in the 5′ untranslated portion of a neuronally expressed gene. Each expanded repeat sequence exhibits similar properties of instability, methylation and extinction of gene expression resulting in a loss of function of the associated genes. Advances in understanding the role of the repeat expansions in *FRAXE* and *FRAXF* and associated genomics and models are presented.

1 History of the Xq27-q28 Fragile Sites

The identification (Lubs 1969) and characterization (Sutherland and Ashforth 1979) of a cytogenetically visible fragile site located at the distal end of the long arm of the X chromosome in families with X-linked mental retardation led to extensive use of cytogenetics to characterize individuals with reduced cognitive capacity. Many additional families with cytogenetically defined Fragile X syndrome were identified and studied for their clinical features throughout the 1970s and 1980s. These analyses led to a broad clinical picture of this common disorder, suggesting reduced penetrance for many of the common features of the syndrome. With the identification of the *FMR1* gene and CGG repeat expansion leading to the *FRAXA* fragile site in 1991 (Fu et al. 1991; Kremer et al. 1991; Oberlé et al. 1991; Pieretti et al. 1991; Verkerk et al. 1991; Yu et al. 1991), it became apparent that some individuals with cytogenetically identical fragile sites in Xq27.3-q28 did not have the expanded d(CGG) repeats at *FMR1*. Two additional fragile sites were found nearby, and these were termed *FRAXE* and *FRAXF* for the fifth and sixth fragile sites described on the X chromosome.

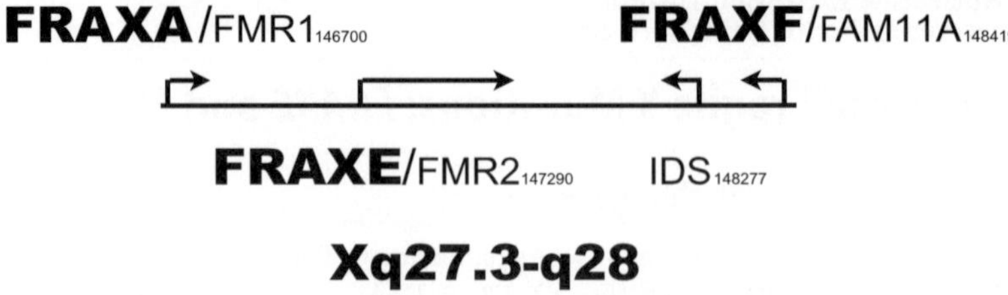

Fig. 1 Map of the Xq27.3-q28 region containing the three folate-sensitive fragile sites, *FRAXA*, *FRAXE* and *FRAXF*, showing relevant genes in the vicinity (*FMR1*, *FMR2* and *FAM11A*) are associated with the three fragile sites, and IDS is the iduronate sulfatase gene, defective in Hunter syndrome. *Numbers* indicate position (in thousands) from the *P*-terminus of the reference X chromosome sequence

The *FRAXE* site was subsequently identified (Flynn et al. 1993; Knight et al. 1994), and found to be 600 kb distal to the *FRAXA/FMR1* site. Similar to the *FRAXA* site, a d(CCG) repeat sequence was found to have increased in length at the *FRAXE* locus in patients with mental handicap, and to have caused a loss-of-function mutation in its associated gene (*FMR2*) (Gecz et al. 1996; Gu et al. 1996). A third fragile site (*FRAXF*) likewise showed d(CGG) repeat expansion in rare individuals who had been identified by cytogenetics (Hirst et al. 1993; Parrish et al. 1994; Ritchie et al. 1994). *FRAXF* is located 1.72 Mb distal to *FRAXA*, and is also associated with a gene, *FAM11A* (Shaw et al. 2002). This tight cluster of folate-sensitive fragile sites is so far unique in the human genome. No adequate explanation has been proposed for the presence of three loci capable of undergoing repeat expansion to form fragile sites in such a short interval. The complication with clinical definition of patients with these three distinct fragile sites was solved with the development of DNA-based tests for each of the repeat expansions.

The recognition that a subset of patients diagnosed with Fragile X syndrome actually carried repeat expansions at *FRAXE* and *FRAXF* provided a more consistent clinical description of the disorder caused by *FRAXA*. Since Fragile X syndrome resulting from loss of *FMR1* function was by far the most prevalent form of the disorder, it has retained the name of the clinical entity. The disorder caused by expansion at *FRAXE* is now termed *FRAXE* disease or *FRAXE* mental retardation. Since it is unclear whether expansions of *FRAXF* lead to disease, there is not yet a name for this clinical entity. Fragile X syndrome is reviewed elsewhere in this volume. Below, we discuss the current state of knowledge for repeat expansions and disease at *FRAXE* and *FRAXF*. A graphical representation of the region is provided in Fig. 1.

2
Clinical features of *FRAXE* disease

Patients with the expanded *FRAXE* repeats show mild to borderline mental retardation, with delays in language development a common problem. Some *FRAXE* patients also exhibit behavioral abnormalities, such as attention deficit, hyperactivity, autistic-like behavior, even schizophrenia and obsessive-compulsive disorder (OCD) (Gecz 2000b; Wang et al. 2003). Most patients with *FRAXE* are not easily distinguished from the general population as there are no consistent physical features in these patients, and *FRAXE* is considered to be a "non-syndromic" form of mental retardation. However, among *FRAXE* patients, reports of a long, narrow face, mild facial hypoplasia, a high-arched palate, irregular teeth, hair abnormalities, angiomata, clinodactyly, thick lips and nasal abnormalities can be found (Barnicoat et al. 1997; Biancalana et al. 1996; Carbonell et al. 1996; Hamel et al. 1994; Knight et al. 1996; Mulley et al. 1995; Russo et al. 1998). In addition, in some families, the *FRAXE* fragile site does not clearly segregate with mental retardation (IQ < 70) and some cytogenetically positive *FRAXE* males carrying an expanded and fully methylated allele have been reported to not display intellectual disability (Lo Nigro et al. 2000; Murgia et al. 1996; Sutherland and Baker 1992). These observations have made it difficult to demonstrate a clear causal relationship between *FRAXE* and non-syndromic mental retardation, leading some to term the consequences of the disorder as a mild mental handicap (Knight et al. 1994).

FRAXE repeat expansion is much less frequent than *FRAXA*, with estimates of the frequency in the 1/50 000 to 1/100 000 range (Crawford et al. 1999; Holden et al. 1996a; Murray et al. 1996; Youings et al. 2000). Recent data from a mouse model of loss of *Fmr2* function suggest that the lesion in the *FMR2* gene is likely to be involved in the MR found in *FRAXE* patients, and that variation in background genetics of families, and/or in the timing of loss of gene expression play significant roles in the phenotypic effect (Gu et al. 2002). These findings are discussed in more detail below.

In addition, a patient with an *FMR2* loss-of-function mutation that does not involve repeat expansion has been described (Gecz et al. 1996). He has mild learning disabilities and speech delay, supporting a role for *FMR2* in these functions, but his phenotype does not exclude the possibility that the repeat expansion mutation can affect additional genes or functions of the locus. In this respect, a transcript has been identified that is expressed in the opposite orientation from *FMR2* from the same promoter (Gecz 2000a). Gecz and colleagues termed this transcript *FMR3*, and suggested that it may also play a role in the *FRAXE* phenotype. It is not clear what the function of the *FMR3* transcript might be, but with the growing appreciation of the importance of RNA molecules in the control of numerous cell functions, the possibility for involvement must be considered.

3 Repeat Dynamics in Patients and Families

As with *FRAXA*, the *FRAXE* triplet repeat is polymorphic in the general population. Lengths ranging from as few as three to as many as 40 d(GCC) · d(CCG) repeats have been reported in studies of either the general population or in targeted populations (learning disabled or retarded individuals). The most common alleles in European individuals range from six to 25 repeats, with 15 repeats being predominant (Hamel et al. 1994; Knight et al. 1993, 1994). Affected individuals may have repeats ranging from ~ 130 to over 1000 triplets, but there is significant variability both between patients and within any individual patient with a full mutation. Methylation can also be variable, but appears to become likely once the repeat reaches the 100–130 d(GCC) range (Gecz 2000b). This contrasts with the situation at *FRAXA*, where the methylation threshold is between 200 and 230 repeats. Large repeats with only partial methylation have also been reported (Gecz 2000b).

Premutation alleles have been less well characterized in *FRAXE*, in part due to the scarcity of the mutation, and lack of extended pedigrees. Thus, the dynamics of this repeat are not as clearly defined as they are in *FRAXA*. However, some family transmission data have been reported, and these findings generally support the repeat's similarity to *FRAXA*. There appears to be a maternal bias for expansion from the premutation to the full mutation, as in *FRAXA* (Hamel et al. 1994; Mulley et al. 1995), yet full mutations appear to be transmitted by males as full mutations to their daughters (Hamel et al. 1994); a phenomenon that is never observed for *FRAXA* full mutations.

For *FRAXF*, the repeat dynamics are even less well known. In the general population, the repeat ranges from six to 38 triplets, with 14 found to be the most common length (Holden et al. 1996b; Parrish et al. 1994; Ritchie et al. 1994). Only four families with large expansions have been characterized. The expansions are large, similar to *FRAXA* and *FRAXE*, and there appears to be an intermediate expanded length that escapes methylation in parents of individuals with the fully methylated and expanded version. Since it appears unlikely that repeat expansion at *FRAXF* results in any pathology, and that the repeat expansion appears rare, collection of individuals with large repeats and studies of the behavior of the repeats have not been carried out.

An extensive survey of mammals demonstrated that the *FRAXA* d(CGG) repeat was conserved in the *FMR1* gene, with significant variation amongst species (Eichler et al. 1995). In the mouse, the *FRAXE* repeat is not conserved as a block of triplet repeat, although there are d(CCG) elements with interruptions in the same region of the 5′ untranslated (UTR) portion of the *Fmr2* gene (Chakrabarti et al. 1998). The *FRAXF* locus has recently demonstrated a gene (*FAM11A*) in humans that contains the d(CGG) repeat in its 5′ UTR (Shaw et al. 2002). The *FAM11A* gene is highly conserved between human and mouse at the amino acid level, but the d(CGG) repeat is not well conserved, although a hexamer sequence d(GCCGTC) that is typically repeated three times in humans and d(CGG) triplets can be found in the same position.

4 The *FMR2* Locus

FRAXE expansion and methylation eliminate expression of *FMR2*, named as the second fragile X-associated mental retardation gene in the region. Identification of the full gene was complicated by the presence of a very large first intron, which extends some 140 kb toward the Xq telomere; the fragile site was found rapidly after *FRAXA*, but determination of the full extent of the affected gene was slowed by the gene's size (500 kb). From its sequence the *FMR2* gene predicts a large protein (up to 1311 amino acids) with a long 3′ untranslated section of the cDNA. Message length varies, with forms ranging between 8.7 and 9.5 kb. There is also a shorter form (approximately 1/3 the full length) that appears to be developmentally regulated (Chakrabarti et al. 1996).

FMR2 is not similar to *FMR1*, and the predicted protein is very rich in proline, serine and threonine residues, with weak similarity ($\sim 25\%$ identity) to three other proteins, forming a small family. The other members of the family include *AF4*, *LAF4* and *AF5q31*. *AF4* was initially described as a frequent chromosome fusion partner in mixed lymphocytic leukemia rearrangements involving MLL, a homolog of the *Drosophila trithorax* gene. *AF4* has been studied in some detail, and appears to be a nuclear protein with features of a transcriptional activator. *FMR2* has proven to be similar, with nuclear localization and transcriptional activation properties (Hillman and Gecz 2001). With the completion of the human genome sequence, it is now clear that the *AF4*/*FMR2* gene family has four members, and nomenclature has been adjusted to reflect this. The new gene name for *FMR2* is *AFF2*, for *AF4*/*FMR2* family, member 2. *AF4* is now referred to as *AFF1*. For this review, the initial name of *FMR2* will be used to reduce confusion.

With the four members completely determined, it is now possible to define more highly conserved domains among the members. As depicted in Fig. 2, members share a conserved *N*-terminal domain of 90 amino acids, a 175 amino acid ALF domain, a 64 amino acid serine rich activation domain, a 20 amino acid consensus bipartite nuclear localization sequence and a conserved *C*-terminal domain that is 240 amino acids in length. The family also has similarity to a protein found in *Drosophila* known as *Lilliputian*, which has a similar activation domain, nuclear localization signal and *C*-terminal sequence, but lacks the *N*-terminal and ALF domains (Fig. 2). Effects of *Lilliputian* mutations are discussed below.

FMR2 expression can be found in abundance in brain and placenta of adults, but is not easily detected in other tissues (Chakrabarti et al. 1998;

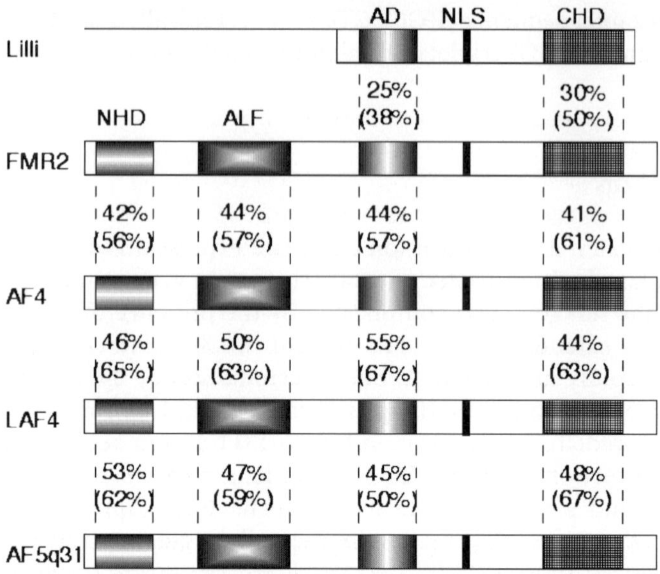

Fig. 2 Graphical representation of the protein sequences of the four paralogous genes in the *AF4/FMR2* family, along with the *Drosophila* orthologue *Lilli*. *Numbers between each protein* indicate percentage of identical amino acids conserved in each domain between the two proteins compared pairwise. *Numbers in parentheses* indicate percentage of similar amino acids conserved in each domain

Gecz et al. 1996; Gecz and Mulley 1999; Gu et al. 1996). This led to difficulties in determining the effects of *FRAXE* expansion, since transformed lymphoblasts do not express the gene normally. However, fibroblasts express sufficient RNA, and this tissue was used to demonstrate that repeat expansion and methylation reduce or eliminate expression of the gene (Gecz et al. 1996; Gu et al. 1996). Within the brain, the structures with the highest levels of expression are the amygdala and hippocampus.

Investigation of the *FMR2* gene product has been hampered by the absence of sensitive and specific antibodies (Miller et al. 2000). As a result, most investigation has relied on expression patterns of the *FMR2* RNA, and these studies have largely been carried out in mouse. The mouse *Fmr2* gene is highly similar to the human gene, with ∼ 77% identity at the nucleotide sequence level and 86% identity at the amino acid level (Chakrabarti et al. 1998). The mouse gene is slightly larger than the human (510 kb), and is also near the mouse *Fmr1* gene, but at a similar distance (658 kb). The human exon 5 is not found in the mouse genome, otherwise, the exons are well conserved, as are their boundaries. The two large introns found in the human gene (introns 1 and 3) are similarly large in the mouse. Thus, it is likely that the mouse and human genes overlap in expression and function, and suggest that the mouse may be a reasonable model for study of the human gene and disease.

5
Mouse Model of *Fmr2* Deficiency and *Drosophila Lilli*

Gu et al. (2002) reported a mouse model for *Fmr2* deficiency where the gene had been disrupted in the first exon with an expression cassette that allowed the *E. coli lacZ* gene to be expressed under control of the *Fmr2* promoter. This provided the ability to follow expression of the gene through staining for beta-galactosidase activity, and was used to define the expression pattern of *Fmr2* in early embryos. *Fmr2* is expressed in neurons early in development; expression can be seen as early as 10.5 days *post coitum* (pc), and is found at the ganglionic eminences of the telencephalon, in regions where the first groups of neuroblast cells differentiate. By 12.5 days pc, expression is found in early differentiating and migrating neuroblasts that form the primitive plexiform layer. The highest levels of expression are found at 15.5 days pc, with the cerebral cortex staining intensely along with other parts of the developing CNS and other regions of the developing fetus (bone, cartilage, hair follicles, lung, cardiac muscle and others). These expression data suggest a role for *Fmr2* in neuronal differentiation, as expression correlates with regions where the neurons are "born" and begin to assume their ultimate identities.

Absence of *Fmr2* in male mice leads to a small increase in the rate of spontaneous death, with 15% of knockout mice dead in the first 13 months compared with none of the normal littermates. This excess death is unexplained, and is not found in heterozygous females. No evidence for loss of embryos was found after genotyping offspring, and the animals that died early had no obvious pathology after histological examination.

Behavioral testing did reveal differences between *Fmr2* knockout mice and their control littermates. *Fmr2* deficient mice showed reduced fear conditioning in both contextual and conditioned fear paradigms that was delay dependent suggesting a memory consolidation defect. They were also more sensitive to painful stimulus in a hot plate test, suggesting a role for *Fmr2* in development of nociception. In the Morris water maze, the *Fmr2* knockout mice were slower to find the hidden platform, but were found to use a spatially biased search pattern in the probe test. Other measures of behavior were not found to be significantly different between the knockouts and their normal littermates.

Electrophysiological studies of the hippocampus of the *Fmr2* knockout mice revealed an enhancement of long-term potentiation (LTP) when measured in the CA1 region. While it is more common to find reduced LTP in mouse models of mental retardation, enhancement of LTP has also been described in mouse knockouts, and it is likely that disturbance of the balance of LTP in either direction can be debilitating to neuronal function.

Overall, the *Fmr2* mouse model shows significant defects in both behavioral and electrophysiological profiles, confirming the likely role for loss of function of this gene in the phenotypes found in patients with *FRAXE* expansion. Curiously, mice missing *Fmr2* appear to be more debilitated than knockouts of *Fmr1* in both behavioral and electrophysiological profiles (Kooy 2003). This is not the result expected from comparison of human patients with the two disorders—*FRAXA* patients are typically more affected than those with *FRAXE* deficiency. One possible explanation for this finding may lie in the mechanisms of mutation. The repeat expansion mutations at *FRAXA* and *FRAXE* ablate gene expression through methylation and down-regulation of the respective loci. While there are no data regarding the timing of these events in *FRAXE* expansion, for *FRAXA*, it is clear that early in fetal development, there is a substantial fraction of expanded (full mutation) alleles that are not methylated. This complicates diagnostic testing using amniocentesis since methylation status is not a reliable predictor of affected status at this time in development (Kallinen et al. 2000). From the mouse data, the majority of *FMR2* expression is found early in fetal development, suggesting that for human *FRAXE* patients, expression may be quasi-normal until methylation is established. In this case, by the time the gene has been turned off by methylation, it may have accomplished much of its early work. In contrast, since *FMR1* is expressed robustly throughout life, loss of function later in development may be more deleterious. Of course in both mouse models, the genes are turned off from conception, and this might exaggerate the phenotype beyond that found in the human repeat-expansion disorders.

Another confounding feature for both *FMR1* and *FMR2* models is the presence of related gene products that could compensate for loss of function. For *FMR1*, two paralogs, *FXR1* and *FXR2* are being studied for overlapping function (Bontekoe et al. 2002; Mientjes et al. 2004; Zhang et al. 1995). For *FMR2*, the other AFF family members may compensate for some of the *FMR2* function in early development. For example, double knockout mice missing *Fmr2* and *Af5q31* show significantly increased mortality compared to either single knockout (Gu and Nelson, 2006, personal communication).

Some significant insight into *FMR2* function has come from study of the fruit fly. In 2001, three groups independently identified an ortholog of the *AF4/FMR2* family in *Drosophila melanogaster*. *Lilliputian (Lilli)* is the only member of the *FMR2/AF4* gene family found in the *Drosophila* genome (Su et al. 2001; Tang et al. 2001; Wittwer et al. 2001). The *Lilli* protein has a C-terminal homology region (CHD) (Tang et al. 2001; Wittwer et al. 2001) that is 31–37% identical to the corresponding sequences of the human AFF proteins. Remarkably, the position of intron/exon boundaries within this region is conserved between *Lilli* and the human gene family. Like other family members, *Lilli* is rich in proline and serine residues (9.0% and 12.7% of all amino acids, respectively) and serines are found in *Lilli* at the same relative positions in *FMR2/AF4* family members. This indicates that the putative transactivation domain is conserved in *Lilli* although some domains in humans (NTD and ALF domains) are not present in *Lilli*. These data suggest that the function of *Lilli* should be similar to the function of the *FMR2/AF4* gene family and that elucidating its function in *Drosophila* should help us understand the function of this family in humans.

Lilli is essential for proper cellularization, gastrulation and segmentation during *Drosophila* embryogenesis (Tang et al. 2001). Most embryos deficient in *Lilli*, fail to hatch and subsequently die. A small percentage of embryos hatch and die as first or second instar larvae. Examination of these dead embryos shows two classes of phenotypes. One is involved with defects in segmentation and germband extension. The other defect is failure to secrete cuticle properly. These phenotypes are variable among individuals. Considering that genes known as pair-rule genes are required for segmental patterning and germband extension, Tang et al. (2001) examined the expression patterns of several of these genes in *Lilli* mutant embryos. Two genes: *fushi tarazu* and *huckebein* display a change in expression patterns in *Lilli* mutant embryos. Failure to secrete cuticle properly is associated with specific defects in the maintenance of the actin network and defects in transport of organelles during cellularization. In the mutants, actin filaments showed uneven distribution in nuclei, ranging from abnormally large bundles to multinucleated cells lacking filaments. These findings suggest potential targets for *FMR2* and family members in vertebrates, and studies are currently underway to define potential mis-regulated targets in the mouse models (Gu and Nelson 2003).

6
The *FMR3* Transcript

In 2000, Gecz reported the identification of a 3.8 kilobase transcript expressed in the opposite orientation from the *FMR2* gene, using the same promoter region, with a 5′ end a mere 3 nucleotides from the transcriptional start of *FMR2* (Gecz 2000a). The transcript is spliced, composed of two exons, but without a lengthy open reading frame, suggesting it does not encode a protein. Expression of *FMR3* is extinguished by repeat expansion and methylation at *FRAXE*, thus absence of its expression could contribute to the *FRAXE* phenotype, but a mechanism for this awaits further analysis of potential function for the *FMR3* transcript.

7
FRAXF and *FAM11A*

The *FRAXF* repeat expansion is rare, but has not been surveyed in detail. Individuals with expansions were identified in cytogenetic fragile site studies among patients with MR prior to molecular testing for *FRAXA* and *FRAXE*. Since such studies are much less frequent now, and since the phenotype caused by *FRAXF* expansion (if any) does not appear to encompass cognitive disabilities, no surveys for *FRAXF* expansions have been carried out. In the general population, the repeat is found to be composed of a variable d(CGG) triplet along with a d(GCCGTC) hexamer repeat that can vary. The most common allele is d(GCCGTC)$_3$d(GCC)$_8$, and the total length varies from 36 to 114 base pairs in the general population (12 to 38 triplet equivalents) (Holden et al. 1996b; Ritchie et al. 1997).

The scarcity of repeat expansion at *FRAXF*, coupled with the apparent absence of phenotype in individuals with expanded and methylated repeats led to little interest in the locus among the medical genetics community. With the

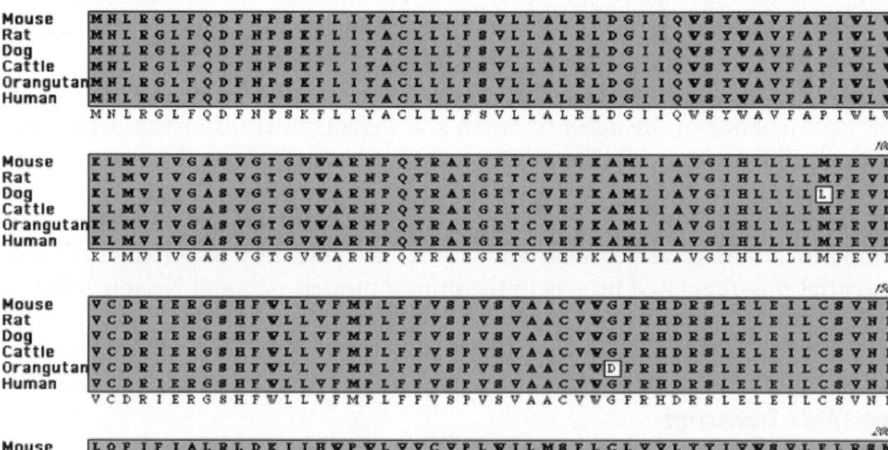

Fig. 3 Alignment of the amino acid sequences of mammalian *FAM11A* proteins predicted from mouse, rat, dog, bovine, orangutan and human. *Shading* indicates identical amino acids, and the *bottom line* represents a consensus sequence

completion of the sequence of the human genome, however, it has become clear that the *FRAXF* repeat, like those at *FRAXA* and *FRAXE*, is embedded in the 5′ untranslated region of a gene. Shaw et al. reported the sequence of *FAM11A*, a gene of unknown function that is transcribed from the *FRAXF* CpG island in a telomere to centromere direction and contains the *FRAXF* repeat in its mRNA (Shaw et al. 2002). Moreover, the gene's expression is extinguished by repeat expansion at *FRAXF*. *FAM11A* is most abundantly ex-

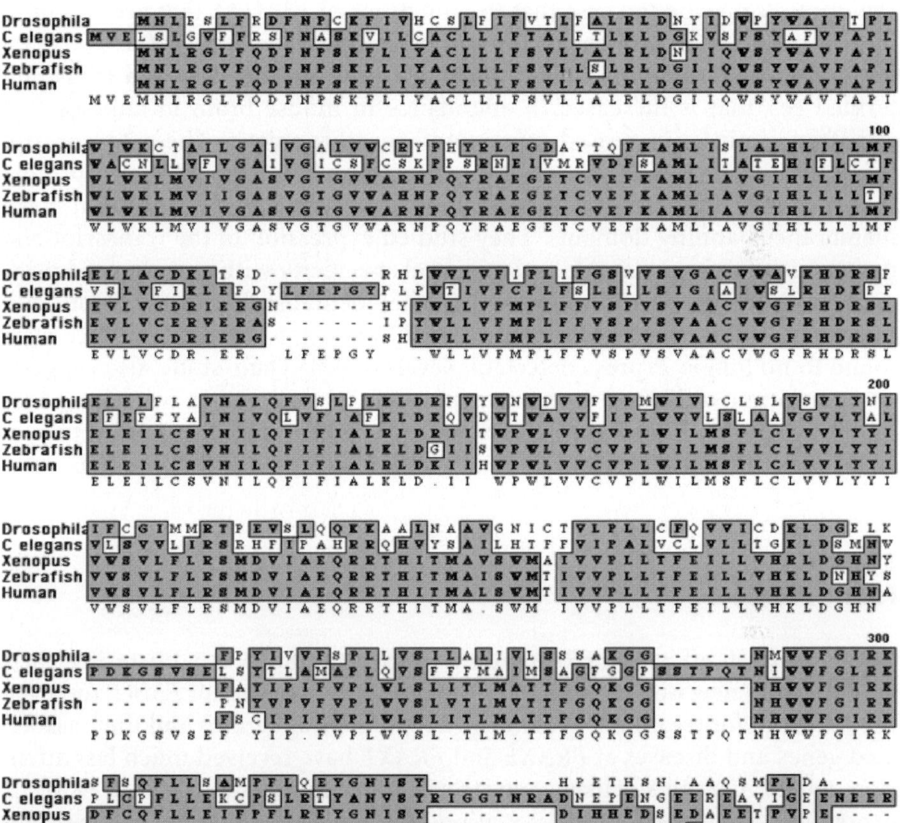

Fig. 4 Alignment of the human *FAM11A* protein with the most similar proteins identified from *Drosophila melanogaster*, *C. elegans*, *Xenopus laevis*, and Zebrafish (*Danio rerio*). *Shading* indicates identical amino acids, and the *bottom line* represents a consensus sequence

pressed in heart and skeletal muscle, along with placenta, while other tissues, such as brain and kidney, show reduced levels of expression.

Remarkably, the predicted amino acid sequence of *FAM11A* (350 amino acids) is 99.7% identical between the human and mouse proteins. Other mammals show similarly high identity (dog, cow, orangutan, and rat), *Xenopus* is 94% identical, chicken is 95% identical and orthologs can be found in both *Drosophila* and *C. elegans* (Figs. 3 and 4). An intronless autosomal paralog (*FAM11B*) that is found in humans (2q14) and mice (and other mammals) is 88% identical to *FAM11A* in the human, and is very highly conserved in evolution. It may be the case that the functions of *FAM11A* that are absent in individuals with *FRAXF* expansion can be compensated by *FAM11B*.

Maurer and coauthors (2004) described the identification of a transcript termed ee3 that is increased in abundance in mouse brain in animals that were transgenic for increased expression of erythropoietin. The ee3 transcript is identical to FAM11A. These authors recognized that the predicted protein sequence was similar to a G-protein coupled receptor, with the typical seven membrane spanning domains. They studied expression of the transcript and protein in adult brain, and found neuronal specificity with an enhanced presence in dendrites. Yeast two-hybrid studies demonstrated association with microtubule associated protein 1b (*Map1b*) and *Map1b* knockout mice were found to no longer express detectable levels of ee3. Their study also suggests an association with the 5-hydroxytryptamine 2a receptor in neurons. These findings suggest a possible role for the *FAM11A* gene and could point to possible areas of clinical investigation into individuals with *FRAXF* expansions.

8 Future Prospects

While research efforts at *FRAXA*/*FMR1* have continued at a blistering pace, with exciting new understanding of the potential function of *FMR1* and even the potential for treatment (Bear et al. 2004), the fragile sites and their associated genes and diseases at *FRAXE* and *FRAXF* have received much less attention. For *FRAXF*, this is likely appropriate since individuals lacking *FAM11A* expression due to *FRAXF* full mutations are apparently normal and this mutation is therefore not known to be pathogenic. *FRAXE* has received more attention, and numerous studies demonstrated that the repeat expansion mutation is rare in individuals with mental retardation or learning disabilities, with estimates of the frequency in the general population in the 1/50 000 to 1/100 000 range. While this is vastly less common than the *FRAXA*/*FMR1* Fragile X syndrome ($\sim$ 1/3500), it is not significantly less common than some of the other triplet repeat disorders. In addition, the size of the *FMR2* gene suggests it may be a reasonable target for mutations that do not involve triplet repeat expansion. Thus far, no systematic surveys of mutation at *FMR2* have been carried out, but as sequencing technology improves, it should be possible to contemplate a more comprehensive screen for *FMR2* mutations in learning disabled populations. It may also be worthwhile to contemplate other phenotypes as candidates for *FMR2* mutations. Association of *FRAXE* with obsessive-compulsive disorder suggests that there may be additional areas for understanding the role of this gene in neuronal development and function (Wang et al. 2003).

Models for *FMR2* are suggesting new avenues for determining function of the *FMR2* gene family. Additional mouse knockouts for the family should allow determination of functions that overlap and those that are unique to each member. Identification of gene targets for *FMR2* will allow a more complete catalog of the genetic pathways in which it participates, and will suggest possible alterations in neuronal function that could lead to the clinical picture of *FRAXE* patients. This knowledge will in turn lead to possible therapeutic interventions that may lead to improved lives for these rare but fascinating patients.

Acknowledgements The authors thank our many collaborators for useful discussions and advice, along with the families and patients who have contributed to these studies.

References

Bagni C, Greenough WT (2005) From mRNP trafficking to spine dysmorphogenesis: the roots of fragile X syndrome. Nat Rev Neurosci 6:376–387

Barnicoat A, Wang Q, Turk J, Green E, Mathew CG, Flynn G, Buckle V, Hirst M, Davies K, Bobrow M (1997) Clinical, cytogenetic, and molecular analysis of three families with FRAXE. J Med Genet 34:13–17

Bear MF, Huber KM, Warren ST (2004) The mGluR theory of fragile X mental retardation. Trends Neurosci 27:370–377

Biancalana V, Taine L, Bouix JC, Finck S, Chauvin A, De Verneuil H, Knight SJ, Stoll C, Lacombe D, Mandel JL (1996) Expansion and methylation status at FRAXE can be detected on EcoRI blots used for FRAXA diagnosis: analysis of four FRAXE families with mild mental retardation in males. Am J Hum Genet 59:847–854

Bontekoe CJ, McIlwain KL, Nieuwenhuizen IM, Yuva-Paylor LA, Nellis A, Willemsen R, Fang Z, Kirkpatrick L, Bakker CE, McAninch R, Cheng NC, Merriweather M, Hoogeveen AT, Nelson D, Paylor R, Oostra BA (2002) Knockout mouse model for Fxr2: a model for mental retardation. Hum Mol Genet 11:487–498

Carbonell P, Lopez I, Gabarron J, Bernabe MJ, Lucas JM, Guitart M, Gabau E, Glover G (1996) FRAXE mutation analysis in three Spanish families. Am J Med Genet 64:434–440

Chakrabarti L, Bristulf J, Foss GS, Davies KE (1998) Expression of the murine homologue of FMR2 in mouse brain and during development. Hum Mol Genet 7:441–448

Chakrabarti L, Knight SJ, Flannery AV, Davies KE (1996) A candidate gene for mild mental handicap at the FRAXE fragile site. Hum Mol Genet 5:275–282

Crawford DC, Meadows KL, Newman JL, Taft LF, Pettay DL, Gold LB, Hersey SJ, Hinkle EF, Stanfield ML, Holmgreen P, Yeargin-Allsopp M, Boyle C, Sherman SL (1999)

Prevalence and phenotype consequence of FRAXA and FRAXE alleles in a large, ethnically diverse, special education-needs population. Am J Hum Genet 64:495–507

Eichler EE, Kunst CB, Lugenbeel KA, Ryder OA, Davison D, Warren ST, Nelson DL (1995) Evolution of the cryptic *FMR1* CGG repeat. Nature Genet 11:301–308

Flynn GA, Hirst MC, Knight SJ, Macpherson JN, Barber JC, Flannery AV, Davies KE, Buckle VJ (1993) Identification of the FRAXE fragile site in two families ascertained for X linked mental retardation. J Med Genet 30:97–100

Fu YH, Kuhl DP, Pizzuti A, Pieretti M, Sutcliffe JS, Richards S, Verkerk AJ, Holden JJ, Fenwick RG Jr, Warren ST, Oostra BA, Nelson DL, Caskey CT (1991) Variation of the CGG repeat at the fragile X site results in genetic instability: resolution of the Sherman paradox. Cell 67:1047–1058

Gecz J (2000a) *FMR3* is a novel gene associated with FRAXE CpG island and transcriptionally silent in FRAXE full mutations. J Med Genet 37:782–784

Gecz J (2000b) The FMR2 gene, FRAXE and non-specific X-linked mental retardation: clinical and molecular aspects. Ann Hum Genet 64:95–106

Gecz J, Gedeon AK, Sutherland GR, Mulley JC (1996) Identification of the gene FMR2, associated with FRAXE mental retardation. Nat Genet 13:105–108

Gecz J, Mulley JC (1999) Characterisation and expression of a large, 13.7 kb FMR2 isoform. Eur J Hum Genet 7:157–162

Gu Y, McIlwain KL, Weeber EJ, Yamagata T, Xu B, Antalffy BA, Reyes C, Yuva-Paylor L, Armstrong D, Zoghbi H, Sweatt JD, Paylor R, Nelson DL (2002) Impaired conditioned fear and enhanced long-term potentiation in Fmr2 knock-out mice. J Neurosci 22:2753–2763

Gu Y, Nelson DL (2003) FMR2 function: insight from a mouse knockout model. Cytogenet Genome Res 100:129–139

Gu Y, Shen Y, Gibbs RA, Nelson DL (1996) Identification of FMR2, a novel gene associated with the FRAXE CCG repeat and CpG island. Nat Genet 13:109–113

Hamel B, Smits AP, de Graaff E, Smeets DF, Schoute F, Eussen BH, Knight SJ, Davies KE, Assman-Hulsmans CF, Oostra BA (1994) Segregation of FRAXE in a large family: clinical, psychometric, cytogenetic, and molecular data. Am J Hum Genet 55:923–931

Hillman MA Gecz J (2001) Fragile XE-associated familial mental retardation protein 2 (FMR2) acts as a potent transcription activator. J Hum Genet 46:251–259

Hirst MC, Barnicoat A, Flynn G, Wang Q, Daker M, Buckle VJ, Davies KE, Bobrow M (1993) The identification of a third fragile site, FRAXF, in Xq27–q28 distal to both FRAXA and FRAXE. Hum Mol Genet 2:197–200

Holden JJ, Julien-Inalsingh C, Chalifoux M, Wing M, Scott E, Fidler K, Swift I, Maidment B, Knight SJ, Davies KE, White BN (1996a) Trinucleotide repeat expansion in the FRAXE locus is not common among institutionalized individuals with non-specific developmental disabilities. Am J Med Genet 64:420–423

Holden JJ, Walker M, Chalifoux M, White BN (1996b) Trinucleotide repeats at the FRAXF locus: frequency and distribution in the general population. Am J Med Genet 64:424–427

Kallinen J, Heinonen S, Mannermaa A, Ryynanen M (2000) Prenatal diagnosis of fragile X syndrome and the risk of expansion of a premutation. Clin Genet 58:111–115

Knight S, Ritchie RJ, Chakrabarti L, Cross G, Taylor GR, Mueller RF, Hurst J, Paterson J, Yates JR, Dow DJ, Davies KE (1996) A study of FRAXE in mentally retarded individuals referred for fragile X syndrome (FRAXA) testing in the United Kingdom. Am J Hum Genet 58:906–913

Knight SJ, Flannery AV, Hirst MC, Campbell L, Christodoulou Z, Phelps SR, Pointon J, Middleton-Price HR, Barnicoat A, Pembrey ME, Holland J, Oostra BA, Bobrow M,

Davies KE (1993) Trinucleotide repeat amplification and hypermethylation of a CpG island in FRAXE mental retardation. Cell 74:127–134

Knight SJ, Voelckel MA, Hirst MC, Flannery AV, Moncla A, Davies KE (1994) Triplet repeat expansion at the FRAXE locus and X-linked mild mental handicap. Am J Hum Genet 55:81–86

Kooy RF (2003) Of mice and the fragile X syndrome. Trends Genet 19:148–154

Kremer EJ, Pritchard M, Lynch M, Yu S, Holman K, Baker E, Warren ST, Schlessinger D, Sutherland GR, Richards RI (1991) Mapping of DNA instability at the fragile X to a trinucleotide repeat sequence $p(CCG)_n$. Science 252:1711–1714

Lo Nigro C, Faravelli F, Cavani S, Perroni L, Novello P, Vitali M, Bricarelli FD, Grasso M (2000) FRAXE mutation in a mentally retarded subject and in his phenotypically normal twin brother. Eur J Hum Genet 8:157–162

Lubs H (1969) A marker X chromosome. Am J Hum Genet 21:231–244

Maurer MH, Grunewald S, Gassler N, Rossner M, Propst F, Wurz R, Weber D, Kuner T, Kuschinsky W, Schneider A (2004) Cloning of a novel neuronally expressed orphan G-protein-coupled receptor which is up-regulated by erythropoietin, interacts with microtubule-associated protein 1b and colocalizes with the 5-hydroxytryptamine 2a receptor. J Neurochem 91:1007–1017

Mientjes EJ, Willemsen R, Kirkpatrick LL, Nieuwenhuizen IM, Hoogeveen-Westerveld M, Verweij M, Reis S, Bardoni B, Hoogeveen AT, Oostra BA, Nelson DL (2004) *Fxr1* knockout mice show a striated muscle phenotype: implications for Fxr1p function in vivo. Hum Mol Genet 13:1291–1302

Miller WJ, Skinner JA, Foss GS, Davies KE (2000) Localization of the fragile X mental retardation 2 (FMR2) protein in mammalian brain. Eur J Neurosci 12:381–384

Mulley JC, Yu S, Loesch DZ, Hay DA, Donnelly A, Gedeon AK, Carbonell P, Lopez I, Glover G, Gabarron I, Yu PWL, Baker E, Haan EA, Hockey A, Knight SJL, Davis KE, Richards RI, Sutherland GR (1995) FRAXE and mental retardation. J Med Genet 32:162–169

Murgia A, Polli R, Vinanzi C, Salis M, Drigo P, Artifoni L, Zacchello F (1996) Amplification of the Xq28 FRAXE repeats: extreme phenotype variability? Am J Med Genet 64:441–444

Murray A, Youings S, Dennis N, Latsky L, Linehan P, McKechnie N, Macpherson J, Pound M, Jacobs P (1996) Population screening at the FRAXA and FRAXE loci: molecular analyses of boys with learning difficulties and their mothers. Hum Mol Genet 5:727–735

Oberlé I, Rousseau F, Heitz D, Kretz C, Devys D, Hanauer A, Boue J, Bertheas MF, Mandel JL (1991) Instability of a 550-base pair DNA segment and abnormal methylation in fragile X syndrome. Science 252:1097–1102

Parrish JE, Oostra BA, Verkerk AJ, Richards CS, Reynolds J, Spikes AS, Shaffer LG, Nelson DL (1994) Isolation of a GCC repeat showing expansion in FRAXF, a fragile site distal to FRAXA and FRAXE. Nat Genet 8:229–235

Pieretti M, Zhang FP, Fu, YH, Warren ST, Oostra BA, Caskey CT, Nelson DL (1991) Absence of expression of the FMR-1 gene in fragile X syndrome. Cell 66:817–822

Ritchie RJ, Chakrabarti L, Knight SJ, Harding RM, Davies KE (1997) Population genetics of the FRAXE and FRAXF GCC repeats, and a novel CGG repeat, in Xq28. Am J Med Genet 73:463–469

Ritchie RJ, Knight SJ, Hirst MC, Grewal PK, Bobrow M, Cross GS, Davies KE (1994) The cloning of FRAXF: trinucleotide repeat expansion and methylation at a third fragile site in distal Xqter. Hum Mol Genet 3:2115–2121

Russo S, Selicorni A, Bedeschi MF, Natacci F, Viziello P, Fortuna R, Pagani G, Dalpra L, Larizza L (1998) Molecular characterization of FRAXE-positive subjects with mental impairement in two unrelated Italian families. Am J Med Genet 75:304–308

Shaw MA, Chiurazzi P, Romain DR, Neri G, Gecz J (2002) A novel gene, FAM11A, associated with the FRAXF CpG island is transcriptionally silent in FRAXF full mutation. Eur J Hum Genet 10:767–772

Su M, Wisotzkey RG, Newfeld SJ (2001) A screen for modifiers of decapentaplegic mutant phenotypes identifies lilliputian, the only member of the Fragile-X/Burkitt's Lymphoma family of transcription factors in *Drosophila melanogaster*. Genetics 157:717–725

Sutherland G, Baker E (1992) Characterization of a new rare fragile site easily confused with the fragile X. Hum Mol Genet 1:111–113

Sutherland GR, Ashforth PLC (1979) X-linked mental retardation with macroorchidism and the fragile site at Xq27 or 28. Hum Genet 48:117–120

Tang A, Neufeld TP, Rubin GM, Muller HA (2001) Transcriptional regulation of cytoskeletal functions and segmentation by a novel maternal pair-rule gene, lilliputian. Development 128:801–813

Verkerk AJ, Pieretti M, Sutcliffe JS, Fu YH, Kuhl DP, Pizzuti A, Reiner O, Richards S, Victoria MF, Zhang FP, Eussen BE, van Ommen GJB, Blonden LAJ, Riggins GJ, Chastain JL, Kunst CB, Galjaard H, Caskey CT, Nelson DL, Oostra BA, Warren ST (1991) Identification of a gene (FMR-1) containing a CGG repeat coincident with a breakpoint cluster region exhibiting length variation in fragile X syndrome. Cell 65:905–914

Wang Q, Gu Y, Ferguson JM, Chen Q, Boatwright S, Gardiner J, Below C, Espinosa J, Nelson DL, Shaffer LG (2003) Cytogenetic analysis of obsessive-compulsive disorder (OCD): identification of a FRAXE fragile site. Am J Med Genet A 118:25–28

Wittwer F, van der Straten A, Keleman K, Dickson BJ, Hafen E (2001) Lilliputian: an AF4/FMR2-related protein that controls cell identity and cell growth. Development 128:791–800

Youings, SA, Murray A, Dennis N, Ennis S, Lewis C, McKechnie N, Pound M, Sharrock, A, Jacobs P (2000) FRAXA and FRAXE: the results of a five year survey. J Med Genet 37:415–421

Yu S, Pritchard M, Kremer E, Lynch M, Nancarrow J, Baker E, Holman K, Mulley JC, Warren ST, Schlessinger D, Sutherland GR, Richards RI (1991) Fragile X genotype characterized by an unstable region of DNA. Science 252:1179–1181

Zhang Y, Oconnor JP, Siomi MC, Srinivasan S, Dutra A, Nussbaum RL, Dreyfuss G (1995) The fragile X mental retardation syndrome protein interacts with novel homologs *FXR1* and *FXR2*. EMBO J 14:5358–5366

Nucleic Acids and Molecular Biology, Vol. 19
M. Fry/K. Usdin (Eds.)
Human Nucleotide Expansion Disorders

Friedreich Ataxia

Massimo Pandolfo

Service de Neurologie, Université Libre de Bruxelles-Hôpital Erasme,
Route de Lennik 808, 1070 Brussels, Belgium
massimo.pandolfo@ulb.ac.be

1
Clinical Features and Pathology

Friedreich ataxia (FA) is the most common of the early-onset hereditary ataxias in Indo-European and North African populations. The disease was first described in 1863 by Nickolaus Friedreich, Professor of Medicine at Heidelberg. Although Friedreich's papers described the essential clinical and pathological features of the disease as "degenerative atrophy of the posterior columns of the spinal cord" leading to progressive ataxia, sensory loss and muscle weakness, often associated with scoliosis, foot deformity and heart disease, subsequent descriptions of atypical cases and of clinically similar diseases clouded classification for many years. Diagnostic criteria were established in the late 1970s, after a renewed interest in the disease prompted several rigorous clinical studies (Geoffroy et al. 1976; Harding 1981). The following clinical features were considered essential to establish the diagnosis: (1) autosomal recessive inheritance, (2) onset before 25 years of age, (3) progressive limb and gait ataxia, (4) absent tendon reflexes in the legs, (5) electrophysiologic evidence of axonal sensory neuropathy, followed within 5 years of onset by (6) dysarthria, (7) areflexia at all four limbs, (8) distal loss of position and vibration sense, (9) extensor plantar responses, and (10) pyramidal weakness of the legs. The associated neuropathology is characterized by atrophy of the sensory pathways, with early loss of large neurons in the dorsal root ganglia (DRG), sensory axonal neuropathy, and degeneration of the posterior columns of the spinal cord. The cerebellum shows atrophy of the deep dentate nucleus, but its cortex is relatively preserved (Koeppen 2003).

The eventual identification of the FA gene (*FRDA*) and its most common mutation, the unstable hyperexpansion of a d(GAA) triplet repeat sequence (Campuzano et al. 1996), has allowed the clinical and pathological spectrum of the disease to be defined better. While the aforementioned criteria certainly identify the typical cases of FA, it is now clear that the disease shows a remarkable clinical variability, sometimes even within the same sibship, a rather uncommon finding for recessive disorders. Variability involves age of onset, rate of progression, and overall severity (Dürr et al. 1996; Montermini et al. 1997a). Cardiomyopathy, kyphoscoliosis, pes cavus, optic atrophy, hearing loss and diabetes mellitus only occur in some patients. Atypical cases with an overall FA-like phenotype but missing some of the essential diagnostic features can be identified. These include late-onset FA, which develops after the age of 25, sometimes as late as the sixth decade, and FA with retained tendon reflexes. The molecular basis for such a variability is only partially understood. Germ-line and somatic instability of the d(GAA) triplet repeat sequence certainly plays a role (Montermini et al. 1997a), but additional genetic and environmental factors are clearly involved. One example of a possible modifier genetic factor is the effect of mitochondrial DNA haplogroups (Giacchetti et al. 2004).

2
Gene Structure and Expression

The *FRDA* locus is in the proximal long arm of chromosome 9 (Chamberlain et al. 1988). The gene contains seven exons spanning 95 kb of genomic DNA. It is transcribed in the cen-tel direction. The major, and probably only functionally relevant messenger RNA (mRNA), has a size of 1.3 kb, corresponding to the first five exons, numbered 1–5a. The encoded protein, predicted to contain 210 amino acids, was designated frataxin (Campuzano et al. 1996).

The gene is expressed in all cells, but at variable levels in different tissues and at different times during development (Campuzano et al. 1996; Jiralerspong et al. 1997; Koutnikova et al. 1997). In adult humans, frataxin mRNA and protein are most abundant in the heart, brain and spinal cord, followed by liver, skeletal muscle, and pancreas. In mouse embryos, expression starts in the neuroepithelium at embryonic day 10.5 (E10.5), then reaches its highest level at E14.5 and into the postnatal period (Jiralerspong et al. 1997; Koutnikova et al. 1997). In developing mice, the highest levels of frataxin mRNA are found in the spinal cord, particularly at the thoracolumbar level, and in the DRG. The developing brain is also very rich in frataxin mRNA, which is abundant in the proliferating neural cells in the periventricular zone, in the cortical plates, and in the ganglionic eminence. Robust expression is also detected in the heart, in the axial skeleton, and in some epithelial (skin, teeth) and mesenchymal (brown fat) tissues (Jiralerspong et al. 1997; Koutnikova et al. 1997).

Overall, frataxin expression is generally higher in mitochondria-rich cells, such as cardiomyocytes and neurons. There is, however, a still-unexplained additional cell specificity, which in the nervous system is reflected in a higher abundance of frataxin in specific neuronal types, such as primary sensory neurons.

3 The d(GAA) Triplet Repeat Mutation

The most common mutation causing FA (98%) is the hyperexpansion of a d(GAA) triplet repeat in the first intron of the *FRDA* gene (Campuzano et al. 1996). FA is the only known disease to be caused by an expansion of d(GAA) triplets. Repeats in normal chromosomes contain up to approximately 40 triplets; disease-associated repeats contain from approximately 70 to more than 1000 triplets, most commonly 600–900 (Campuzano et al. 1996; Montermini et al. 1997b). Because of the recessive nature of the disease, affected individuals have expansions in both homologues of chromosome 9. Heterozygous carriers are clinically normal. This is the most common disease-causing triplet repeat expansion identified so far with 1 : 90 Europeans being a carrier. A small minority of patients (approximately 5%) are heterozygous for a d(GAA) expansion and a missense or nonsense point mutation disrupting the frataxin coding sequence (Campuzano et al. 1996; Cossée et al. 1999). No patients have been identified so far that carry point mutations in both copies of the frataxin gene.

3.1 Instability of Expanded Repeats

The FA-associated expansion shows instability when transmitted from parent to child (Campuzano et al. 1996; Dürr et al. 1996; Filla et al. 1996; Montermini et al. 1997a). Expansions and contractions of expanded d(GAA) repeats can both be observed. Expanded repeats are equally likely to further expand or contract during maternal transmission, but they most often contract during paternal transmission (Pianese et al. 1997; Monros et al. 1997), a result also supported by sperm analysis (Pianese et al. 1997). In this regard, FA resembles the other diseases associated with very large expansions in noncoding regions, such as fragile X syndrome and myotonic dystrophy, while smaller expansions of d(CAG) repeats in coding regions, such as those found in dominant ataxias and Huntington disease, are more likely to undergo size increases during paternal transmission.

Mitotic instability, leading to somatic mosaicism for expansion sizes, can be observed in FA (Montermini et al. 1997c). Analysis of d(GAA) expansions reveals ample variations in different cell types or tissues from the same patient. Furthermore, heterogeneity among cells occurs to a variable degree in different tissues. For instance, cultured fibroblasts and cerebellar cortex show very little heterogeneity of expansion size among individual cells, lymphocytes are more heterogeneous, and most brain regions show a rather complex pattern of allele sizes, indicating extensive cellular heterogeneity (Montermini et al. 1997c). While some of these differences could be accounted for by a major period of instability during the first weeks of embryonic development, expanded d(GAA) repeats may be inherently more stable in some cell types (Montermini et al. 1997c). In general, it is clear that determining the size of a patient's expansions in peripheral blood lymphocytes, from which DNA is usually obtained, only provides a single sample of the overall repeat size distribution occurring within that patient, and therefore only an approximate estimate of expansion sizes in affected tissues.

3.2 Origin and Mechanisms of Expansion of the Repeat

The d(GAA) repeat associated with FA is localized within an Alu sequence (GAA – Alu). Alu sequences are a heterogeneous group of primate-specific interspersed repetitive DNA elements with an estimated frequency of $5 \times 10^5$5–1×10^6 copies per genome. They may serve as functional polIII-transcribed genes and are probably derived from 7SL genes. Their pervasiveness and variability are the result of constant amplification and retrotranposon-mediated reinsertion throughout the genome over 65 million years of primate evolution. Despite their diversity, Alu sequences can be grouped into subfamilies whose members share a few, common diagnostic base changes. By comparing differences between these sequences, Alu elements can be used as molecular clocks to estimate the age of a particular subfamily or member of a subfamily. GAA – Alu is assigned to the AluSx subfamily. Identity between GAA – Alu and the AluSx consensus sequence is 89%, in agreement with the overall $92 \pm 3\%$ identity between individual AluSx subfamily sequences and the consensus sequence. On the basis of sequence similarity, the average age of the AluSx subfamily has been estimated to be 37 million years (Kapitonov and Jurka 1996). The FA-associated d(GAA) repeat is situated in the middle of GAA – Alu, preceded by a stretch of an average of 16 adenine residues, apparently derived from an expansion of the canonical A_5TACA_6 sequence linking the two halves of Alu sequences. GAA – Alu is flanked by a 13-bp perfect direct repeat d(AAAATGGATTTCC), suggesting a recent Alu retroposition/insertion event, an idea supported by the estimated age of the AluSx subfamily.

Alleles at the d(GAA) repeat site can be subdivided into three classes depending on their length: short normal alleles (SN, approximately 82% in Europeans), long normal alleles (LN, approximately 17% in Europeans), and pathological expanded alleles (E, approximately 1% in Europeans) (Cossée et al. 1997; Montermini et al. 1997b). The length polymorphism of the d(GAA) repeat in normal alleles suggests that it was generated by two types of events. Small changes, plus or minus one trinucleotide, may have caused limited size heterogeneity. Such small changes were likely to be the consequence of occasional events of polymerase "stuttering" during DNA replication, i.e., slippage followed by misrealignment of the newly synthesized strand by one or, rarely, a few repeat units (Richards and Sutherland 1994).

This basic polymorphism-generating mechanism has been postulated for all simple-sequence repeats (Wells 1996). By comparison, the jump from the SN to the LN group was probably a singular event. Linkage disequilibrium studies were carried out in European and also in Yemenite and North African families, with single nucleotide polymorphisms spanning the *FRDA* gene and with polymorphisms of the polyalanine sequence adjacent to the d(GAA) repeat. These studies indicate that E and LN alleles appear genetically homogeneous and likely related, while SN represents a more heterogeneous class of alleles (Monticelli et al. 2004).

Possibly, the event that created LN alleles was the sudden duplication of an SN allele containing eight or nine d(GAA) triplets, creating an LN allele with 16 or 18 d(GAA) triplets. This occurred presumably in Africa, leading to a population of chromosomes with LN alleles sharing the same background haplotype. Single repeat insertion/deletions, resulting from DNA polymerase stuttering, gave rise to the spectrum of stable d(GAA) repeats ranging from 12 to about 25 triplets. One or a few of these chromosomes subsequently migrated to Europe and/or to the Middle East, but not to East Asia, where no LN (or E) alleles are found. It is hard to speculate about the mechanism leading to such a sudden doubling of the repeat; however, similar events have been shown to occur in triplet repeats cloned into bacterial plasmids (Pluciennik et al. 2000). Recombination-based mechanisms such as unequal sister-chromatid exchange and gene conversion have been proposed as generators of variability in tandem repeats (Wells 1996) and in microsatellites (Jakupciak and Wells 2000), but alternative hypotheses such as the occurrence of an exceptionally large slippage event cannot be excluded. The passage from LN to E alleles probably involved a second genetic event of the same kind, which generated "very long" LN alleles containing 32–36d(GAA)triplets still on the same haplotype background as the "shorter" LN alleles from which they derived. By reaching the instability threshold, estimated as 34 d(GAA)triplets (Montermini et al. 1997b), they form a reservoir for expansions. The occurrence of a second duplication event is suggested by the lack of both E and LN alleles with more than 21 d(GAA) triplets alleles in Africans. The ethnic–geographic distribution of FA could be explained if the second event occurred prior to the divergence of Indo-Europeans and Afro-Asiatic speakers. According to the previously described scenario, the extent of linkage disequilibrium between LN alleles and linked marker loci on chromosomes of African descent is expected to be lower than that between LN and E alleles and the same marker in Europeans (Labuda et al. 1997; Harpending et al. 1998), as is in fact observed (Labuda et al. 2000). Accordingly, LN chromosomes in Africa appear to be 3.2 times older than the LN chromosomes in Europe, and these appear to be 1.27 times older than E chromosomes. Assuming the age of LN African chromosomes to be about 100 000 years, one would date the origin of European LN chromosomes at about 30 000 years ago and that of the

E chromosomes at about 25 000 years ago, i.e., following the Upper Paleolithic population expansion.

It was possible to directly observe the hyperexpansion of "very long" LN premutation alleles containing more than 34 d(GAA) triplets. This length is close to the instability threshold for other triplet repeat associated disorders, such as those involving d(CGG) and d(CAG) repeats (Eichler et al. 1994).

Strand displacement during DNA replication is thought to be the mechanism that leads to reiterative synthesis and expansion. For this phenomenon to occur, the displaced strand has to form some kind of secondary structure (Parniewski and Staczek 2002). A single DNA strand containing a d(GAA) repeat is able to form different types of secondary structure (LeProust et al. 2000), which may be involved in instability. A single d(CTT) strand seems structureless (LeProust et al. 2000), and this difference may play a role in determining whether deletions or expansions are favored according to the direction of the replicating fork. Strand displacement is promoted by stalling of DNA polymerase caused by an alternate DNA structure, or by tightly bound proteins, or both (Wells 1996). The triplex-forming ability of long FA d(GAA) repeats, discussed next, may be involved in repeat instability by causing DNA polymerase stalling as well as by forming a target for protein binding.

4 Pathogenic Mechanisms: Triplexes and Sticky DNA

FA is due to frataxin deficiency. Frataxin levels in pathology specimens and in cultured cells from FA patients are markedly lower than in normal controls (Campuzano et al. 1997). The decrease in frataxin protein and mRNA is proportional to the size of the expanded d(GAA) repeats, particularly the smaller one, indicating a direct etiologic role of these repeats in suppressing frataxin gene expression (Campuzano et al. 1997). How can an intron-located repetitive sequence have such an effect? The currently prevailing hypothesis is that the d(GAA) repeat adopts an unusual DNA structure that interferes with transcription of the frataxin gene. An effect on transcription was first suggested by transfection experiments in which the expression of a two-exon reporter gene was inhibited by the insertion of a d(GAA) repeat of pathological length in the intron. Those experiments did not reveal splicing abnormalities and provided evidence in favor of a transcription block between the two exons (Ohshima et al. 1998). The frataxin repeat is a tract of oligopurines (R) and oligopyrimidines (Y). It has been proposed that the pathological structure adopted by disease-causing lengths of this repeat is a triplex (Ohshima et al. 1996, 1998; Bidichandani et al. 1998; Filla et al. 1996; Grabczyk and Usdin 1998; Gacy et al. 1997). Triplexes are three-stranded nucleic acid structures that can form at such RY sequences (Wells 1996). The third strand occupies the major grove of the DNA double helix, forming Hoogsteen pairs between R or Y bases of the Watson–Crick base pairs. In intramolecular triplexes, as can be observed in vitro in supercoiled plasmid DNA, the RY DNA folds back onto itself to form the triple-helical structure. Four different isomers may form, two based on RRY and two on YRY structures. Intermolecular triplexes are formed between oligonucleotides or polynucleotides (DNA or RNA) and target RY sequences on duplex DNA. Extensive investigations of triplexescon-ducted in the 1980s and 1990s provided substantial information on the type of sequence required, the effects of pH and methylation of the cytosine residues, the effect of interposing non-RY sequences, the influence of environmental factors on the stabilization of the four triplex isomers, and stabilization by factors like intercalating agents (Wells et al. 1988; Frank-Kamenetskii and Mirkin 1995; Soyfer and Potaman 1996; Sinden 1994; Guieysse et al. 1996; Bacolla et al. 1995; Xu and Goodridge 1996; Hanvey et al. 1988, 1989a, b; Shimizu et al. 1989, 1990; Kang et al. 1992a, b; Ohshima et al. 1996). RRY triplexes are more versatile than YRY triplexes since they tolerate more diverse pairing schemes and their stability does not depend on lower pH for hemiprotonation of the cytosine residues, but rather requires the presence of divalent metal ions; however, in vitro short repeat tracts formed by the FA d(GAA) repeat at neutral pH are of the YRY type. Bimolecular YRY triplexes form when the repeat tract approaches the premutation range and the amount of negative supercoiling is higher (Potaman et al. 2004). A new type of DNA structure, consistent with intramolecular triplex formation, was shown to be adopted by lengths of d(GAA) as found in FA. This structure, called "sticky DNA", was first demonstrated in plasmids containing long tracts of d(GAA) where it appeared as an anomalously retarded band in agarose gels in which linearized plasmids containing d(GAA) repeats were separated (Sakamoto et al. 1999). Such slow-migrating bands were shown to have a number of physicochemical properties that are typical of intramolecular RRY triplexes. In particular, the retarded band appeared only if the plasmid was negatively supercoiled prior to linearization, and it was sensitive to divalent ion concentration and temperature as is typical for RRY triplexes. An excellent correlation was found between the lengths of d(GAA) and the formation of this novel conformation: FA patients have 66 or more repeats; sticky DNA was found only for repeats longer than 59 units. In vitro transcription studies of $d(GAA)_n$ repeats ($n = 9–150$) using T7 or SP6 RNA polymerase showed that, when a gel-isolated sticky DNA template was transcribed, the amount of full-length RNA synthesized was significantly reduced compared with the amount synthesized by transcription of the linear template. Surprisingly, transcriptional inhibition was observed not only for the sticky DNA template but also for another DNA molecule used as an internal control in an orientation-independent manner. The molecular mechanism of transcriptional inhibition by sticky DNA was a sequestration of the RNA polymerases by direct binding to the complex DNA structure (Sakamoto et al. 2001a). A $d(GAAGGA)_{65}$ sequence, also found in intron 1 of the frataxin gene, does not form sticky

DNA nor does it inhibit transcription in vivo and in vitro and it does not associate with the FA disease state (Ohshima et al. 1999). This finding suggests that interruptions in the d(GAA) sequence may destabilize its structure and facilitate transcription. Systematic analysis of the effects of introducing interruptions into a d(GAATTC)$_{150}$ repeat by substituting an increasing number of adenines with guanines has confirmed that the sticky DNA/triplex structure is progressively destabilized and fails to form when the sequence becomes d(GAAGGA)$_{75}$. As the tendency to form a sticky DNA/triplex structure decreases, less and less inhibition of transcription is observed in vivo and in vitro (Sakamoto et al. 2001b).

5 Genotype–Phenotype Correlation

5.1 d(GAA) Triplet Repeat Expansion

As expected by the experimental finding that smaller expansions allow a higher residual gene expression (Campuzano et al. 1997), expansion sizes have an influence on the severity of the phenotype. A direct correlation has been firmly established between the size of d(GAA) repeats and an earlier age of onset, earlier age of confinement to a wheelchair, more rapid rate of disease progression, and presence of nonobligatory disease manifestations indicative of more widespread degeneration (Dürr et al. 1996; Montermini et al. 1997b). However, differences in d(GAA) expansions account for only about 50% of the variability in the age of onset, indicating that other factors influence the phenotype. These may include somatic mosaicism for expansion sizes, variations in the frataxin gene itself, modifier genes, and environmental factors.

5.2 Point Mutations

About 2% of the FA chromosomes carry d(GAA) repeat tracts of normal length, but have missense, nonsense, or splice-site mutations ultimately affecting the frataxin coding sequence (Campuzano et al. 1996; Cossée et al. 1999). All affected individuals with a point mutation identified so far are heterozygous for an expanded d(GAA) repeat on the other homologue of chromosome 9. It is possible that homozygotes for point mutations have not yet been found just because point mutations are rare, but it is more likely that homozygosity for frataxin point mutations would cause a lethal phenotype, as suggested by the recent observation that frataxin knockout mice (Cossée et al. 2000) and mice homozygous for a frataxin missense mutation (P. Ioannu, personal communication) die during embryonic development.

A few missense mutations are associated with milder atypical phenotypes with slow progression, suggesting that the mutated proteins preserve some residual function. Patients carrying the G130V mutation have early onset but slow progression, no dysarthria, mild limb ataxia, and retained reflexes. A similar phenotype occurs in individuals with the mutations D122Y and R165P. For reasons that are not yet clear, individuals with frataxin point mutations have a much higher frequency of optic atrophy (50%) than individuals with repeat expansions (Cossée et al. 1999).

6 Frataxin Structure and Function

The FA gene (*FRDA*) (Chamberlain et al. 1988; Campuzano et al. 1996) encodes a small mitochondrial matrix protein, frataxin that is highly conserved in evolution. A single frataxin gene is found in all eukaryotes, including fungi and plants. A homologue, CyaY, is present in Gram-negative bacteria and in other prokaryotes like *Rickettsia prowazeckii*, thought to be related to the hypothetical mitochondrial precursor. *FRDA* is expressed in all cells, but at variable levels in different tissues and during development (Koutnikova et al. 1997; Jiralerspong et al. 1997). In adult humans, frataxin mRNA is most abundant in the heart, brain, and spinal cord, followed by liver, skeletal muscle, and pancreas.

FA patients have a profound but not complete frataxin deficiency, with a small residual amount of normal protein as a result of the d(GAA) triplet repeat expansion.

Structural studies have been carried out on frataxin (Dhe-Paganon et al. 2000; Musco et al. 2000) and its bacterial homologue, CyaY (Cho et al. 2000) by nuclear magnetic resonance and by crystallography. The structure is compact, overall globular, containing an *N*-terminal α-helix, a middle β-sheet region composed of seven β-strands, a second α-helix, and a *C*-terminal coil. On the outside, a ridge of negatively charged residues and a patch of hydrophobic residues are highly conserved.

Knockout of the yeast frataxin homologous gene (*YFH1*) in yeast (*Δyfh1*) causes the loss of oxidative phosphorylation and of mitochondrial DNA (Babcock et al. 1997; Wilson and Roof 1997). Iron accumulates in mitochondria of *Δyfh1* to more then tenfold its level in wild-type yeast. Loss of respiratory competence requires the presence of iron in the culture medium, and occurs more rapidly as the iron concentration in the medium is increased, suggesting that permanent mitochondrial damage is the consequence of iron toxicity (Radisky et al. 1999). Formation of the highly toxic hydroxyl radical through the Fenton reaction is suggested by the enhanced sensitivity of *Δyfh1* to H_2O_2 (Babcock et al. 1997). In *Δyfh1* yeast, there is a marked induction (tenfold to 50-fold) of the high-affinity iron transport system on the cell membrane, normally not expressed in yeast cells that are iron-replete (Babcock et al. 1997). This induction has been recently related to a deficit in mitochondrial synthesis of iron–sulfur clusters (ISCs), rather than cytosolic iron depletion as previously thought (Chen et al. 2004). ISC-containing enzymes, such as respiratory chain complexes I, II, and III, and aconitase, are impaired in *Δyfh1* yeast (Rötig et al. 1997). Frataxin appears to be involved in an early step of ISC synthesis (Muhlenhoff et al. 2003), through its interaction with the scaffold protein Isu1, where the first ISC assembly takes place, probably facilitating iron incorporation (Yoon and Cowan 2003). This finding suggests that frataxin may be a mitochondrial iron chaperone, protecting this metal from reactive oxygen species and making it bioavailable. Recent data support this view, suggesting that frataxin also acts as an iron chaperone in heme synthesis (Yoon and Cowan 2004), and in the modulation of aconitase activity (Bulteau et al. 2004). A much higher affinity of frataxin for the heme-synthesis enzyme ferrochelatase than for Isu1 (Yoon and Cowan 2004) would explain why heme synthesis is resistant to low frataxin levels and is essentially unaffected in FA patients.

7
Animal Models

A mouse model of FA has been difficult to generate because a complete loss of frataxin, such as in frataxin knockout mice, causes early embryonic lethality (Cossée et al. 2000). Viable mouse models have been obtained so far only through a conditional gene targeting approach. The first two models utilized Cre transgenes under the control of the muscle creatine kinase (*MCK*) and of the neuron-specific enolase (*NSE*) promoters to induce striated muscle- and neuron-restricted exon deletion, respectively. *NSE* mutants have a low birth weight and develop a progressive neurological phenotype with an average onset of ataxia at 12 days, hunched stance, and loss of proprioception (Puccio et al. 2001). *MCK* mutants show cardiac hypertrophy with thickening of the walls of the left ventricle, and show myocardial degeneration with cytoplasmic vacuolization in the myocytes, evidence of necrosis, and postnecrotic fibrosis (Puccio et al. 2001). Loss of activity of ISC-containing enzymes is an early finding in these models. The *MCK* mutants accumulate iron in heart mitochondria at later stages. Using a similar conditional knockout approach, but with a tamoxifen-inducible Cre recombinase under the control of a neuron-specific prion protein promoter, Simon et al. (2004) developed two different lines developed which exhibit a progressive neurological phenotype with slow evolution that recreates the neurological features of the human disease. An autophagic process was detected in the DRG, leading to removal of mitochondrial debris and the appearance of lipofuscin deposits.

8
Pathogenic Mechanisms in Friedreich Ataxia

Altered iron metabolism, free-radical damage, and mitochondrial dysfunction all occur in FA patients, suggesting that information derived from investigations on frataxin function and from the yeast and animal models is relevant for the pathogenesis of the human disease. Oxidative stress is revealed by increased plasma levels of malondialdheyde, a lipid peroxidation product (Emond et al. 2000), increased urinary 8-hydroxy-2′-deoxyguanosine, a marker of oxidative DNA damage (Schulz et al. 2000), decreased plasma-free glutathione (Piemonte et al. 2001) and elevated plasma glutathione S-transferase activity (Tozzi et al. 2002). Increased free-radical production could be directly demonstrated in cultured cells engineered to produce reduced levels of frataxin (Santos et al. 2001). In addition, H_2O_2 induces apoptosis in patients' fibroblasts at lower doses than in control fibroblasts (Wong et al. 1999), suggesting that even nonaffected cells are at risk for oxidative stress as a consequence of the primary genetic defect. FA fibroblasts also show abnormal antioxidant responses, in particular a blunted increase in mitochondrial superoxide dismutase triggered by iron and by oxidants in control cells (Jiralerspong et al. 2001). Mitochondrial dysfunction has been proven to occur in vivo in FA patients. Phosphorus magnetic resonance spectroscopy analysis of skeletal muscle and heart shows a reduced rate of ATP synthesis (Lodi et al. 1999). Finally and most importantly, the same multiple ISC-containing enzyme dysfunctions found in *Δyfh1* yeast and in mouse models are also found in affected tissues from FA patients (Rötig et al. 1997).

Activation of stress pathways, triggered by mitochondrial dysfunction, occurs in FA and is likely to play an important role in cell atrophy and death. Studies on cultured PC12 cells, rat pheochromocytoma cells that can be differentiated into neurons by adding nerve growth factor, showed in particular an increased expression and activity of the MKK4-JNK kinase pathway, which may be at first a protective response but eventually triggers apoptosis. Different vulnerable cell types may activate different pathways, as suggested by the observation of the specific occurrence of autophagic vacuoles only in primary sensory neurons in the inducible conditional knockout mouse model (Simon et al. 2004).

9
Perspectives for Treatment

All FA patients carry at least one allele with an expanded d(GAA) repeat and therefore make an insufficient amount of otherwise normal frataxin. If it were possible to increase their frataxin production to levels that are similar to those of healthy carriers, one could possibly stop the course of the disease and maybe even induce some improvement. Increased frataxin production could be obtained:

1. Through gene replacement therapy, i.e., by introducing a frataxin gene without the d(GAA) expansion into the patient cells.
2. By giving frataxin directly. The protein should, however, be modified such that it will be able to reach the nerve cells affected by the disease and the mitochondria within these cells.
3. By using molecules that can destabilize the triple-helical structure formed by the d(GAA) repeat and shift the equilibrium toward the physiological double helix that allows frataxin expression.

Though still in their infancy, all these approaches are under study. Recently, encouraging results have been obtained for gene replacement therapy, with partial correction of the oxidative stress hypersensitivity of FA fibroblasts by frataxin-encoding adeno-associated virus and lentivirus vectors (Fleming et al. 2005).

Additional ways to treat the disease may become apparent from studies on the function of frataxin. On the basis of these findings, therapeutic approaches aimed at controlling the levels of free radicals and regulating respiratory chain activation may be proposed. Concerning antioxidant molecules and respiratory chain stimulants, some coenzyme Q derivatives (idebenone, CoQ-10) have already yielded promising results, not only in experimental models (Seznec et al. 2004), but also in clinical trials, at least with respect to FA cardiomyopathy (Buyse et al. 2003; Mariotti et al. 2003). Automated high-throughput tests to evaluate a large number of molecules for their ability to correct the functional consequences of frataxin deficiency are under way. An intriguing possibility would be the identification of small molecules capable of effectively replacing frataxin by binding mitochondrial iron and increasing its bioavailability.

Last, cellular therapies, in particular the use of stem cells, could be useful in the treatment of FA. However, the widespread nature of neurodegeneration in FA is a major obstacle to this approach since it would require the widespread delivery of cells in the central nervous system of the patients.

Remarkable progress has been made in understanding the pathogenesis of FA since the gene responsible was discovered in 1996. In addition, investigating the pathogenesis of FA has stimulated research on numerous basic areas of biology, from DNA structure and biochemistry to iron metabolism. However, most excitingly perhaps is the now realistic perspective of developing a treatment for this so far incurable neurodegenerative disease.

References

Babcock M, de Silva D, Oaks R, Davis-Kaplan S, Jiralerspong S, Montermini L, Pandolfo M, Kaplan J (1997) Regulation of mitochondrial iron accumulation by Yfh1, a putative homolog of frataxin. Science 276:1709–1712

Bacolla A, Ulrich MJ, Larson JE, Ley TJ, Wells RD (1995) An intramolecular triplex in the human gamma-globin 5′-flanking region is altered by point mutations associated with hereditary persistence of fetal hemoglobin. J Biol Chem 270:24556–24563

Bidichandani SI, Ashizawa T, Patel PI (1998) The GAA triplet-repeat expansion in Friedreich's ataxia interferes with transcription and may be associated with an unusual DNA structure. Am J Hum Genet 62:111–121

Bulteau AL, O'Neill HA, Kennedy MC, Ikeda-Saito M, Isaya G, Szweda LI (2004) Frataxin acts as an iron chaperone protein to modulate mitochondrial aconitase activity. Science 305:242–245

Buyse G, Mertens L, Di Salvo G, Matthijs I, Weidemann F, Eyskens B, Goossens W, Goemans N, Sutherland GR, Van Hove JL (2003) Idebenone treatment in Friedreich's ataxia: neurological, cardiac, and biochemical monitoring. Neurology 60:1679–1681

Campuzano V, Montermini L, Moltó MD, Pianese L, Cossée M, Cavalcanti F, Monros E, Rodius F, Duclos F, Monticelli A, Zara F, Canizares J, Koutnikova H, Bidichandani SI, Gellera C, Brice A, Trouillas P, De Michele G, Filla A, De Frutos R, Palau F, Patel PI, Di Donato S, Mandel JL, Cocozza S, Koenig M, Pandolfo M (1996) Friedreich ataxia: autosomal recessive disease caused by an intronic GAA triplet repeat expansion. Science 271:1423–1427

Campuzano V, Montermini L, Lutz Y, Cova L, Hindelang C, Jiralerspong S, Trottier Y, Kish SJ, Faucheux B, Trouillas P, Authier FJ, Durr A, Mandel JL, Vescovi A, Pandolfo M, Koenig M (1997) Frataxin is reduced in Friedreich ataxia patients and is associated with mitochondrial membranes. Hum Mol Genet 6:1771–1780

Chamberlain S, Shaw J, Rowland A, Wallis J, South S, Nakamura Y, von Gabain A, Farrall M, Williamson R (1988) Mapping of mutation causing Friedreich' ataxia to human chromosome 9. Nature 334:248–250

Chen OS, Crisp RJ, Valachovic M, Bard M, Winge DR, Kaplan J (2004) Transcription of the yeast iron regulon does not respond directly to iron but rather to iron-sulfur cluster biosynthesis. J Biol Chem 279:29513–29518

Cho SJ, Lee MG, Yang JK, Lee JY, Song HK, Suh SW (2000) Crystal structure of Escherichia coli CyaY protein reveals a previously unidentified fold for the evolutionarily conserved frataxin family. Proc Natl Acad Sci USA 97:8932–8937

Cossée M, Schmitt M, Campuzano V, Dahl N, Trouillas P, Allinson P, Kostrzewa M, Nivelon-Chevallier A, Gustavson KH, Kohlschutter A, Muller U, Mandel JL, Brice A, Koenig M, Cavalcanti F, Tammaro A, De Michele G, Filla A, Cocozza S, Labuda M, Montermini L, Poirier J, Pandolfo M (1997) Evolution of the Friedreich ataxia trinucleotide repeat expansion: founder effect and premutations. Proc Natl Acad Sci USA 94:7452–7457

Cossée M, Dürr A, Schmitt M, Dahl N, Trouillas P, Allinson P, Kostrzewa M, Nivelon-Chevallier A, Gustavson KH, Kohlschutter A, Muller U, Mandel JL, Brice A, Koenig M, Cavalcanti F, Tammaro A, De Michele G, Filla A, Cocozza S, Labuda M, Montermini L, Poirier J, Pandolfo M (1999) Frataxin point mutations and clinical presentation of compound heterozygous Friedreich ataxia patients. Ann Neurol 45:200–206

Cossée M, Puccio H, Gansmuller A, Koutnikova H, Dierich A, LeMeur M, Fischbeck K, Dolle P, Koenig M (2000) Inactivation of the Friedreich ataxia mouse gene leads to early embryonic lethality without iron accumulation. Hum Mol Genet 9:1219–1226

Dhe-Paganon S, Shigeta R, Chi YI, Ristow M, Shoelson SE (2000) Crystal structure of human frataxin. J Biol Chem 275:30753–30756

Dürr A, Cossée M, Agid Y, Campuzano V, Mignard C, Penet C, Mandel JL, Brice A, Koenig M (1996) Clinical and genetic abnormalities in patients with Friedreich ataxia. N Engl J Med 335:1169–1175

Eichler EE, Holden JJA, Popovich BW, Reiss AL, Snow K, Thibodeau SN, Richards CS, Ward PA, Nelson DL (1994) Length of uninterrupted CGG repeats determines instability in the FMR1 gene. Nat Genet 8:88–94

Emond M, Lepage G, Vanasse M, Pandolfo M (2000) Increased levels of plasma malondialdehyde in Friedreich ataxia. Neurology 55:1752–1753

Filla A, De Michele G, Cavalcanti F, Pianese L, Monticelli A, Campanella G, Cocozza S (1996) The relationship between trinucleotide (GAA) repeat length and clinical features in Friedreich ataxia. Am J Hum Genet 59:554–560

Foury F (1999) Low iron concentration and aconitase deficiency in a yeast frataxin homologue deficient strain. FEBS Lett 456:281–284

Frank-Kamenetskii MD, Mirkin SM (1995) Triplex DNA Structures. Annu Rev Biochem 64:65–95

Geoffroy G, Barbeau A, Breton G, Lemieux B, Aube M, Leger C, Bouchard JP (1976) Clinical description and roentgenologic evaluation of patients with Friedreich ataxia. Can J Neurol Sci 3:279–286

Geschwind DH, Perlman S, Grody W et al. (1997) The Friedreich's ataxia GAA repeat expansion in patients with recessive or sporadic ataxia. Neurology 49:1004–1009

Giacchetti M, Monticelli A, De Biase I, Pianese L, Turano M, Filla A, De Michele G, Cocozza S (2004) Mitochondrial DNA haplogroups influence the Friedreich's ataxia phenotype. J Med Genet 41:293–295

Grabczyk E, Usdin K (2000) The GAA*TTC triplet repeat expanded in Friedreich's ataxia impedes transcription elongation by T7 RNA polymerase in a length and supercoil dependent manner. Nucleic Acids Res 28:2815–2822

Guieysse A-L, Praseuth D, Grigoriev M, Harel-Bellan A, Helene C (1996) Detection of covalent triplex with human cells. Nucleic Acids Res 24:4210–4216

Hanvey JC, Klysik J, Wells RD (1988a) Influence of DNA sequence on the formation of non-B right-handed helices in oligopurine · oligopyrimidine inserts in plasmids. J Biol Chem 263:7386–7396

Hanvey JC, Shimizu M, Wells RD (1988b) Intramolecular DNA triplexes in supercoiled plasmids. Proc Natl Acad Sci USA 85:6292–6296

Hanvey JC, Shimizu M, Wells RD (1989a) Intramolecular DNA triplexes in supercoiled plasmids: II. Effect of base composition and non-central interruptions on formation and stability. J Biol Chem 264:5950–5956

Hanvey JC, Shimizu M, Wells RD (1989b) Site-specific inhibition of EcoRI restriction/modification enzymes via DNA triple helix. Nucleic Acids Res 18:157–161

Harding AE (1981) Friedreich ataxia: a clinical and genetic study of 90 families with an analysis of early diagnosis criteria and intrafamilial clustering of clinical features. Brain 104:589–620

Jakupciak JP, Wells RD (2000) Gene conversion (recombination) mediates expansions of CTG · CAG repeats. J Biol Chem 275:40003–40013

Jiralerspong S, Liu Y, Montermini L, Stifani S, Pandolfo M (1997) Frataxin shows developmentally regulated tissue-specific expression in the mouse embryo. Neurobiol Dis 4:103–113

Jiralerspong S, Ge B, Hudson TJ, Pandolfo M (2001) Manganese superoxide dismutase induction by iron is impaired in Friedreich ataxia cells. FEBS Lett 509:101–105

Kang S, Wohlrab F, Wells RD (1992a) Metal ions cause the isomerization of certain intramolecular triplexes. J Biol Chem 267:1259–1264

Kang S, Wohlrab F, Wells RD (1992b) GC rich flanking tracts decrease the kinetics of intramolecular DNA triplex formation. J Biol Chem 267:19435–19442

Kapitonov V, Jurka J (1996) The age of Alu subfamilies. J Mol Evol 42:59–65

Koeppen A (2002) The neuropathology of inherited ataxias. In: Manto M, Pandolfo M (eds) The cerebellum and its disorders. Cambridge University Press, Cambridge, pp 387–408

Koutnikova H, Campuzano V, Foury F, Dollé P, Cazzalini O, Koenig M (1997) Studies of human, mouse and yeast homologues indicate a mitochondrial function for frataxin. Nat Genet 16:345–351

Labuda M, Labuda D, Miranda C, Poirier J, Soong B, Barucha NE, Pandolfo M (2000) Unique origin and specific ethnic distribution of the Friedreich ataxia GAA expansion. Neurology 54:2322–2324

Lamarche JB, Côté M, Lemieux B (1980) The cardiomyopathy of Friedreich ataxia morphological observations in 3 cases. Can J Neurol Sci 7:389–396

LeProust EM, Pearso CE, Sinden RR, Gao X (2000) Unexpected formation of parallel duplex in GAA, TTC trinucleotide repeats of Friedreich's ataxia. J Mol Biol 302:1063–1080

Lodi R, Cooper JM, Bradley JL, Manners D, Styles P, Taylor DJ, Schapira AH (1999) Deficit of in vivo mitochondrial ATP production in patients with Friedreich ataxia. Proc Natl Acad Sci USA 96:11492–11495

Mariotti C, Solari A, Torta D, Marano L, Fiorentini C, Di Donato S (2003) Idebenone treatment in Friedreich patients: one-year-long randomized placebo-controlled trial. Neurology 60:1676–1679

Montermini L, Andermann E, Labuda M, Richter A, Pandolfo M, Cavalcanti F, Pianese L, Iodice L, Farina G, Monticelli A, Turano M, Filla A, De Michele G, Cocozza S (1997a) The Friedreich ataxia GAA triplet repeat: premutation and normal alleles. Hum Mol Genet 6:1261–1266

Montermini L, Richter A, Morgan K, Justice CM, Julien D, Castellotti B, Mercier J, Poirier J, Capozzoli F, Bouchard JP, Lemieux B, Mathieu J, Vanasse M, Seni MH, Graham G, Andermann F, Andermann E, Melancon SB, Keats BJ, Di Donato S, Pandolfo M (1997b) Phenotypic variability in Friedreich ataxia: role of the associated GAA triplet repeat expansion. Ann Neurol 41:675–682

Montermini L, Kish SJ, Jiralerspong S, Lamarche JB, Pandolfo M (1997c) Somatic mosaicism for the Friedreich's ataxia GAA triplet repeat expansions in the central nervous system. Neurology 49:606–610

Monticelli A, Giacchetti M, De Biase I, Pianese L, Turano M, Pandolfo M, Cocozza S (2004) New clues on the origin of the Friedreich ataxia expanded alleles from the analysis of new polymorphisms closely linked to the mutation. Hum Genet 114:458–463

Muhlenhoff U, Gerber J, Richhardt N, Lill R (2003) Components involved in assembly and dislocation of iron-sulfur clusters on the scaffold protein Isu1p. EMBO J 22:4815–4825

Musco G, Stier G, Kolmerer B, Adinolfi S, Martin S, Frenkiel T, Gibson T, Pastore A (2000) Towards a structural understanding of Friedreich's ataxia: the solution structure of frataxin. Struct Fold Des 8:695–707

Ohshima K, Kang S, Larson JE, Wells RD (1996) Cloning, characterization, and properties of seven triplet repeat DNA sequences. J Biol Chem 271:16773–16783

Ohshima K, Montermini L, Wells RD, Pandolfo M (1998) Inhibitory effects of expanded GAATTC triplet repeats from intron 1 of Friedreich's ataxia gene on transcription and replication in vivo. J Biol Chem 273:14588–14595

Ohshima K, Sakamoto N, Labuda M, Poirier J, Moseley ML, Montermini L, Ranum LP, Wells RD, Pandolfo M (1999) A nonpathogenic GAAGGA repeat in the Friedreich gene: implications for pathogenesis. Neurol 53:1854–1857

Parniewski P, Staczek P (2002) Molecular mechanisms of TRS instability. Adv Exp Med Biol 516:1–25

Piemonte F, Pastore A, Tozzi G, Tagliacozzi D, Santorelli FM, Carrozzo R, Casali C, Damiano M, Federici G, Bertini E (2001) Glutathione in blood of patients with Friedreich's ataxia. Eur J Clin Invest 31:1007–1011

Pluciennik A, Iyer RR, Parniewski P, Wells RD (2000) Tandem duplication. A novel type of triplet repeat instability. J Biol Chem 275:28386–28397

Potaman VN, Oussatcheva EA, Lyubchenko YL, Shlyakhtenko LS, Bidichandani SI, Ashizawa T, Sinden RR (2004) Length-dependent structure formation in Friedreich ataxia $(GAA)_n^*(TTC)_n$ repeats at neutral pH. Nucleic Acids Res 32:1224–1231

Puccio H, Simon D, Cossée M, Criqui-Filipe P, Tiziano F, Melki J, Hindelang C, Matyas R, Rustin P, Koenig M (2001) Mouse models for Friedreich ataxia exhibit cardiomyopathy, sensory nerve defect, Fe – S enzyme deficiency followed by intramitochondrial iron deposits. Nat Genet 27:181–618

Radisky DC, Babcock MC, Kaplan J (1999) The yeast frataxin homologue mediates mitochondrial iron efflux. Evidence for a mitochondrial iron cycle. J Biol Chem 274:4497–4499

Reaban ME, Griffin JA (1990) Induction of RNA-stabilized DNA conformers by transcription of an immunoglobulin switch region. Nature 348:342–344

Richards RI, Sutherland GR (1994) Simple repeat DNA is not replicated simply. Nat Genet 6:114–116

Rötig A, deLonlay P, Chretien D, Foury F, Koenig M, Sidi D, Munnich A, Rustin P (1997) Frataxin gene expansion causes aconitase and mitochondrial iron-sulfur protein deficiency in Friedreich ataxia. Nat Genet 17:215–217

Rustin P, von Kleist-Retzow JC, Chantrel-Groussard K, Sidi D, Munnich A, Rötig A (1999) Effect of idebenone on cardiomyopathy in Friedreich's ataxia: a preliminary study. Lancet 354:477–479

Sakamoto N, Chastain PD, Parniewski P, Ohshima K, Pandolfo M, Griffith JD, Wells RD (1999) Sticky DNA: self-association properties of long GAATTC repeats in RRY triplex structures from Friedreich ataxia. Mol Cell 3:465–475

Sakamoto N, Ohshima K, Montermini L, Pandolfo M, Wells RD (2001) Sticky DNA, a self-associated complex formed at long GAATTC repeats in intron 1 of the Frataxin gene, inhibits transcription. J Biol Chem 276:27171–27177

Santos M, Ohshima K, Pandolfo M (2001) Frataxin deficiency enhances apoptosis in cells differentiating into neuroectoderm. Hum Mol Genet 10:1935–1944

Schulz JB, Dehmer T, Schöls L, Mende H, Hardt C, Vorgerd M, Bürk K, Matson W, Dichgans J, Beal MF, Bogdanov MB (2000) Oxidative stress in patients with Friedreich ataxia. Neurology 55:1719–1721

Seznec H, Simon D, Monassier L, Criqui-Filipe P, Gansmuller A, Rustin P, Koenig M, Puccio H (2004) Idebenone delays the onset of cardiac functional alteration without correction of Fe – S enzymes deficit in a mouse model for Friedreich ataxia. Hum Mol Genet 13:1017–1024

Shimizu M, Hanvey JC, Wells RD (1989) Intramolecular DNA Triplexes in supercoiled plasmids: I. Effect of loop size on formation and stability. J Biol Chem 264:5944–5949

Shimizu M, Hanvey JC, Wells RD (1990) Multiple Non-B-DNA Conformations of polypurine · polypyrimidine sequences in plasmids. Biochemistry 29:4704–4713

Simon D, Seznec H, Gansmuller A, Carelle N, Weber P, Metzger D, Rustin P, Koenig M, Puccio H (2004) Friedreich ataxia mouse models with progressive cerebellar and sensory ataxia reveal autophagic neurodegeneration in dorsal root ganglia. J Neurosci 24:1987–1995

Sinden RR (1994) DNA structure, function. Academic Press, San Diego

Soyfer VN, Potaman VN (1996) Triple-helical nucleic acids. Springer, Berlin Heidelberg New York

Tozzi G, Nuccetelli M, Lo Bello M, Bernardini S, Bellincampi L, Ballerini S, Gaeta LM, Casali C, Pastore A, Federici G, Bertini E, Piemonte F (2002) Antioxidant enzymes in blood of patients with Friedreich's ataxia. Arch Dis Child 86:376–379

Wells RD (1996) Molecular basis of genetic instability of triplet repeats. J Biol Chem 271:2875–2878

Wells RD, Collier DA, Hanvey JC, Shimizu M, Wohlrab F (1988) The chemistry and biology of unusual DNA structures adopted by oligopurine-oligopyrimidine sequences. FASEB J 2:2939–2949

Wilson RB, Roof DM (1997) Respiratory deficiency due to loss of mitochondrial DNA in yeast lacking the frataxin homologue. Nat Genet 16:352–357

Wong A, Yang J, Cavadini P, Gellera C, Lonnerdal B, Taroni F, Cortopassi G (1999) The Friedreich ataxia mutation confers cellular sensitivity to oxidant stress which is rescued by chelators of iron and calcium and inhibitors of apoptosis. Hum Mol Genet 8:425–430

Xu G, Goodridge AG (1996) Characterization of a polypyrimidine/polypurine tract in the promoter of the gene for chicken malic enzyme. J Biol Chem 271:16008–16019

Yoon T, Cowan JA (2004) Frataxin-mediated iron delivery to ferrochelatase in the final step of heme biosynthesis. J Biol Chem 279:25943–25946

Yoon T, Cowan JA (2003) Iron-sulfur cluster biosynthesis. Characterization of frataxin as an iron donor for assembly of [2Fe – 2S] clusters in ISU-type proteins. J Am Chem Soc 125:6078–6084

Nucleic Acids and Molecular Biology, Vol. 19
M. Fry/K. Usdin (Eds.)
Human Nucleotide Expansion Disorders

Dodecamer Repeat Expansion in Progressive Myoclonus Epilepsy 1

Maria D. Lalioti[1] · Stylianos E. Antonarakis[2] (✉) · Hamish S. Scott[3]

[1]Department of Ob/Gyn, Yale School of Medicine, 333, Cedar Street, New Haven, CT 06510, USA
maria.lalioti@yale.edu

[2]Department of Genetic Medicine and Development, University of Geneva Medical School, and Cantonal Hospital of Geneva, 1 Rue Michel Servet, 1205 Geneva, Switzerland
stylianos.antonarakis@medecine.unige.ch

[3]Genetics and Bioinformatics Division, The Walter and Eliza Hall Institute of Medical Research, 1G Royal Parade, 3050 Parkville,Victoria, Australia
hscott@wehi.edu.au

1 Introduction and Disease Features

Progressive myoclonus epilepsy of the Unverricht–Lundborg type (EPM1; OMIM no. 254800) is an autosomal recessive disorder originally described by Unverricht (1891) and Lundborg (1903). EPM1 is a rare disorder in the general population but is relatively more common in Finland (1 : 20000) and the western Mediterranean, and is thus also known as Baltic and Mediterranean myoclonus (Norio and Koskiniemi 1979; Eldridge et al. 1983; Genton et al. 1990; Labauge et al. 1997). However, affected families have been described in various areas around the world, e.g., the USA or the eastern Mediterranean (Eldridge et al. 1983; Mazarib et al. 2001).

The disease is characterized by severe stimulus-sensitive myoclonus, generalized tonic–clonic seizures, and a characteristic electroencephalogram (Berkovic et al. 1991; Canafoglia et al. 2004). The onset of the disease is between 6 and 18 years of age, and the progression, the severity, and survival vary between and within families (Koskiniemi et al. 1974a, b; Norio and Koskiniemi 1979; Lehesjoki 2002). Mental deterioration, dementia, and cerebellar ataxia develop late in the course of the disease, which is usually 10–20 years in duration (Koskiniemi et al. 1974a; Eldridge et al. 1983). Cognitive functions are only mildly if at all affected and the patients do not present psychotic symptoms; they are emotionally labile, however, and show a higher rate of suicide.

The symptoms are now efficiently managed using antiepileptic drugs, mainly valproic acid alone or in combination with clonazepam and piracetam.

Histologically, the brain shows degenerative changes which can be easily distinguished from Lafora bodies, the characteristic acid mucopolysaccharide inclusions found in EPM2 (Carpenter and Karpati 1981). The most consistent finding in EPM1 is the marked loss of Purkinje cells in the cerebellum, neuronal loss in medial thalamus and spinal cord, and most likely cell loss in the granular cell layer (Haltia et al. 1969; Koskiniemi et al. 1974a, b; Eldridge et al. 1983; Meldium and Bruton 1992; Mascalchi et al. 2002; Takuma et al. 2003). Extracellular and occasionally intracellular periodic acid–Schiff positive granulations were found in neurons and glial cells and in the liver, spleen, heart, lungs, renal tubules, and posterior lobes of the pituitary gland in a large consanguineous Swiss family (Klein et al. 1968; Klein and Rabinowicz 1980). Membrane-bound vacuoles with clear contents in eccrine cells have also been reported in some EPM1 patients (Cochius et al. 1994).

The molecular genetics findings described herein have allowed EPM1 to be distinguished from other epilepsies using genetic diagnosis, and have shed light on the cause and pathophysiology of the disease.

2 Linkage Analysis and Positional Cloning of the EPM1 Gene

EPM1 was originally mapped to the terminal band of human chromosome 21q22.3 in Finnish patients (Lehesjoki et al. 1991). An identical linkage mapping was confirmed for EPM1 patients from other parts of the world (Malafosse et al. 1992). With the refinement of the genetic maps, as a result of the human genome project, the availability of more efficient linkage methods, and the recruitment of more families, the disease was finally mapped to a 175-kb interval on chromosome 21 (Cochius et al. 1993; Lehesjoki et al. 1993a, b; Lalioti et al. 1995; Virtaneva et al. 1996). The Finnish population showed strong linkage disequilibrium, predicting a founder effect and a common mutation (Lehesjoki et al. 1993b; Virtaneva et al. 1996). A founder effect was also described in other parts of the world (Parmeggiani et al. 1997; Moulard et al. 2002, 2003).

Since the biochemical defect in EPM1 remained undiscovered, positional cloning approaches were undertaken. Complete arrays of overlapping clones covering the area were constructed (Lafreniere et al. 1995; Stone et al. 1996), and were extensively used to identify candidate genes through complementary DNA selection and exon trapping. Candidate genes were tested in a trial-and-error fashion, and excluded if no mutation was found. This approach led to the identification of point mutations on the cystatin B (also named stefin B or CSTB) gene on a small subgroup of patients with EPM1 (Pennacchio et al. 1996).

3 Cystatin B

Cystatin B is a member of the cysteine proteinase inhibitor superfamily (PROSITE PDOC00259; reviewed in Turk and Bode 1991; Turk et al. 1997, 2002b). This name was first used to describe an inhibitor of papain and related endopeptidases isolated from chicken egg white (Barrett 1981).

Cysteine proteinases constitute a group of proteolytic enzymes that cleave peptide bonds using a catalytic cysteine residue. Their inhibitors are classified into three distinct groups:

1. Type 1 cystatins (also named stefins) are molecules of approximately 100 amino acid residues and molecular size of 11 kDa, with no disulfide bonds or carbohydrate groups. Cystatins A and B belong in this category. They are stable at neutral and alkaline pH and resist heat. These proteins are potent and reversible competitive inhibitors of cysteine proteinases, with highest inhibition constants for papain, and cathepsins B, H, and L. In addition to animal cystatins, two cystatins have been described in rice that were found to inhibit insect digestive cysteine proteinases, therefore acting as plant-resistance mechanisms (Liang et al. 1991).
2. Type 2 cystatins are slightly larger molecules of 115 amino acids and a molecular size of 13 kDa. In contrast to type 1 cystatins, proteins of this group contain one or two disulfide loops near their C-terminus. Cystatin C is the best-characterized member of this category. It was isolated from serum of patients with autoimmune diseases (Brzin et al. 1984). A mutant cystatin C, with a Q68L substitution is a major constituent of amyloid fibrils in patients with hereditary cerebral hemorrhage with amyloidosis, described in Ireland (Ghiso et al. 1986).
3. Type 3 cystatins or kininogens are much larger plasma glycoproteins of 68–120 kDa.

Residues 46–50 of cystatin B constitute the so-called QVVAG domain, which is highly conserved in type 1 and plant cystatins. This region is deleted in some EPM1 patients with splice site mutations as discussed later. In type 2 cystatins only Q46 and G50 are consistently present. Kininogens contain three QVVAG domains and are predicted to have arisen from type 1 cystatins by gene triplication (Muller-Esterl et al. 1985; Rawlings and Barrett 1990).

The other highly conserved residue among all cystatins (except rice II) is Gly4. One EPM1 patient was found to be homozygous for a G4R mutation (*vide infra*).

Human cystatin B (NM_000100) is a small gene of three exons and a total genomic size of 3 kb (Fig. 1a). The coding region which is 297-bp long encodes a protein of 98 amino acids (Swiss-Prot P04080). In human and mouse, cystatin B is expressed in all tissues tested (Pennacchio et al. 1996; Pennacchio and Myers 1997). In the brain, cystatin B is present in neural stem cells and in mature neurons and glial cells. However, there are some differences in the subcellular localization: in stem cells it is localized in the nucleus and in astrocytes in nucleus and cytoplasm (Riccio et al. 2001; Brannvall et al. 2003). In the cytoplasm, cystatin B is present in lysosomes. The Purkinje cells of the cerebellum, which are affected in EPM1, express cystatin B and its distribution is developmentally regulated (Riccio et al. 2005). Cystatin B is present in the nucleus of proliferating primary myoblasts and COS-1 cells (Alakurtti et al. 2005). In differentiated myotubes cystatin B is excluded from the nucleus and is detected in punctate cytoplasmic structures, some of which are lysosomes. In embryonic liver cells, cystatin B is diffusely distributed throughout the cytoplasm (Calkins et al. 1998). Although mainly an intracellular protein, cystatin B has also been isolated from extracellular fluids (Abrahamson et al. 1986).

Human cystatin B forms inactive disulfide-linked dimers of 25 kDa. Under reducing conditions, these dimers are converted into active monomers (Wakamatsu et al. 1984). The crystal structure of human cystatin B in stoichiometric complex with papain has been determined to 2.4-Å resolution (Stubbs et al. 1990), (Fig. 2a). X-ray crystallography revealed that the molecule consists of five-stranded β-sheets wrapped around a five-turn α-helix. Crystallography also verified, as is the case in general with cystatins, that the main interactions with papain are provided by the amino terminal

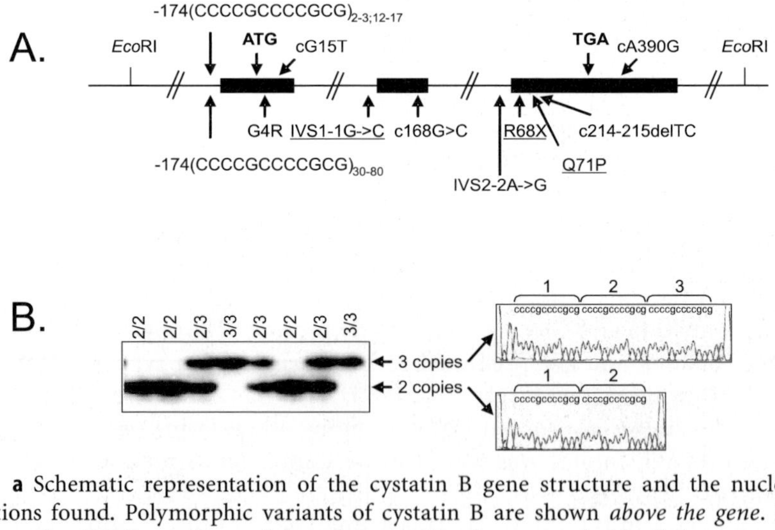

Fig. 1 **a** Schematic representation of the cystatin B gene structure and the nucleotide variations found. Polymorphic variants of cystatin B are shown *above the gene*. Mutations causing EPM1 are shown *below the gene*. Numbers preceded by "c" correspond to position in the complementary DNA (*cDNA*), A of the ATG being 1. The *underlined mutations* are the first ones described by Pennacchio et al. (1996). **b** Agarose gel and sequence of PCR-amplified normal alleles containing two or three copies of the dodecamer. The genotype is showing *above each lane*

domain and the first hairpin loop, containing the highly conserved QVVAG motif, with minor contributions from the second hairpin loop. The carboxyl terminus of cystatin B is an additional site of interaction, dominated by hydrophobic contacts. Gly4 is close to but not in direct contact with the Cys25 in the active site of papain. Amino-terminal deletions of recombinant chicken cystatin provided functional support to the crystallographic data. In particular, a protein starting at Gly9 (equivalent to Gly4 of human cystatin B) exhibited 5000-fold to 10 000-fold weaker inhibition of papain (Machleidt et al. 1989, 1991). Directed mutagenesis, deletions, and elongations identified regions of cystatin B involved in its biological activity that are consistent with data for chicken cystatin (Abrahamson et al. 1987; Thiele et al. 1990; Jerala et al. 1994; Pol and Bjork 2003).

In vitro, cystatin B is a tight-binding reversible inhibitor of papain and cathepsins B, H, L, and S (Green et al. 1984; Popovic et al. 1988; Bromme et al. 1991). Except for cathepsin B, which is also present at the cytoplasm, the others are lysosomal proteinases responsible for protein degradation (Schwartz and Barrett 1980; Barrett and Kirschke 1981; Bohley and Seglen

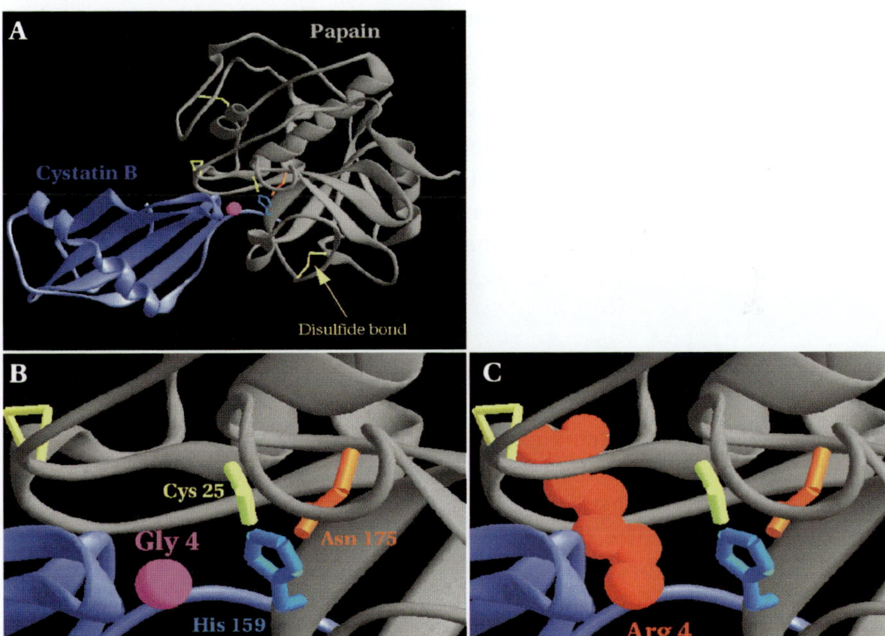

Fig. 2 **a** Ribbon representation of the cystatin B (blue)–papain (*gray*) complex. **b, c** Magnification of the active site in the wild type and the G4R mutant cystatin C. The location of Gly4 in the wild-type protein is shown with a *magenta sphere*, and that of Arg4 with a group of *red spheres*. This side chain is in steric conflict with the binding site on papain, and is likely to reduce the inhibitory activity of the G4R cystatin B mutant

1992). Some cathepsins conduct nonselective protein degradation, while others have tissue- and substrate-specific functions.

4
Point Mutations and Polymorphisms of the Cystatin B Gene

Only seven point mutations in the coding region of the cystatin B gene have been identified to date, accounting for approximately 10% of the EPM1 alleles examined (Pennacchio et al. 1996; Bespalova et al. 1997a, b; Lalioti et al. 1997a; Kagitani-Shimono et al. 2002; de Haan et al. 2004), (Fig. 1, Table 1). These alterations include two splice-site variants (IVS1-1 G > C, and IVS2-2 A > G), one glycine-to-arginine change (G4R), a silent c168G > C change at the last nucleotide of exon 2, an arginine to termination (R68X), a glutamine to termination (Q71X), and a deletion of two nucleotides resulting in protein truncation (214-215delTC).

A few polymorphisms have also been described in cystatin B (Fig. 1A, Table 1). These include a dodecamer repeat [d(CCCCGCCCCGCG)$_n$] in the 5′ upstream region of the cystatin B gene, located approximately 70 bp upstream of the transcription initiation site, and 174 nucleotides upstream of the ATG translation initiation codon (Lalioti et al. 1997b). Normal individuals contain two to three copies of the repeat. The frequency of the two-copy allele was 34–47% and that of the thee-copy allele was 66–53% in different populations (Lalioti et al. 1997b). While alleles with four to 11 repeats have not been described to date, two related CEPH families (102 and 104) had alleles containing 12 to 17 repeats. Since the carriers of these alleles are unaffected and EPM1 is a fully penetrant disease, these alleles were considered normal.

5
Effect of Point Mutations on Cystatin B Expression

Normal and mutant proteins expressed in vitro provided evidence for the effect of the mutations in the normal protein function. The splice-site mutations result in abnormal splicing of the affected exon (Bespalova et al. 1997b; Lalioti et al. 1997b). The truncated R68X cystatin B protein is very rapidly degraded and therefore unable to function (Alakurtti et al. 2005). This is consistent with the lack of cystatin B staining in cells of patients with this mutation.

The amino acid substitution G4R occurs in a highly conserved residue of all cystatins. The crystal structure of the cystatin B–papain complex has been resolved and shown that the amino terminal of cystatin B, including Gly4, interacted with papain (Bode et al. 1988; Stubbs et al. 1990). Three-dimensional modeling suggested that the G4R mutation is a large and charged side chain

Table 1 Mutations and polymorphisms of the cystatin B gene

Position	Nucleotide change[a]	Amino acid change	Inheritance	References
Mutations				
Promoter	c-174 d(ccccgccccgcg)$_{30-80}$ g207	None	Homozygosity Heterozygosity	Lalioti et al. (1997b)
Exon 1	c10G > C g426G > C	G4R	Homozygosity	Lalioti et al. (1997a)
Intron 1	IVS1-1G > C g1925G > C	Exon 2 Skipping[b]	Heterozygosity	Pennacchio et al. (1996) Bespalova et al. (1997b) Lafreniere et al. (1997) Lalioti et al. (1997a)
Exon 2	c168G > C g2027G > C	Aberrant Splicing?[c]	Heterozygosity	Kagitani-Shimono et al. (2002)
Intron 2	IVS2-2A > G g2353A > G	Aberrant splicing[d]	Heterozygosity	Lalioti et al. 1997a
Exon 3	c202C > T g2388C > T	R68X	Heterozygosity	Pennacchio et al. (1996) Lafreniere et al. 1997
Exon 3	c212A > C g2398A > C	Q71P	Heterozygosity	deHaan et al. (2004)
Exon 3	c214-215delTC gdel2400-2401TC	K73fsX2 Truncation	Heterozygosity	Bespalova et al. (1997b) Lafreniere et al. (1997) Lalioti et al. 1997a
Polymorphisms				
Promoter	c-174 d(ccccgccccgcg)$_{2-3;12-17}$ g207	None	Homozygosity Heterozygosity	Lalioti et al. (1997a,b)
Exon 1	c15G > T g431G > T	None	Unknown	Lalioti et al. (1997a)
3′UTR	c371T > C g2557T > C	None	Homozygosity heterozygocity	dbSNP/HapMap rs6385
3′UTR	c390A > G g2576A > G	None	Unknown	Lalioti et al. (1997a)

UTR untranslated region

[a]Mutation nomenclature is according to Dunnen and Antonarakis (2000). "g" in front of a nucleotide position indicates the position in the genomic sequence U46692. "c" indicates the position in the complementary DNA sequence NM_000100, assuming the A of the ATG (start codon) as 1.

[b]Complete exon skipping demonstrated by Bespalova et al. (1997b). Aberrant splicing utilizing cryptic splice sites flanking the exon 2 acceptor splice site also detected using RNase protection (Lalioti et al. 1997b)

[c]c168G is the last nucleotide of exon 2. This mutation was found in heterozygocity in a patient with a dodecamer repeat expansion in the other allele. The splicing of the cystatin B RNA carrying the point mutation was not examined, but it is possible to be affected.

[d]Splicing is likely to be affected because of the nonconservative nucleotide change within the critical nucleotides of the splice site. However, it has not been tested yet.

and is, therefore, likely to jeopardize and reduce or abolish the interaction of the two proteins (Fig. 2) (Lalioti et al. 1997a). The 214-215delTC mutation results in frameshift and subsequent truncation of the COOH terminal of the cystatin B protein, two amino acids downstream of the mutation (K73fsX2). When expressed in vitro in cell lines, the normal protein is located in the nucleus, in cytoplasmic granular structures, and in lysosomes (Alakurtti et al. 2005). In contrast, the G4R, K73fsX2, and Q71P mutants show a diffuse distribution in the nucleus and the cytoplasm but fail to colocalize with a specific marker of the lysosomes. However, it is currently unknown whether the loss of lysosomal association is related to loss of inhibitory function of cystatin B. In summary, the point mutations eliminate cystatin B function by deleting important domains, perturbing subcellular localization, or shortening protein half-life.

6
A Dodecamer Repeat Expansion is the Most Common Mutation in EPM1

Southern blot analysis revealed that all EPM1 alleles devoid of a point mutation were larger than normal alleles or alleles with a point mutation (Lafreniere et al. 1997; Lalioti et al. 1997b; Virtaneva et al. 1997) (Fig. 3a). Sequencing of these larger fragments revealed large uninterrupted expansion of the dodecamer repeat 5′-d(CCCCGCCCCGCG)-3′ to more than 50 copies (Lalioti et al. 1997b) (Fig. 3c). Development of optimized PCR methods have led to the accurate measurement of the expansion size in all patients (Lalioti et al. 1998). Normal alleles contain two to three copies of the dodecamer. EPM1 alleles contain 30 to 80 repeats (Lalioti et al. 1998; Larson et al. 1999). More recently, a method combining DNA deamination and PCR greatly improved the accuracy of amplification of large alleles (Weinhaeusel et al. 2003).

The dodecamer repeat expansion is the most common mutation found in *EPM1*, accounting for 90% of the *EPM1* alleles (Fig. 3c).

A similar unit, which contains two identical dodecamers d(gccgccccccgc) and a third differing by a T, is expanded in canine *EPM2b* (Lohi et al. 2005). The histology of the dogs' brains shows degenerations similar that of to Lafora bodies. Unaffected dogs contain two dodecamers, while affected dogs contain 19 to 26 dodecamers. Unlike the dodecamer repeat in *EPM1*, the canine repeat is within the coding region of the gene. In human and mouse, the nucleotide sequence of this region is different from that of the dog.

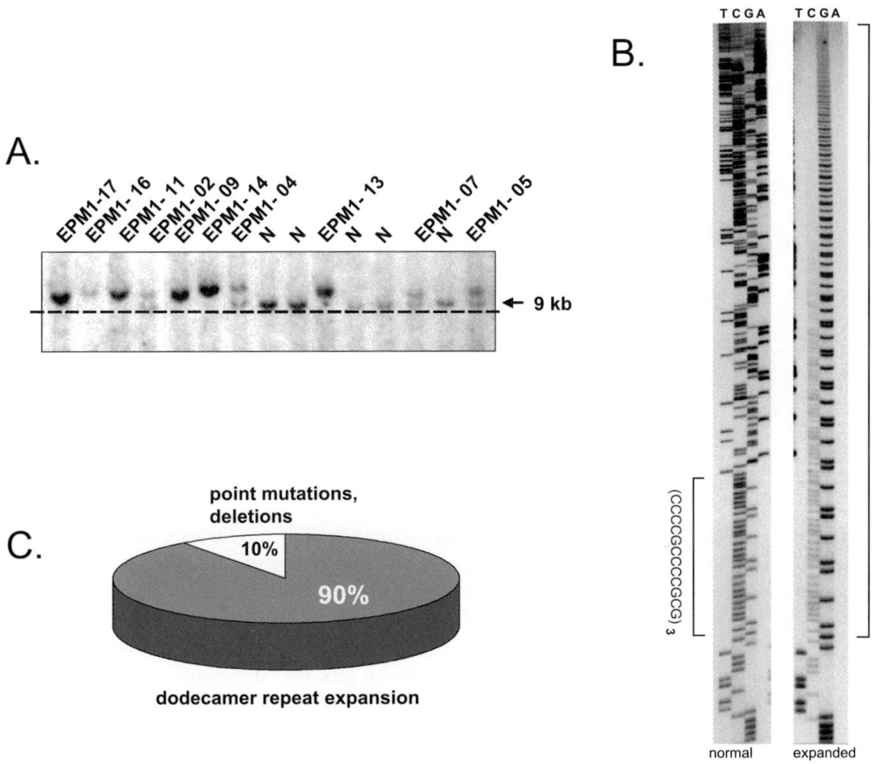

Fig. 3 **a** Southern blot analysis of patients homozygous or compound heterozygous for the repeat and another mutation. DNA is digested with *Eco*RI (see Fig. 1a) and probed with the cystatin B cDNA. Compound heterozygotes carry a normal-size and a larger-size allele. The large allele varies between patients. **b** Sequence of dodecamer repeat area from a normal individual with three repeats and from a patient with more than 50 uninterrupted repeats of the dodecamer. **c** The point mutations account for only a small percentage of the mutant alleles (10%). The repeat expansion in the promoter is the most common defect in progressive myoclonus epilepsy of the Unverricht–Lundborg type (*EPM1*) even in patients with different haplotypes. *N* normal individual

7 Instability of the Dodecamer Repeat

Meiotic instability is one of the characteristic features of repeat expansions. In *EPM1*, the common two-copy to three-copy alleles show no mitotic or meiotic instability. The large *EPM1* alleles containing 30 to 80 copies of the repeat unit show meiotic instability, including both expansions and contractions (Fig. 4) (Lalioti et al. 1998; Larson et al. 1999; Mazarib et al. 2001). These are both maternally and paternally transmitted. Patients with the same haplotype share different allele sizes, providing further evidence of meiotic instability. The intermediate, no-EPM1 causing alleles containing 12 to 17 repeats are also

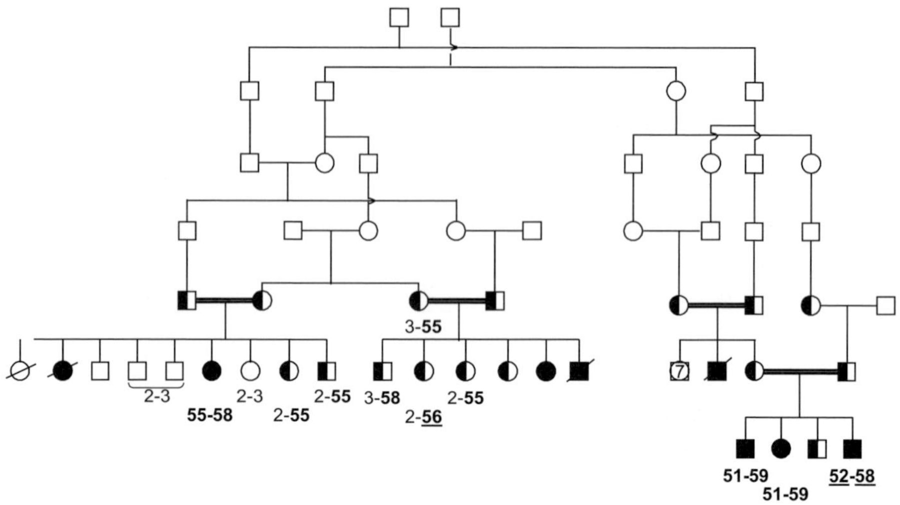

Fig. 4 Meiotic instability of the dodecamer repeat pedigree of the Swiss EPM1 family described by Klein et al. (1968). Carrier and affected status is indicated for the last three generations. All EPM1 alleles share the same haplotype around cystatin B. The number of repeats in the two alleles of tested individuals is shown. Disease alleles are shown in *bold*. *Underlined alleles* were subject to meiotic instability. For example, one of the affected children in the far-right branch of the family has two different expanded alleles from both of his siblings

unstable during meiosis (Lalioti et al. 1997b). In particular, from the 21 expansions of 12-copy and 13-copy alleles, there are six expansions to alleles with 13, 14, 15, and 17 copies, with the largest expansion being four dodecamers. Although these alleles do not cause *EPM1*, they can be considered premutation alleles on the basis of their instability. The largest addition observed during transmission of *EPM1* alleles is three repeats (Larson et al. 1999). Therefore, large expansions do not seem to be more prone to size increase than premutation alleles.

There is no apparent somatic mosaicism of the repeat in blood DNA from EPM1 patients, excluding extensive mitotic instability. In contrast, three out of six lymphoblastoid cell lines from EPM1 patients showed different alleles from blood of the corresponding patients (Larson et al. 1999).

The dodecamer repeat expansion forms more secondary structures than any of the other triplet repeat expansions, including hairpins, tetraplexes and I-motifs (Jithesh et al. 2001; Pataskar et al. 2001a,b; Saha and Usdin 2001). Tetraplexes, which seem to be the predominant structures of the EPM1 repeat, are stable at physiological temperatures, pH, and ionic strength (Saha and Usdin 2001). These secondary structures formed by intrastrand folding are likely to affect DNA replication and repair and to contribute to the instability of these sequences (Usdin and Grabczyk 2000).

8
Size of the Repeat and Age of Onset of the Epilepsy

Unlike many other repeat expansion disorders, the size of the EPM1 repeat does not correlate with the age of onset of the disease (Lalioti et al. 1998). This suggests that once the repeat extends beyond a critical threshold, cystatin B expression is repressed to pathological levels. Thus, the age of onset and the severity of the disease must depend on modifier genes and/or environmental factors. The critical threshold for cystatin B repression due to repeat expansion is at the 12–29 copy range, because an individual with 12/17 copies in his two alleles was unaffected and an individual with 30 copies in his smallest allele was affected (Lalioti et al. 1997b, 1998). Reporter gene assays have shed some light on the cystatin B expression under the control of the expanded dodecamer repeat, and are described in the next section.

9
Effect of Expansion on Cystatin B Expression

The cystatin B gene is ubiquitously expressed with high levels of expression comparable to housekeeping genes in both human and mouse (Pennacchio et al. 1996; Lalioti et al. 1997b; Pennacchio and Myers 1997; Hsiao et al. 2001). Sensitive RNase protection experiments showed that the expression of cystatin B is greatly reduced in blood leukocytes from patients homozygous for the repeat expansion (Lalioti et al. 1997b) (Fig. 5). Antibody staining for cystatin B in the brain of EPM1 patients also showed reduced expression (Kinne et al. 2002). Some lymphoblastoid and fibroblast cell lines from EPM1 patients display reduced cystatin B expression, whereas in others normal cystatin B expression was restored following growth in culture (Pennacchio et al. 1996; Bespalova et al. 1997a; Lafreniere et al. 1997; Lalioti et al. 1997b). It is possible that the dodecamer repeat downregulates cystatin B expression in primary cells but occasionally fails to do so after the cells are transformed and/or cultured. Culture conditions and other unknown factors cause cell lines to acquire phenotypes different from those of the cells of origin. In these cell types, cystatin B expression, as a response to dodecamer expansion regulation, may vary. This is supported by further in vitro promoter assays showing that the dodecamer repeat shuts down expression of reporter genes in some cell types but not in others (Lalioti et al. 1999). Lymphoblastoid cell lines from the CEPH family, carrying the intermediate dodecamer expansions with 12 to 17 copies of the repeat also show reduced cystatin B expression (Alakurtti et al. 2000). The expression of cystatin B in primary cells of these unaffected individuals has not been examined; therefore, it is not known whether the expression in their cell lines is at all modified by growth in culture. As a result, it

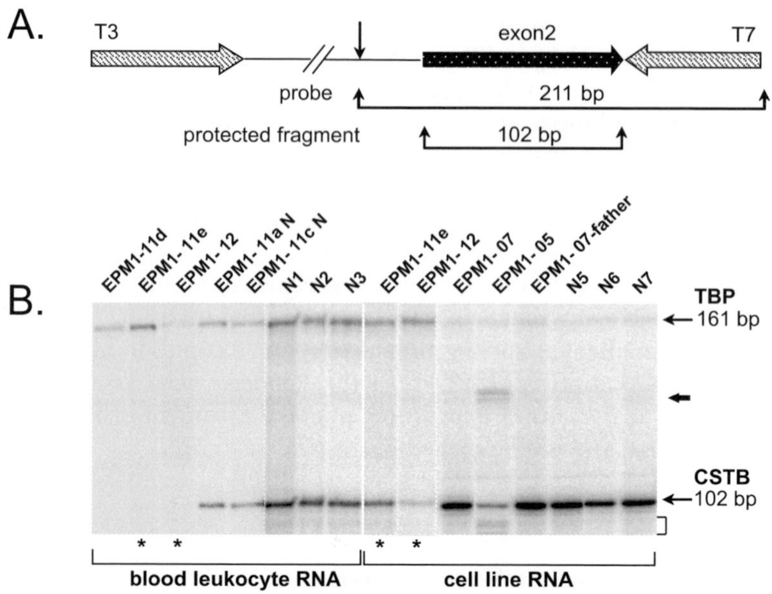

Fig. 5 Quantitation of cystatin B expression in EPM1 and normal samples. **a** Schematic representation of the plasmid used RNase protection experiment to produce the riboprobe. The transcribed (probe) and protected fragments are shown with *arrows under the plasmid*. **b** Autoradiogram showing the cystatin B (*CSTB*) expression relative to that of TATA binding protein (*TBP*, control probe) in blood leukocyte RNA from patients and controls (*N*), or in lymphoblastoid cell lines. *EPM1-11d*, *EPM1-11e*, *EPM1-11a*, and *EPM1-11c* are siblings; EPM1-11d and EPM1-11e are homozygous for the expansion; EPM1-11a and EPM1-11c are unaffected. In blood leukocytes, there is a marked reduction of cystatin B RNA in patients, compared with that in controls. In cell lines the expression is either normal or slightly reduced. For patients 11d, and 12, there are both blood and lymphoblastoid cell lines data (*asterisks*). The *bold arrow* and *bracket on the side* highlight the aberrant protected products in patient EPM1-05, who is a compound heterozygote for a repeat expansion and a splicing mutation (IVS1-1G > C)

has not yet been determined beyond what threshold of cystatin B expression EPM1 symptoms are initiated.

Sensitive and quantitative in vitro reporter assays have shown that a repeat as short as 19 copies results in tenfold downregulation of the gene (Alakurtti et al. 2000).

10 Mechanism of Transcriptional Repression

The experiments already outlined provide evidence that the dodecamer repeat expansion has a direct effect on cystatin B transcription. Possible explanations for the reduced transcription include altered spacing of promoter elements, hypermethylation, altered chromatin structure, and recruitment of transcription repressors to the repeat sequence.

The cystatin B minimal promoter was mapped using reporter gene assays and serial deletions of upstream sequences (Lalioti et al. 1999; Alakurtti et al. 2000). This approach demonstrated that there are important transcription factor binding sites or other regulatory sequences upstream of the dodecamer repeat. It was hypothesized that the insertion of a large DNA fragment could alter the spacing of transcription factor binding sites and/or the transcription initiation complex and result in gene suppression (Lalioti et al. 1999) (Fig. 6). This hypothesis was supported by the finding that a different sequence of the

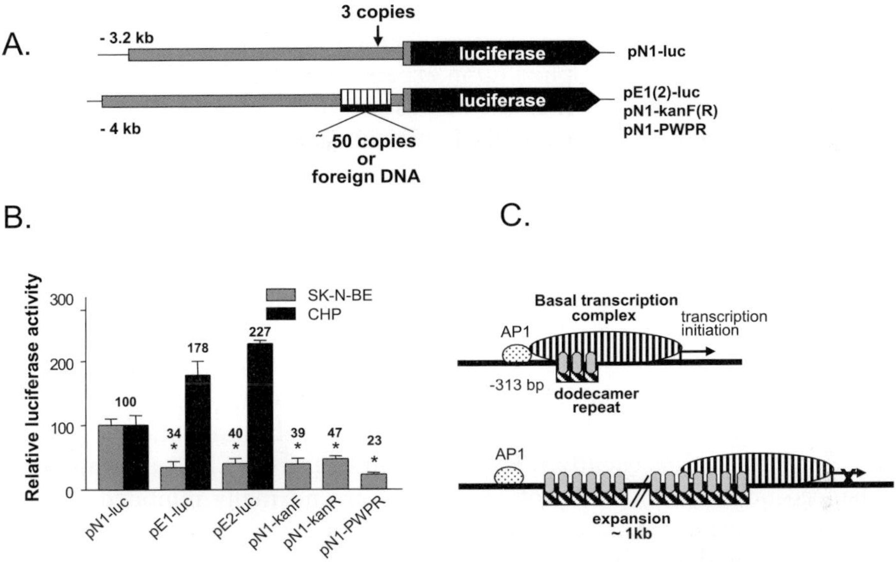

Fig. 6 Transcriptional repression of the expanded cystatin B promoter. **a** Schematic representation of the constructs. A 3.1-kb fragment of the cystatin B promoter containing three dodecamer repeats was cloned upstream of the luciferase reporter gene. A similar construct containing 50 copies of the repeat or heterologous DNA insertions instead of the repeat was also engineered. **b** Relative promoter activity in two different neuroblastoma cell lines. The activity of the wild-type promoter is set at 100%. *Asterisks* indicate loss of promoter activity in the SK-N-BE cell line due to expansion or introduction of heterologous DNA. **c** Model of cystatin B transcriptional repression due to spacing of promoter elements or recruitment of repressors. For simplicity, all other regulators are referred as a *basal transcription complex* and are shown as *vertically striped ovals.* The dodecamers are shown as *hatched boxes.* An "activator" is shown as a *dotted ball*, and can normally interact with the complex. When the distance is increased the activator is no longer able to interact with the complex and activate transcription. Alternatively, the repeat may be able to bind transcriptional repressors *gray octagons.* The position of the critical AP1 binding site (Lalioti et al. 1999; Alakurtti et al. 2000) is shown

same size inserted in the cystatin B promoter instead of the dodecamer repeat could suppress expression in a similar manner (Lalioti et al. 1999).

Methylation of CpG sites is a characteristic feature of the fragile X repeat (Oberle et al. 1991; Knight et al. 1993); however, both the *Hpa*II/*Msp*I CpG islands throughout the cystatin B genomic area and the dodecamer repeat are unmethylated (Lalioti et al. 1997b; Weinhaeusel et al. 2003).

Secondary structures such as hairpins, tetraplexes, and I-motif structures of the expanded dodecamer repeat (Pataskar et al. 2001a,b; Saha and Usdin 2001) are likely to play a role in repeat instability. It is plausible that such structures also modify chromatin, making it inaccessible to transcription factors or other regulatory proteins, thus diminishing transcription (Li et al. 2004).

Recruitment of transcriptional repressors to the expanded repeat is another appealing, yet untested possibility. The repeat sequence contains several binding sites for the SP1 transcription factor (Lalioti et al. 1999; Alakurtti et al. 2000), which are multiplied upon expansion.

While spacing of promoter elements was shown to downregulate cystatin B expression, the reduction seen in EPM1 patients may be due to the synergistic effect of more than one mechanism.

11 Loss of Cystatin B Function and Disease Pathophysiology

As discussed already, despite the different types of EPM1 mutations, they all have as a consequence the loss of cystatin B function through three apparent mechanisms: lack of the protein, abnormal localization, and deletion of critical residues. Consistently, lymphoblastoid lines from EPM1 patients show enhanced activity of those proteinases that are normally inhibited by cystatin B: cathepsins B, L, and S (Kinne et al. 2002).

A mouse model of EPM1 with a deletion of the cystatin B gene was engineered to mimic the human condition (Pennacchio et al. 1998). Like patients with the disease, the mice show no cystatin B expression and develop progressive ataxia, myoclonus, and seizures. In addition, these mice have a pronounced granule cell loss in the cerebellum due to apoptosis (Pennacchio et al. 1998). Neuronal atrophy, gliosis, and apoptosis are present also outside the cerebellum (Shannon et al. 2002).

As expected with loss of cystatin B inhibitory activity, brains from *Cstb–/–* mice show an upregulation of genes involved in proteolysis, apoptosis, and glial activation (Lieuallen et al. 2001). To identify which symptoms of EPM1 were due to excessive proteolysis of the normally inhibited proteinases, cathepsins B, L, and S were selectively deleted from the *Cstb* knockout mice. Applying this approach it was shown that cathepsins L and S have no contribution to the phenotype. In contrast, the double-knockout mouse for cathepsin B and cystatin B has a greatly reduced neuronal cell death owing to apoptosis (Houseweart et al. 2003a). Surprisingly, the ataxia and seizures are not ameliorated in these mice, suggesting that these symptoms are not a consequence of neuronal apoptosis and that cathepsin B is not the only downstream affector leading to the epilepsy. The exact mechanism by which increased cathepsin activity leads to apoptosis has not yet been identified (Houseweart et al. 2003b). It is possible that increased unspecific proteolysis might initiate cellular apoptotic pathways (Williams and Henkart 1994; Turk et al. 2002a). Alternatively, cathepsins might directly cleave caspases, the proteins responsible for cellular degradation and apoptosis, and thus play a direct role in this well-defined cellular pathway.

In the EPM1 mouse, ataxia and seizures do not seem to be the consequence of neuronal cell death. It is possible that a different mechanism leads to the neurological symptoms. Whether this is mediated through a different function of cystatin B is unknown. For example, in rats, seizures have been shown to upregulate cystatin B (D'Amato et al. 2000), implying that cystatin B may play some role in neuronal protection. The localization of cystatin B in the nucleus of certain cell types is intriguing and supports the possibility of an additional, yet unknown function. Moreover, cystatin B in brain cells can interact and form a complex with other proteins such as RACK-1, β-spectrin and NF-L, none of which is a proteinase (Di Giaimo et al. 2002).

In contrast to humans, the *Cstb* knockout mice did not develop tonic–clonic seizures, showed no photosensitivity, and had seizures only during sleep. Whether these differences indicate additional contributing disease factors in humans, or reflect the differences between human and mouse biology, brain development, or a background strain effect, remains to be investigated. For example, cerebellar granule cells in humans are produced during embryonic life and in mice only postnatally. Interestingly, the symptoms of the EPM1 mouse model depended on the genetic background, implicating modifier genes in the development or severity of the phenotype.

The *Cstb* mutant mouse has similar a phenotype to the human patients lacking cystatin B protein and has been used to elucidate the pathophysiology and explore the progressive nature of the disease. It is a unique tool for validating in vitro data, and most importantly for testing new therapeutic approaches.

Acknowledgements We thank EPM1 patients and their families for donation of precious samples. We thank our previous collaborators and members for the Antonarakis laboratory for help, discussions, and suggestions.

References

Abrahamson M, Barrett AJ, Salvesen G, Grubb A (1986) Isolation of six cysteine proteinase inhibitors from human urine. Their physicochemical and enzyme kinetic properties and concentrations in biological fluids. J Biol Chem 261:11282–11289

Abrahamson M, Ritonja A, Brown MA, Grubb A, Machleidt W, Barrett AJ (1987) Identification of the probable inhibitory reactive sites of the cysteine proteinase inhibitors human cystatin C and chicken cystatin. J Biol Chem 262:9688–9694

Alakurtti K, Virtaneva K, Joensuu T, Palvimo JJ, Lehesjoki AE (2000) Characterization of the cystatin B gene promoter harboring the dodecamer repeat expanded in progressive myoclonus epilepsy, EPM1. Gene 242:65–73

Alakurtti K, Weber E, Rinne R, Theil G, de Haan GJ, Lindhout D, Salmikangas P, Saukko P, Lahtinen U, Lehesjoki AE (2005) Loss of lysosomal association of cystatin B proteins representing progressive myoclonus epilepsy, EPM1, mutations. Eur J Hum Genet 13:208–215

Barrett AJ (1981) Cystatin, the egg white inhibitor of cystein proteinases. Methods Enzymol 80:771–778

Barrett AJ, Kirschke H (1981) Cathepsin B, cathepsin H, and cathepsin L. Methods Enzymol 80:535–561

Berkovic SF, So NK, Andermann F (1991) Progressive myoclonus epilepsies: clinical and neurophysiological diagnosis. J Clin Neurophysiol 8:261–274

Bespalova IN, Adkins S, Pranzatelli M, Burmeister M (1997a) Novel cystatin B mutation and diagnostic PCR assay in an Unverricht-Lundborg progressive myoclonus epilepsy patient. Am J Med Genet 74:467–471

Bespalova IN, Pranzatelli M, Burmeister M (1997b) G to C transversion at a splice acceptor site causes exon skipping in the cystatin B gene. Mutat Res 382:67–74

Bode W, Engh R, Musil D, Thiele U, Huber R, Karshikov A, Brzin J, Kos J, Turk V (1988) The 2.0 A X-ray crystal structure of chicken egg white cystatin and its possible mode of interaction with cysteine proteinases. EMBO J 7:2593–2599

Bohley P, Seglen PO (1992) Proteases and proteolysis in the lysosome. Experientia 48:151–157

Brannvall K, Hjelm H, Korhonen L, Lahtinen U, Lehesjoki AE, Lindholm D (2003) Cystatin-B is expressed by neural stem cells and by differentiated neurons and astrocytes. Biochem Biophys Res Commun 308:369–374

Bromme D, Rinne R, Kirschke H (1991) Tight-binding inhibition of cathepsin S by cystatins. Biomed Biochim Acta 50:631–635

Brzin J, Popovic T, Turk V, Borchart U, Machleidt W (1984) Human cystatin, a new protein inhibitor of cysteine proteinases. Biochem Biophys Res Commun 118:103–109

Calkins CC, Sameni M, Koblinski J, Sloane BF, Moin K (1998) Differential localization of cysteine protease inhibitors and a target cysteine protease, cathepsin B, by immunoconfocal microscopy. J Histochem Cytochem 46:745–751

Canafoglia L, Ciano C, Panzica F, Scaioli V, Zucca C, Agazzi P, Visani E, Avanzini G, Franceschetti S (2004) Sensorimotor cortex excitability in Unverricht-Lundborg disease and Lafora body disease. Neurology 63:2309–2315

Carpenter S, Karpati G (1981) Sweat gland duct cells in Lafora disease: diagnosis by skin biopsy. Neurology 31:1564–1568

Cochius J, Carpenter S, Andermann E, Rouleau G, Nousiainen U, Kalviainen R, Farrell K, Andermann F (1994) Sweat gland vacuoles in Unverricht-Lundborg disease: a clue to diagnosis? Neurology 44:2372–2375

Cochius JI, Figlewicz DA, Kalviainen R, Nousiainen U, Farrell K, Patry G, Soderfeldt B, Frydman M, Lerman P, Andermann F (1993) Unverricht-Lundborg disease: absence of nonallelic genetic heterogeneity. Ann Neurol 34:739–741

D'Amato E, Kokaia Z, Nanobashvili A, Reeben M, Lehesjoki AE, Saarma M, Lindvall O (2000) Seizures induce widespread upregulation of cystatin B, the gene mutated in progressive myoclonus epilepsy, in rat forebrain neurons. Eur J Neurosci 12:1687–1695

de Haan GJ, Halley DJ, Doelman JC, Geesink HH, Augustijn PB, Jager-Jongkind AD, Majoie M, Bader AJ, Leliefeld-Ten Doeschate LA, Deelen WH, Bertram E, Lehesjoki AE, Lindhout D (2004) Unverricht-Lundborg disease: underdiagnosed in the Netherlands. Epilepsia 45:1061–1063

Di Giaimo R, Riccio M, Santi S, Galeotti C, Ambrosetti DC, Melli M (2002) New insights into the molecular basis of progressive myoclonus epilepsy: a multiprotein complex with cystatin B. Hum Mol Genet 11:2941–2950

Dunnen JT, Antonarakis SE (2000) Mutation nomenclature extensions and suggestions to describe complex mutations: a discussion. Hum Mutat 15:7–12

Eldridge R, Iivanainen M, Stern R, Koerber T, Wilder BJ (1983) Baltic myoclonus epilepsy: hereditary disorder of childhood made worse by phenytoin. Lancet 2:838–842

Genton P, Michelucci R, Tassinari CA, Roger J (1990) The Ramsay Hunt syndrome revisited: Mediterranean myoclonus versus mitochondrial encephalomyopathy with ragged-red fibers and Baltic myoclonus. Acta Neurol Scand 81:8–15

Ghiso J, Pons-Estel B, Frangione B (1986) Hereditary cerebral amyloid angiopathy: the amyloid fibrils contain a protein which is a variant of cystatin C, an inhibitor of lysosomal cysteine proteases. Biochem Biophys Res Commun 136:548–554

Green GD, Kembhavi AA, Davies ME, Barrett AJ (1984) Cystatin-like cysteine proteinase inhibitors from human liver. Biochem J 218:939–946

Haltia M, Kristensson K, Sourander P (1969) Neuropathological studies in three Scandinavian cases of progressive myoclonus epilepsy. Acta Neurol Scand 45:63–77

Houseweart MK, Pennacchio LA, Vilaythong A, Peters C, Noebels JL, Myers RM (2003a) Cathepsin B but not cathepsins L or S contributes to the pathogenesis of Unverricht-Lundborg progressive myoclonus epilepsy (EPM1). J Neurobiol 56:315–327

Houseweart MK, Vilaythong A, Yin XM, Turk B, Noebels JL, Myers RM (2003b) Apoptosis caused by cathepsins does not require Bid signaling in an in vivo model of progressive myoclonus epilepsy (EPM1). Cell Death Differ 10:1329–1335

Hsiao LL, Dangond F, Yoshida T, Hong R, Jensen RV, Misra J, Dillon W, Lee KF, Clark KE, Haverty P, Weng Z, Mutter GL, Frosch MP, Macdonald ME, Milford EL, Crum CP, Bueno R, Pratt RE, Mahadevappa M, Warrington JA, Stephanopoulos G, Gullans SR (2001) A compendium of gene expression in normal human tissues. Physiol Genomics 7:97–104

Jerala R, Kroon-Zitko L, Popovic T, Turk V (1994) Elongation on the amino-terminal part of stefin B decreases inhibition of cathepsin H. Eur J Biochem 224:797–802

Jithesh PV, Singh P, Joshi R (2001) Molecular dynamics studies of trinucleotide repeat DNA involved in neurodegenerative disorders. J Biomol Struct Dyn 19:479–495

Kagitani-Shimono K, Imai K, Okamoto N, Ono J, Okada S (2002) Unverricht-Lundborg disease with cystatin B gene abnormalities. Pediatr Neurol 26:55–60

Kinne R, Saukko P, Jarvinen M, Lehesjoki AE (2002) Reduced cystatin B activity correlates with enhanced cathepsin activity in progressive myoclonus epilepsy. Ann Med 34:380–385

Klein D, Mumenthaler M, Kraus-Ruppert R, Rallo E (1968) A large family of Valais affected with progressive myoclonic epilepsy and retinitis pigmentosa. A clinical, genetical and anatomo-pathological study. Humangenetik 6:237–252

Klein D, Rabinowicz T (1980) Clinical and genetic aspects of certain neuro-opthalmological diseases in Swiss isolates. In: Eriksson A (ed) Population structure and genetic disorders, vol 1. Academic, London, pp 367–382

Knight SJ, Flannery AV, Hirst MC, Campbell L, Christodoulou Z, Phelps SR, Pointon J, Middleton-Price HR, Barnicoat A, Pembrey ME et al (1993) Trinucleotide repeat amplification and hypermethylation of a CpG island in FRAXE mental retardation. Cell 74:127–134

Koskiniemi M, Donner M, Majuri H, Haltia M, Norio R (1974a) Progressive myoclonus epilepsy. A clinical and histopathological study. Acta Neurol Scand 50:307–332

Koskiniemi M, Toivakka E, Donner M (1974b) Progressive myoclonus epilepsy. Electroencephalographical findings. Acta Neurol Scand 50:333–359

Labauge P, Ouazzani R, M'Rabet A, Grid D, Genton P, Dravet C, Chkili T, Beck C, Buresi C, Baldy-Moulinier M, Malafosse A (1997) Allelic heterogeneity of Mediterranean myoclonus and the cystatin B gene. Ann Neurol 41:686–689

Lafreniere RG, de Jong PJ, Rouleau GA (1995) A 405-kb cosmid contig and HindIII restriction map of the progressive myoclonus epilepsy type 1 (EPM1) candidate region in 21q22.3. Genomics 29:288–290

Lafreniere RG, Rochefort DL, Chretien N, Rommens JM, Cochius JI, Kalviainen R, Nousiainen U, Patry G, Farrell K, Soderfeldt B, Federico A, Hale BR, Cossio OH, Sorensen T, Pouliot MA, Kmiec T, Uldall P, Janszky J, Pranzatelli MR, Andermann F, Andermann E, Rouleau GA (1997) Unstable insertion in the 5′ flanking region of the cystatin B gene is the most common mutation in progressive myoclonus epilepsy type 1, EPM1. Nat Genet 15:298–302

Lalioti MD, Bottani A, Morris MA, Antonarakis SE (1995) The Swiss type of progressive myoclonous epilepsy (EPM1) maps tp 21q22.3. Med Genet 2:249

Lalioti MD, Mirotsou M, Buresi C, Peitsch MC, Rossier C, Ouazzani R, Baldy-Moulinier M, Bottani A, Malafosse A, Antonarakis SE (1997a) Identification of mutations in cystatin B, the gene responsible for the Unverricht-Lundborg type of progressive myoclonus epilepsy (EPM1). Am J Hum Genet 60:342–351

Lalioti MD, Scott HS, Buresi C, Rossier C, Bottani A, Morris MA, Malafosse A, Antonarakis SE (1997b) Dodecamer repeat expansion in cystatin B gene in progressive myoclonus epilepsy. Nature 386:847–851

Lalioti MD, Scott HS, Genton P, Grid D, Ouazzani R, M'Rabet A, Ibrahim S, Gouider R, Dravet C, Chkili T, Bottani A, Buresi C, Malafosse A, Antonarakis SE (1998) A PCR amplification method reveals instability of the dodecamer repeat in progressive myoclonus epilepsy (EPM1) and no correlation between the size of the repeat and age at onset. Am J Hum Genet 62:842–847

Lalioti MD, Scott HS, Antonarakis SE (1999) Altered spacing of promoter elements due to the dodecamer repeat expansion contributes to reduced expression of the cystatin B gene in EPM1. Hum Mol Genet 8:1791–1798

Larson GP, Ding S, Lafreniere RG, Rouleau GA, Krontiris TG (1999) Instability of the EPM1 minisatellite. Hum Mol Genet 8:1985–1988

Lehesjoki AE (2002) Clinical features and genetics of Unverricht-Lundborg disease. Adv Neurol 89:193–197

Lehesjoki AE, Eldridge R, Eldridge J, Wilder BJ, de la Chapelle A (1993a) Progressive myoclonus epilepsy of Unverricht-Lundborg type: a clinical and molecular genetic study of a family from the United States with four affected sibs. Neurology 43:2384–2386

Lehesjoki AE, Koskiniemi M, Norio R, Tirrito S, Sistonen P, Lander E, de la Chapelle A (1993b) Localization of the EPM1 gene for progressive myoclonus epilepsy on chromosome 21: linkage disequilibrium allows high resolution mapping. Hum Mol Genet 2:1229–1234

Lehesjoki AE, Koskiniemi M, Sistonen P, Miao J, Hastbacka J, Norio R, de la Chapelle A (1991) Localization of a gene for progressive myoclonus epilepsy to chromosome 21q22. Proc Nat Acad Sci USA 88:3696–3699

Li YJ, Fu XH, Liu DP, Liang CC (2004) Opening the chromatin for transcription. Int J Biochem Cell Biol 36:1411–1423

Liang C, Brookhart G, Feng GH, Reeck GR, Kramer KJ (1991) Inhibition of digestive proteinases of stored grain Coleoptera by oryzacystatin, a cysteine proteinase inhibitor from rice seed. FEBS Lett 278:139–142

Lieuallen K, Pennacchio LA, Park M, Myers RM, Lennon GG (2001) Cystatin B-deficient mice have increased expression of apoptosis and glial activation genes. Hum Mol Genet 10:1867–1871

Lohi H, Young EJ, Fitzmaurice SN, Rusbridge C, Chan EM, Vervoort M, Turnbull J, Zhao XC, Ianzano L, Paterson AD, Sutter NB, Ostrander EA, Andre C, Shelton GD, Ackerley CA, Scherer SW, Minassian BA (2005) Expanded repeat in canine epilepsy. Science 307:81

Lundborg H (1903) Die progressive Myoclonous-Epilepsie (Unverricht's Myoclonie). Almquist and Wiksell, Uppsala

Machleidt W, Thiele U, Assfalg-Machleidt I, Forger D, Auerswald EA (1991) Molecular mechanism of inhibition of cysteine proteinases by their protein inhibitors: kinetic studies with natural and recombinant variants of cystatins and stefins. Biomed Biochim Acta 50:613–620

Machleidt W, Thiele U, Laber B, Assfalg-Machleidt I, Esterl A, Wiegand G, Kos J, Turk V, Bode W (1989) Mechanism of inhibition of papain by chicken egg white cystatin. Inhibition constants of N-terminally truncated forms and cyanogen bromide fragments of the inhibitor. FEBS Lett 243:234–238

Malafosse A, Lehesjoki AE, Genton P, Labauge P, Durand G, Tassinari CA, Dravet C, Michelucci R, de la Chapelle A (1992) Identical genetic locus for Baltic and Mediterranean myoclonus. Lancet 339:1080–1081

Mascalchi M, Michelucci R, Cosottini M, Tessa C, Lolli F, Riguzzi P, Lehesjoki AE, Tosetti M, Villari N, Tassinari CA (2002) Brainstem involvement in Unverricht-Lundborg disease (EPM1): an MRI and (1)H MRS study. Neurology 58:1686–1689

Mazarib A, Xiong L, Neufeld MY, Birnbaum M, Korczyn AD, Pandolfo M, Berkovic SF (2001) Unverricht-Lundborg disease in a five-generation Arab family: instability of dodecamer repeats. Neurology 57:1050–1054

Meldium BS, Bruton CJ (1992) Epilepsy. In: Duchen JHALW (ed) Greenfield's neuropathology. Arnold, London, pp 1246–1283

Moulard B, Darcel F, Mignard D, Jeanpierre M, Genton P, Cartault F, Yaouanq J, Roubertie A, Biraben A, Buresi C, Malafosse A (2003) FOunder effect in patients with Unverricht-Lundborg disease on reunion island. Epilepsia 44:1357–1360

Moulard B, Genton P, Grid D, Jeanpierre M, Ouazzani R, Mrabet A, Morris M, LeGuern E, Dravet C, Mauguiere F, Utermann B, Baldy-Moulinier M, Belaidi H, Bertran F, Biraben A, Ali Cherif A, Chkili T, Crespel A, Darcel F, Dulac O, Geny C, Humbert-Claude V, Kassiotis P, Buresi C, Malafosse A (2002) Haplotype study of West European and North African Unverricht-Lundborg chromosomes: evidence for a few founder mutations. Hum Genet 111:255–262

Muller-Esterl W, Fritz H, Kellermann J, Lottspeich F, Machleidt W, Turk V (1985) Genealogy of mammalian cysteine proteinase inhibitors. Common evolutionary origin of stefins, cystatins and kininogens. FEBS Lett 191:221–226

Norio R, Koskiniemi M (1979) Progressive myoclonus epilepsy: genetic and nosological aspects with special reference to 107 Finnish patients. Clin Genet 15:382–398

Oberle I, Rousseau F, Heitz D, Kretz C, Devys D, Hanauer A, Boue J, Bertheas MF, Mandel JL (1991) Instability of a 550-base pair DNA segment and abnormal methylation in fragile X syndrome. Science 252:1097–1102

Parmeggiani A, Lehesjoki AE, Carelli V, Posar A, Santi A, Santucci M, Gobbi G, Pini A, Rossi PG (1997) Familial Unverricht-Lundborg disease: a clinical, neurophysiologic, and genetic study. Epilepsia 38:637–641

Pataskar SS, Dash D, Brahmachari SK (2001a) Intramolecular i-motif structure at acidic pH for progressive myoclonus epilepsy (EPM1) repeat $d(CCCCGCCCCGCG)_n$. J Biomol Struct Dyn 19:307–313

Pataskar SS, Dash D, Brahmachari SK (2001b) Progressive myoclonus epilepsy [EPM1] repeat $d(CCCCGCCCCGCG)_n$ forms folded hairpin structures at physiological pH. J Biomol Struct Dyn 19:293–305

Pennacchio LA, Bouley DM, Higgins KM, Scott MP, Noebels JL, Myers RM (1998) Progressive ataxia, myoclonic epilepsy and cerebellar apoptosis in cystatin B-deficient mice. Nat Genet 20:251–258

Pennacchio LA, Lehesjoki AE, Stone NE, Willour VL, Virtaneva K, Miao J, D'Amato E, Ramirez L, Faham M, Koskiniemi M, Warrington JA, Norio R, de la Chapelle A, Cox DR, Myers RM (1996) Mutations in the gene encoding cystatin B in progressive myoclonus epilepsy (EPM1). Science 271:1731–1734

Pennacchio LA, Myers RM (1997) Isolation and characterization of the mouse cystatin B gene. Genome Res 6:1103–1109

Pol E, Bjork I (2003) Contributions of individual residues in the N-terminal region of cystatin B (stefin B) to inhibition of cysteine proteinases. Biochim Biophys Acta 1645:105–112

Popovic T, Brzin J, Kos J, Lenarcic B, Machleidt W, Ritonja A, Hanada K, Turk V (1988) A new purification procedure of human kidney cathepsin H, its properties and kinetic data. Biol Chem Hoppe Seyler 369 Suppl:175–183

Rawlings ND, Barrett AJ (1990) Evolution of proteins of the cystatin superfamily. J Mol Evol 30:60–71

Riccio M, Di Giaimo R, Pianetti S, Palmieri PP, Melli M, Santi S (2001) Nuclear localization of cystatin B, the cathepsin inhibitor implicated in myoclonus epilepsy (EPM1). Exp Cell Res 262:84–94

Riccio M, Santi S, Dembic M, Di Giaimo R, Cipollini E, Costantino-Ceccarini E, Ambrosetti D, Maraldi NM, Melli M (2005) Cell-specific expression of the epm1 (cystatin B) gene in developing rat cerebellum. Neurobiol Dis 20:104–114

Saha T, Usdin K (2001) Tetraplex formation by the progressive myoclonus epilepsy type-1 repeat: implications for instability in the repeat expansion diseases. FEBS Lett 491:184–187

Schwartz WN, Barrett AJ (1980) Human cathepsin H. Biochem J 191:487–497

Shannon P, Pennacchio LA, Houseweart MK, Minassian BA, Myers RM (2002) Neuropathological changes in a mouse model of progressive myoclonus epilepsy: cystatin B deficiency and Unverricht-Lundborg disease. J Neuropathol Exp Neurol 61:1085–1091

Stone NE, Fan JB, Willour V, Pennacchio LA, Warrington JA, Hu A, de la Chapelle A, Lehesjoki AE, Cox DR, Myers RM (1996) Construction of a 750-kb bacterial clone contig and restriction map in the region of human chromosome 21 containing the progressive myoclonus epilepsy gene. Genome Res 6:218–225

Stubbs MT, Laber B, Bode W, Huber R, Jerala R, Lenarcic B, Turk V (1990) The refined 2.4 A X-ray crystal structure of recombinant human stefin B in complex with the cysteine proteinase papain: a novel type of proteinase inhibitor interaction. EMBO J 9:1939–1947

Takuma K, Kiriu M, Mori K, Lee E, Enomoto R, Baba A, Matsuda T (2003) Roles of cathepsins in reperfusion-induced apoptosis in cultured astrocytes. Neurochem Int 42:153–159

Thiele U, Assfalg-Machleidt I, Machleidt W, Auerswald EA (1990) N-terminal variants of recombinant stefin B: effect on affinity for papain and cathepsin B. Biol Chem Hoppe-Seyler 371(Suppl):125–136

Turk B, Turk V, Turk D (1997) Structural and functional aspects of papain-like cysteine proteinases and their protein inhibitors. Biol Chem 378:141–150

Turk B, Stoka V, Rozman-Pungercar J, Cirman T, Droga-Mazovec G, Oresic K, Turk V (2002a) Apoptotic pathways: involvement of lysosomal proteases. Biol Chem 383:1035–1044

Turk B, Turk D, Salvesen GS (2002b) Regulating cysteine protease activity: essential role of protease inhibitors as guardians and regulators. Curr Pharm Des 8:1623–1637

Turk V, Bode W (1991) The cystatins: protein inhibitors of cysteine proteinases. FEBS Lett 285:213–219

Unverricht H (1891) Die Myoclonie. Deuticke, Leipzig

Usdin K, Grabczyk E (2000) DNA repeat expansions and human disease. Cell Mol Life Sci 57:914–931

Virtaneva K, D'Amato E, Miao J, Koskiniemi M, Norio R, Avanzini G, Franceschetti S, Michelucci R, Tassinari CA, Omer S, Pennacchio LA, Myers RM, Dieguez-Lucena JL, Krahe R, de la Chapelle A, Lehesjoki AE (1997) Unstable minisatellite expansion causing recessively inherited myoclonus epilepsy, EPM1. Nat Genet 15:393–396

Virtaneva K, Miao J, Traskelin AL, Stone N, Warrington JA, Weissenbach J, Myers RM, Cox DR, Sistonen P, de la Chapelle A (1996) Progressive myoclonus epilepsy EPM1 locus maps to a 175-kb interval in distal 21q. Am J Hum Genet 58:1247–1253

Wakamatsu N, Kominami E, Takio K, Katunuma N (1984) Three forms of thiol proteinase inhibitor from rat liver formed depending on the oxidation–reduction state of a sulfhydryl group. J Biol Chem 259:13832–13838

Weinhaeusel A, Morris MA, Antonarakis SE, Haas OA (2003) DNA deamination enables direct PCR amplification of the cystatin B (CSTB) gene-associated dodecamer repeat expansion in myoclonus epilepsy type Unverricht-Lundborg. Hum Mutat 22:404–408

Williams MS, Henkart PA (1994) Apoptotic cell death induced by intracellular proteolysis. J Immunol 153:4247–4255

Nucleic Acids and Molecular Biology, Vol. 19
M. Fry/K. Usdin (Eds.)
Human Nucleotide Expansion Disorders

Myotonic Dystrophies Types 1 and 2

Patana Teng-umnuay · Maurice S. Swanson (✉)

Department of Molecular Genetics and Microbiology and Powell Gene Therapy Center, College of Medicine, University of Florida, Gainesville, FL 32610-0266, USA
mswanson@ufl.edu

1
The Myotonic Dystrophies: an Overview

Myotonic dystrophy (dystrophia myotonica, DM) is a dominantly inherited neuromuscular disease that is characterized by a distinctive combination of clinical features, including skeletal muscle myotonia and weakness/wasting, cardiac muscle arrhythmias and conduction defects, unusual ocular cataracts, insulin insensitivity, male hypogonadism, balding and hypogammaglobulinemia (Harper 2001). Moreover, the genetic basis of DM is novel because this disease is caused by the expansion of different, but structurally similar, microsatellite repeats in two unrelated genes. Type 1 DM (DM1) is associated with the expansion of a $d(CTG)_n$ repeat [poly r(CUG)] positioned in the 3′-untranslated region (UTR) of the *DMPK* gene, while type 2 disease (DM2) results from a $d(CCTG)_n$ expansion [poly r(CCUG)] in the first intron of *ZNF9* (Brook et al. 1992; Liquori et al. 2001). Congenital DM (CDM), which is the most severe form of this disease, is exclusively associated with very large d(CTG) expansions in the *DMPK* gene.

The focus of this review is to highlight recent studies which examine the mechanistic question of how microsatellite expansions in the noncoding regions of different genes cause the multisystemic DM phenotype. We begin by comparing the genetics and clinical features of DM1 and DM2. The striking similarities between these two disease forms has led to the prevailing view that DM is an RNA-mediated disease in which mutant DMPK and ZNF9 transcripts accumulate in the nucleus and affect the normal activities of precursor messenger RNA (pre-mRNA) splicing factors during postnatal development. This RNA-mediated pathogenesis model predicts that the characteristic constellation of clinical features associated with this neuromuscular disease result from the retention of specific fetal protein isoforms which fail to function properly in adult tissues. While this model provides a reasonable explanation for many of the characteristic features of the juvenile and adult-onset forms of DM, it fails to account for the underlying cause of the congenital disease in which the embryonic development of brain and muscle is severely affected. Therefore, we will also reexamine an earlier disease model which emphasizes a central role for the *DMPK* gene locus in pathogenesis.

2
Genetics and Clinical Presentation of the Myotonic Dystrophies: DM1 Versus DM2 Disease: Many Similarities but Significant Differences

Prior to discussing disease models for DM pathogenesis, it is important to distinguish between the clinical presentations of types 1 and 2. Another related disorder with severe frontotemporal dementia, myotonia and DM-type cataracts, but no genetic linkage to *DMPK* or *ZNF9*, has been suggested as a candidate for DM type 3 (DM3) (Le Ber et al. 2004). However, the molecular basis for this disease, and its relationship to DM1 and DM2, is still obscure and therefore discussion of this disease will be reserved for a future review.

Both DM1 and DM2 are characterized by a distinguishing pattern of multisystemic abnormalities, including myotonia, muscle weakness, distinctive particulate cataracts, cardiac conduction defects and insulin insensitivity (Table 1) (reviewed in Finsterer 2002; Mankodi and Thornton 2002; Day et al. 2003; Meola and Moxley 2004; Day and Ranum 2005; Machuca-Tzili et al. 2005). Nevertheless, there is a consensus that DM1 is a more severe disease with earlier onset, severe neonatal hypotonia and mental retardation in the congenital form, prominent facial weakness and ptosis, more pronounced distal muscle weakness/wasting, readily apparent genetic anticipation, hypersomnia and dysphagia (Table 1). Therefore, pathogenesis models must account for both the similarities and the differences between DM1 and DM2. Interestingly, homozygosity of either the DM1 or the DM2 mutant alleles has little or no effect on disease severity (Schoser et al. 2004). Additional information on DM1 and DM2 clinical presentations, and the discovery of *DMPK* and *ZNF9* as the genes mutated in DM1 and DM2, respectively, may be found in previous reviews (Harper 2001; Finsterer 2002; Nykamp and Swanson 2004; Ranum and Day 2004; Day and Ranum 2005; Machuca-Tzili et al. 2005).

A particularly striking difference between DM1 and DM2 is the age at onset. DM1 has congenital, juvenile and adult-onset forms, while DM2 generally appears in the fourth or fifth decade and may not appear until age 70 with proximal weakness and very mild myotonia (van Engelen et al. 2005). Although early reports indicated that DM1 was primarily, if not exclusively, a disease of adults, subsequent studies revealed a form of the disorder that is present at birth (Harper 2001). CDM is associated with a high (17–41%) neonatal mortality rate and is characterized by profound immobility and hypotonia at birth (floppy baby), bilateral facial weakness and difficulties with suckling/swallowing, moderate to severe respiratory insufficiency, de-

Table 1 Characteristic features associated with myotonic dystrophy type 1 (*DM1*) versus myotonic dystrophy type 2 (*DM2*)

Category	Clinical feature	DM1	DM2
Genetics			
Inheritance		Dominant	Dominant
Congenital onset		Yes	No
Anticipation		Yes	Infrequent
Gene (chromosome)		*DMPK* (19q13.3)	*ZNF9* (3q31.3)
Protein function		Ser-Thr protein kinase	Transcription, translation
Expansion mutation		d(CTG)$_{37\rightarrow>3000}$	d(CCTG)$_{75\rightarrow\sim11\,000}$
Mutation position		3′ untranslated region	Intron 1
Clinical presentation			
Brain	Mental retardation (congenital only)	+	–
	Hypersomnia	+	–
	White matter abnormalities[a]	+	+
Endocrine/other	Frontal balding	+	+/–
	hypogammaglobulinemia	+	+
	Hypogonadism (male)	+	+
	Insulin insensitivity	+	+
Eye	Iridescent cataracts	+	+
Muscle—cardiac	Arrhythmia	+	+/–
	Conduction defect[b]	+	+/–
Muscle—skeletal	Distal weakness/wasting	+	+/–
	Neonatal hypotonia and respiratory insufficiency	+	–
	Myotonia	+	+
	Proximal weakness/wasting	+/–	+
Muscle—smooth	Constipation	+	+/–
	Dysphagia	+	–
Skeleton	Talipes	+	–

Clinical features are graded as + (characteristic/routinely observed), +/– (less common) and – (infrequently/never observed)

[a] Determined by cranial magnetic resonance imaging

[b] Determined by electrocardiography

layed motor development and mental retardation (Campbell et al. 2004). Clinically, CDM infants are extremely floppy with a "tented" or myopathic mouth and often require respiratory support in combination with tube feeding (Johnston 2003). Interestingly, the other major cause of extreme neonatal hypotonia is an imprinting disease, Prader–Willi syndrome (Tsai et al. 1999). In contrast to myotonia congenita, clinical myotonia is not present in CDM although electrical myotonia may be detectable at an early age. CDM is generally transmitted maternally and the mother is often mildly affected and generally unaware that she is a carrier. Male transmission is rare, possibly because very large $d(CTG)_{>1000}$ expansions associated with CDM might impair spermatogenesis and/or sperm viability. If the CDM infant survives, the hypotonia resolves by age 3–4 years. Thus, an important feature of CDM is that the differentiation and maturation of muscle is delayed during embryonic development but can proceed following birth although motor development remains abnormal. The partial respite from these muscle effects is temporary since adult-onset DM1 emerges in the second decade with the development of myotonia and progressive muscle weakness/wasting. The very large d(CTG) repeat expansions in CDM patients are also associated with earlier onset and more severe symptoms at this stage. Smooth muscle is also affected in CDM with colonic dilatation and poor intestinal motility. Additional clinical features unique to CDM include severe mental retardation, talipes (clubfoot) and strabismus (eyeball muscle imbalance). Why CDM is unique to large $d(CTG)_n$ expansions in the *DMPK* gene is unclear but might be due to differences in embryonic expression between *DMPK* and *ZNF9* or to an unusual property of very large d(CTG) repeats.

3
RNA Gain-of-Function Model for Myotonic Dystrophy

3.1
Myotonic Dystrophy Associated Microsatellite Expansions Are Toxic at the RNA Level

Although unstable microsatellite expansions can influence gene expression at multiple levels, several observations suggest that adult-onset DM is an RNA gain-of-function disease. The first clue that mutant DM RNA molecules are unusual was based on computer modeling which predicts that these d(CTG) expansions form stable double-stranded (ds) RNA structures or RNA hairpins (Zuker et al. 1999). Indeed, the existence of these RNA structures was demonstrated using chemical and enzymatic structure probing, thermal denaturation, magic angle spinning solid-state NMR and visualization of rod-like RNA duplexes in the electron microscope (Napierala and Krysysosiak

1997; Michalowski et al. 1999; Tian et al. 2000; Sobczak et al. 2003; Leppert et al. 2004). The r(CCUG) repeats in mutant ZNF9 transcripts also form RNA hairpin structures (Sobczak et al. 2003). However, there are significant differences in the stability of ds r(CCUG) versus ds r(CUG) with tandem C : U and U : C mismatches and a larger terminal loop for ds r(CCUG) hairpins compared with U : U mismatches for ds r(CUG).

Although DM repeat expansions affect both DNA and RNA structures, another distinguishing attribute of *DMPK* and *ZNF9* mutant allele transcripts is that they are retained in nuclear foci while normal transcripts are exported to the cytoplasm. These ribonuclear foci, which were originally detected by RNA fluorescence in situ hybridization (FISH) analysis, do not colocalize with any known nuclear structures, including splicing factor compartments or speckles, Cajal bodies, the perinucleolar compartment and promyelocytic leukemia nuclear bodies (Taneja et al. 1995; Davis et al. 1997). RNA FISH also indicates that DM2 ribonuclear foci in skeletal muscle are more intense and larger than DM1 ribonuclear foci, perhaps reflecting higher *ZNF9* expression levels (Mankodi et al. 2003). The discovery of these novel nuclear structures was particularly striking in light of concurrent studies on coding-region microsatellite expansion diseases, such as Huntington's disease (HD) and the spinocerebellar ataxias (SCAs). In HD and the SCAs, $d(CAG)_n$ expansions result in the synthesis of proteins containing a toxic polyglutamine (polyQ) region which accumulates in intranuclear inclusions (reviewed in Landles and Bates 2004; Taroni and DiDonato 2004).

Do the DM1 and DM2 expansion mutations affect the processing of their host transcripts? In contrast to an earlier report, mutant DMPK transcripts are correctly spliced and polyadenylated (Wang et al. 1995; Davis et al. 1997). FISH analysis indicates that these mRNA molecules remain intact even within RNA foci since hybridization signals using probes against the first seven *DMPK* exons and the $d(CTG)_n$ repeat colocalize to these foci (Taneja et al. 1995). This result is in agreement with the majority of expression studies which have reported only modest changes in DMPK RNA levels while DMPK protein levels decline (reviewed in Nykamp and Swanson 2004). In contrast, recent work suggests that ZNF9 RNA and protein levels are unaffected in DM2 heterozygous and homozygous individuals (L. Ranum, personal communication). The processing of ZNF9 pre-mRNA is probably not adversely influenced by the DM2 expansion mutation because it is positioned in the first intron approximately 850 nucleotides upstream of the 3′ splice site of *ZNF9* exon 2.

Although these observations suggested that poly r(CUG) and poly r(CCUG) exist as dsRNAs which accumulate in ribonuclear foci, transgenic mouse studies were required to confirm RNA-mediated pathogenesis as a viable disease model (reviewed in Wansink and Wieringa 2003). Transgenic mice carrying a greater than 45 kb fragment from the DM1 locus, which contains the *DMWD*, *DMPK* and *SIX5* genes as well as a *DMPK* $d(CTG)_{300}$ expansion, develop myotonia and DM-associated muscle histopathology (Seznec et al. 2001). Interestingly, other effects of transgene expression that are unrelated to DM disease, such as elongated crossed teeth, were also present. The possibility that d(CTG)$_n$ expansions alone are toxic independent of gene context was tested by creating mouse lines carrying a human skeletal actin (HSA) transgene with either a d(CTG)$_5$ (*HSA*^{SR}) or a d(CTG)$_{250}$ (*HSA*^{LR}) repeat tract inserted into the HSA 3′-UTR (Mankodi et al. 2000). While the *HSA*^{SR} mice are indistinguishable from normal sibs, *HSA*^{LR} mice develop skeletal muscle myotonia, centralized myonuclei and split myofibers characteristic of DM disease. Notably, several lines were created which express different levels of the transgene and *HSA*^{LR} mice with no, or relatively low levels of, transgene expression are not affected by myotonia, while high expressers develop robust myotonia. This result, together with the discovery that DM1 and DM2 are caused by structurally related repeat expansions in unrelated and unlinked genes, provides strong support for the conclusion that DM is an RNA gain-of-function disease which results from the expression of pathogenic ds r(CUG) and ds r(CCUG) RNA molecules.

3.2 Toxic RNAs Molecules Sequester Muscleblind-like Proteins

While the minimal microsatellite expansions associated with disease vary between DM1 and DM2, the predicted stability of the respective dsRNAs is remarkably similar [approximately 70 kcal/mol for both r(CUG)$_{50}$ and r(CCUG)$_{75}$]. Why are poly r(CUG) and poly r(CCUG) RNA molecules toxic above a certain repeat length? One possibility is that these RNA molecules are high-affinity binding sites for cellular factors. Binding of these factors might be proportional to the number of repeats and thus they are effectively sequestered above a certain length threshold. As the name implies, CUGBP1 was the first r(CUG)-binding protein identified and it is the founding member of the mammalian CELF family of RNA-binding proteins that contain three RNA recognition motifs (Caskey et al. 1996; Timchenko et al. 1996; Good et al. 2000; Ladd et al. 2001). However, several properties of this protein make it an unlikely candidate for a sequestered factor in DM. While disease-associated r(CUG) and r(CCUG) repeats form RNA hairpins, CUGBP1 is a single-stranded (ss) RNA-binding protein that recognizes r(CUG) trinucleotide and UG dinucleotide repeats (Timchenko et al. 1996; Michalowski et al. 1999; Takahashi et al. 2000). Although r(CUG)$_8$ binding activity and protein levels increase in DM1 cells and skeletal muscle, CUGBP1 does not colocalize with ribonuclear foci. Thus, CUGBP1 activity appears to be indirectly influenced by poly r(CUG) and poly r(CCUG) expression (Timchenko et al. 1996; Savkur et al. 2001; Ho et al. 2004).

In contrast to CUGBP1, considerable evidence now suggests that the muscleblind-like (MBNL) proteins are the sequestered factors in DM (Fig. 1a).

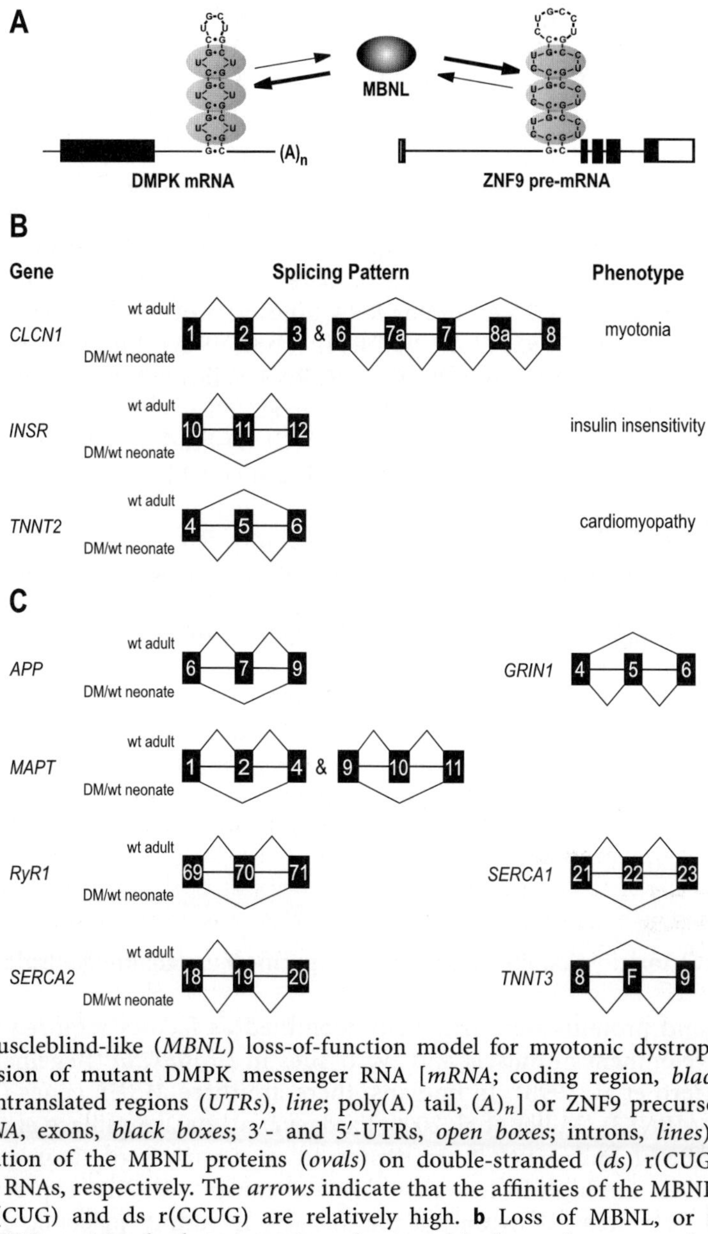

Fig. 1 Muscleblind-like (*MBNL*) loss-of-function model for myotonic dystrophy (*DM*). **a** Expression of mutant DMPK messenger RNA [*mRNA*; coding region, *black box*; 3′ and 5′ untranslated regions (*UTRs*), *line*; poly(A) tail, $(A)_n$] or ZNF9 precursor mRNA (*pre-mRNA*, exons, *black boxes*; 3′- and 5′-UTRs, *open boxes*; introns, *lines*) leads to sequestration of the MBNL proteins (*ovals*) on double-stranded (*ds*) r(CUG) and ds r(CCUG) RNAs, respectively. The *arrows* indicate that the affinities of the MBNL proteins for ds r(CUG) and ds r(CCUG) are relatively high. **b** Loss of MBNL, or upregulation of CELF, proteins leads to retention of neonatal isoforms (exons, *numbered black boxes*; introns, *horizontal lines*; splicing pattern, *angled lines*) in adult tissues, which, in turn, results in distinct pathophysiological effects (e.g., myotonia). The connection between TNNT2 missplicing and DM-associated heart defects (cardiomyopathy, conduction block) has not been established. **c** Missplicing of APP, GRIN1 (NMDA R1), MAPT, RyR1, SERCA1, SERCA2 and TNNT3 in DM tissues results in neonatal exon retention in adults but the phenotypic effects have not been determined

The MBNL proteins were originally identified on the basis of their unusual ability to bind and photo-cross-link to ds r(CUG), but not to ss r(CUG) or other ds RNA molecules [ds r(CAG), HIV TAR ds RNA], in HeLa nuclear extracts (Miller et al. 2000). MBNL binding to ds r(CUG) is proportional to repeat length in vitro and these proteins colocalize with poly r(CUG) and poly r(CCUG) RNA foci in cotransfected cells as well as DM skeletal muscle and cortical neurons (Fardaei et al. 2001, 2002; Mankodi et al. 2001, 2003; Jiang et al. 2004). While several additional nuclear RNA-binding proteins, such as hnRNPs F and H, also accumulate in ribonuclear foci to a much lesser degree, other ss RNA- and ds RNA-binding proteins (2′,5′-OAS, ADAR, CUGBP1, CUGBP2/ETR3, FLAP-1/LRRFIP1, hnRNP A1, hnRNP I, hnRNP M, KSRP, HuR, NF90/ILF3, PACT/RAX, PKR, RNA helicase A) and DNA-binding proteins (Sp1, RARγ) do not (Mankodi et al. 2003; Jiang et al. 2004; Kim et al. 2005). The observation that the ss r(CUG)-binding proteins CUGBP1 and CUGBP2/ETR3 do not colocalize with either DM1 or DM2 ribonuclear foci supports previous suggestions that these nuclear structures contain primarily ds r(CUG) and ds r(CCUG). Significantly, RNA FISH combined with immunocytochemistry indicates that formation of ribonuclear foci in DM1 cortical neurons correlates with a decrease in the diffuse nuclear, or nucleoplasmic, population (Jiang et al. 2004). Interestingly, three proteasome subunits (20Sα, 11Sα, 11Sγ) also colocalize with neuronal ribonuclear foci, suggesting that functional depletion of MBNL might result from targeted protein turnover. Although it is tempting to speculate that the formation of these ribonuclear foci is a primary event in the DM pathogenesis pathway, complexes between MBNL and ds r(CUG) and ds r(CCUG) RNAs which exist outside of these foci might also effectively sequester MBNL proteins (Ho et al. 2005b).

3.3
Poly r(CUG) Toxicity Requires Expression of Specific Muscleblind-like Isoforms

Muscleblind proteins were originally identified as factors required for late-stage development of muscle and eye tissues in *Drosophila* (Begemann et al. 1997; Artero et al. 1998). In humans, there are three *MBNL* genes (*MBNL1*, *MBNL2*, *MBNL3*) (Miller et al. 2000; Fardaei et al. 2002; Squillace et al. 2002). While *MBNL1* and *MBNL2* are expressed in a variety of tissues, MBNL1 mRNA levels are high in heart and skeletal muscle. *MBNL3* expression appears to be restricted to only a few tissues, including placenta. All three MBNL proteins colocalize with r(CUG) repeats in cells cotransfected with d(CTG) repeat and green fluorescent protein (GFP)–MBNL expression plasmids (Fardaei et al. 2002; Ho et al. 2005). Moreover, MBNL1 and MBNL2 accumulate in ribonuclear foci in neurons (Jiang et al. 2004). Intriguingly, MBNL1 proteins may play a fundamental role in ribonuclear foci formation and/or maintenance because small interfering RNA (siRNA)-mediated knockdown of MBNL1 mRNA leads to a substantial loss (approximately 70%) of these foci (Dansithong et al. 2005). Similar reductions of CUGBP1 and MBNL2 mRNA levels resulted in a smaller effect (approximately 20%) on the number of ribonuclear foci.

Are ds r(CUG) and ds r(CCUG) RNA molecules, or ribonuclear foci, inherently toxic to metazoan cells or does toxicity result from MBNL protein sequestration? In support of the latter possibility, a recent study suggests that ds r(CUG) RNA chains are not toxic in *Drosophila* (Houseley et al. 2005). Transgenic flies expressing GFP–DMPK–$(CTG)_{11-162}$ 3′-UTR fusions develop ribonuclear foci in some larval and adult muscles, but not in neurons, only when $d(CTG)_{162}$ is expressed. Although endogenous muscleblind proteins colocalize with these foci, these flies are viable, overtly normal and have extended lifespans. Interestingly, fly muscleblind proteins are not required for foci formation but coexpression of human MBNL1 results in the appearance of neuronal ribonuclear foci. It is important to note that the *Drosophila* muscleblind proteins vary from the vertebrate MBNL homologues since they possess only two of the four CCCH (C_3H) zinc-finger-related motifs which are required for high-affinity ds r(CUG) binding in vitro (Fig. 1) (Miller et al. 2000; Yuan et al., unpublished data). These results support the hypothesis that pathogenesis associated with $d(CTG)_n$ expression is mediated by interactions with specific MBNL proteins which are expressed in vertebrate cells.

If DM disease results from loss of certain MBNL isoforms owing to sequestration by toxic poly r(CUG) and poly r(CCUG) RNA molecules, then disease-associated phenotypes common to DM1 and DM2 should be recapitulated in *Mbnl* knockout mice. This possibility has been tested by generating mice which fail to express the 40–43-kDa isoforms which utilize an initiation codon in exon 3 of the *Mbnl1* gene. These larger isoforms bind, and photo-cross-link, to ds r(CUG), while the shorter 26–36-kDa Mbnl1 proteins do not. Mice carrying a homozygous *Mbnl1* exon 3 deletion ($Mbnl1^{\Delta E3/\Delta E3}$) are viable but develop the most characteristic features of adult-onset DM, including myotonia, dustlike cataracts and heart conduction defects (Kanadia et al. 2003a, and unpublished data). Since the adult-onset disease can be modeled in *Mbnl1* knockout mice in the absence of toxic ds r(CUG) and ds r(CCUG) RNA and ribonuclear foci, the striking conclusion is that DM is an MBNL loss-of-function disease resulting from an RNA gain-of-function mutation.

3.4 The Splicing Connection: DM is Associated with Fetal Exon Retention in Adults

Mutations associated with a large number of inherited diseases result in the perturbation of normal patterns of pre-mRNA splicing (Faustino and Cooper 2003; Garcia-Blanco et al. 2004; Matlin et al. 2005). In humans, multiexon genes are generally alternatively spliced. During postnatal development, fetal tissues are remodeled by a series of alternative splicing events to generate specific isoform ratios. Generally, these splicing decisions are temporally coordinated during the postnatal period so that tissues at various stages of maturation are responsive to the specific physiological demands characteristic of each developmental interval. How are these splicing decisions regulated so precisely so that the correct protein isoforms are synthesized at the proper time? Surprisingly, studies designed to reveal DM pathogenesis have provided fundamental insights into the regulation of pre-mRNA alternative splicing during the fetal-to-adult transition period.

In DM1 and DM2, the processing of *DMPK* and *ZNF9* mutant allele transcripts does not appear to be significantly affected by $r(CUG)_n$ and $r(CCUG)_n$ expansions; however, the alternative splicing of other transcripts is influenced. The discovery of misregulated splicing in DM1 resulted from studies designed to define the RNA sequence elements, and the corresponding *trans*-acting binding partners, which regulated the alternative splicing of exon 5 of chicken cardiac troponin T (TNNT2/cTNT) (Phillips et al. 1998). The splicing of TNNT2 exon 5 is developmentally regulated with inclusion favored in embryonic tissues, while skipping of this exon is the predominant pattern in adults. For chicken cTNT, exon 5 splicing is regulated by four muscle-specific splicing enhancers (MSE1–MSE4) in intron 5 and MSE1 and MSE4 each contain $r(CUG)_2$ repeats. Human cTNT contains an $r[CUG(N)_9(CUG)_2C(CUG)_2]$ repeat motif and exon 5 inclusion is favored in DM1, but not in normal, adult heart muscle (Fig. 1b) (Phillips et al. 1998; Ladd et al. 2001). These results suggest that retention of *TNNT2* exon 5 in adults might contribute to the heart conduction defects characteristically seen in DM. However, the connection between aberrant TNNT2 RNA splicing and DM disease remains tenuous. In DM1 and DM2, progressive cardiac conduction impairment, including atrioventricular (A-V) block, atrial fibrillation and ventricular/supraventricular arrhythmias are the most common heart problems (Pelargonio et al. 2002; Schoser et al. 2004b). Dilated cardiomyopathy has been documented for several DM2 patients and cardiomyopathies also develop in some DM1 patients. Interestingly, *TNNT2* mutations are generally linked to hypertrophic cardiomyopathy and dilated cardiomyopathy (DCM) and abnormal cTNT pre-mRNA splicing also occurs in mammals prone to DCM (Watkins et al. 1995; Biesiadecki et al. 2002; Pelargonio et al. 2002; Schoser et al. 2004b). Thus, TNNT2 pre-mRNA missplicing may be one component of the heart conduction defect common to DM1 but additional contributing factors will probably be uncovered in the future. Splicing of myotubularin-related 1 (MTMR1) is also dysregulated in DM1 adult heart with enhanced retention of the fetal A isoform (Ho et al. 2005a).

A more convincing argument for a direct role of disrupted RNA splicing in DM pathogenesis is provided by the myotonia which is a characteristic feature of DM1 and DM2. Mutations in both the skeletal muscle sodium (SCN5A) and chloride (CLCN1/ClC-1) channels cause myotonia in humans (Chen et al. 1997; Pusch 2002). DM-relevant myotonia has been linked to a defect in the CLCN1 channel but in this disease CLCN1 missplicing is the underlying pathogenic event (Fig. 1b) (Charlet-B et al. 2002; Mankodi et al. 2002). In normal adults, *CLCN1* exons 6, 7 and 8 are spliced together directly to generate functional chloride channels. During the fetal and neonatal periods and in either HSA^{LR} or $Mbnl1^{\Delta E3/\Delta E3}$ adult knockout mice, intron 2 and exons 6b, 7a and 8a are frequently included (Mankodi et al. 2002). These intronic and exonic sequences contain in-frame termination codons which make the resulting mRNA susceptible to turnover by the nonsense-mediated decay (NMD) pathway. For mRNAs that escape NMD, translation of truncated CLCN1 proteins has a dominant-negative effect on chloride channel function (Berg et al. 2004). Additionally, missplicing of skeletal muscle TNNT3 pre-mRNA has been documented in DM1 as well as in the HSA^{LR} and $Mbnl1^{\Delta E3/\Delta E3}$ mouse models but the physiological effects, if any, of adult expression of fetal TNNT3 isoforms are unknown (Kanadia et al. 2003b). Missplicing of the skeletal muscle ryanodine receptor RyR1 and sarcoplasmic/endoplasmic reticulum Ca^{2+}-ATPase (SERCA) 1 and 2 has also been reported and may account for altered calcium homeostasis in DM myotubes (Fig. 1c) (Kimura et al. 2005). The splicing of MTMR1 pre-mRNA is also abnormal in DM1 skeletal muscle (Buj-Bello et al. 2002).

Insulin resistance, which is another characteristic pathological feature of DM, is also caused by abnormal developmental regulation of splicing (Savkur et al. 2001). The fetal splicing pattern for the insulin receptor (IR) is inclusion of exon 11, which generates the lower signaling IR-A isoform. While IR exon 11 is included in normal adults, this exon is skipped in DM1 and DM2 adults (Fig. 1b). Thus, current evidence supports the conclusion that retention of the fetal TNNT2, CLCN1 and IR splicing pattern is responsible for the cardiac, myotonia and insulin resistance characteristic of DM disease.

DM has a significant effect on the function of the central nervous system (CNS), with distinctive behavioral effects and hypersomnia in adults as well as mental retardation in the congenital disease. As described previously, *DMPK* is expressed in the CNS and $r(CUG)_n$ expansions accumulate in neuronal nuclei. These CNS defects might result from abnormal splicing of several pre-mRNAs, including amyloid precursor protein (APP), microtubule-associated protein tau (MAPT) and the glutamate receptor, *N*-methyl-D-aspartate 1 (GRIN1/NMDA R1) (Fig. 1c) (Sergeant et al. 2001; Jiang et al. 2004). In conclusion, expression of the DM1 and DM2 expansion mutations clearly perturbs the regulation of RNA alternative splicing during postnatal development. Nevertheless, the molecular events underlying additional manifestations of DM disease (hypersomnia, mental retardation, muscle weakness/wasting, testicular atrophy, hypogammaglobulinemia, cataracts) have yet to be elucidated.

3.5 MBNL and CELF Proteins are Splicing Antagonists Which Regulate Fetal Exon Splicing

How are the expression of r(CUG) and r(CCUG) repeat expansions and sequestration of the MBNL proteins related to aberrant splicing during development? CUGBP1 was the first RNA-binding protein implicated in DM pathogenesis and was initially characterized as an $r(CUG)_8$-binding protein (Timchenko et al. 1996). Enhanced r(CUG)-binding activity was observed using extracts prepared from DM1 cells and subsequent analysis demonstrated that CUGBP1 is a splicing factor. Cells cotransfected with cTNT minigene reporter and CUGBP1 protein expression plasmids show enhanced cTNT exon 5 inclusion and this effect is abolished by mutation of the r(CUG) repeats to r(CAG) (Philips et al. 1998). Although CUGBP1 activity and steady-state protein levels are elevated in DM1 muscle and myoblasts, the connection between increased CUGBP1 splicing activity and expression of mutant DM1 and DM2 RNA was obscure until recent results became available that linked CUGBP1 and MBNL splicing activities (Savkur et al. 2001; Timchenko et al. 2001a; Dansithong et al. 2005). Intriguingly, CELF and MBNL protein families are antagonistic regulators of fetal exon splicing (Ho et al. 2004). Cotransfection analysis was used to demonstrate that overexpression of MBNL1, MBNL2 or MBNL3 proteins results in either enhanced skipping of TNNT2 exon 5 or increased inclusion of IR exon, which is identical to the normal adult splicing pattern. Alternatively, the DM splicing pattern is seen following siRNA-mediated knockdown of MBNL1 in HeLa cells, which promotes TNNT2 exon 5 inclusion and IR exon 11 skipping. Mutational analysis revealed that several MBNL1 binding sites (consensus is YGCUU/GY) exist immediately upstream of the 3′ splice site of TNNT2 exon 5 and mutation of these sites abolishes the effect of MBNL overexpression on exon 5 splicing (Ho et al. 2004). The similarity in Tnnt2 and Clcn1 pre-mRNA splicing patterns between $Mbnl1^{\Delta E3/\Delta E3}$ knockout and CUGBP1 transgenic (MCKCUG-BP1) mice also indicates that the MBNL and CELF protein families are antagonistic splicing regulators in vivo (Kanadia et al. 2003a; Ho et al. 2005a). CELF–MBNL interactions may also function in additional posttranscriptional regulatory pathways since CELF proteins have been implicated in RNA editing as well as mRNA translation and turnover (Anant et al. 2001; Timchenko et al. 2002, 2005; Mukhopadhyay et al. 2003; Iakova et al. 2004; Baldwin et al. 2004).

3.6 Is Myotonic Dystrophy Caused by MBNL Loss, CUGBP1 Overexpression or Both?

O the basis of the results described in the preceding sections, the original MBNL loss-of function model proposed for DM pathogenesis can be updated (Miller et al. 2000). Certain genes implicated in tissue-specific effects in DM, including CLCN1 and IR, contain fetal exons and inclusion of these exons during pre-mRNA splicing is promoted by CELF activity. In contrast, adult splicing patterns are triggered by activation of MBNL sometime during the neonatal-to-adult transition. Alternatively, CELF protein activity may decline during this transition as suggested by a recent study which demonstrated that CUGBP1 and CUGBP2/ETR-3 protein levels are relatively high in embryos and low in most adult somatic tissues, with the striking exception of brain (Ladd et al. 2005). Loss of MBNL activity, either by sequestration on ds r(CUG) and ds r(CCUG) RNA molecules in DM tissues and HSA^{LR} skeletal muscle, or in all tissues in *Mbnl1* knockout mice, leads to fetal exon retention in adult mRNA molecules because CELF splicing activity is unopposed. This simple model is appealing since it accounts for the increase in CELF splicing activity in DM tissues and cells due to loss of the MBNL splicing antagonist. However, this MBNL1 loss-of-function model fails to explain the observed increase in CUGBP1 steady-state tissue levels in DM1 skeletal muscle and myoblasts as well as the elevated $r(CUG)_8$ RNA-binding activity in vitro (Timchemko et al. 1996; Savkur et al. 2001; Dansithong et al. 2005). Unfortunately, these effects on CUGBP1 protein levels may be specific to humans since they are not reproduced in either HSA^{LR} or $Mbnl1^{\Delta E3/\Delta E3}$ mice. Another puzzling observation argues against the MBNL loss-of-function model. Mutation of the CUGBP1 binding site downstream of TNNT2 exon 5 does not affect MBNL1 splicing regulation since siRNA-mediated knockdown of MBNL1 levels still leads to enhanced exon 5 inclusion in transfected HeLa cells (Ho et al. 2004). In contrast to the prediction of the model, this CUGBP1 mutant binding site minigene no longer responds to expression of an r(CUG) repeat expansion RNA; therefore, siRNA-induced depletion of MBNL1 may not be synonymous with loss due to sequestration by $r(CUG)_n$ expansions. Nevertheless, this conclusion is tentative because we do not know how MBNL proteins interact with either precursor RNAs or ds r(CUG). The $r(CUG)_n$ expansion (CUG960) used in this study consisted of discontinuous $r[(CUG)_{20}CUCGA]_{48}$ repeats so the affinity of MBNL proteins for these repeats might be low relative to that for the continuous DM repeats. An alternative conclusion is that the TNNT2 intron 5 mutation creates a higher-affinity binding site target for MBNL binding which effectively competes with CUG960 binding activity. In summary, current evidence suggests that both MBNL and CELF protein activities in RNA splicing, and potentially other posttranscriptional regulatory steps, are adversely affected in DM.

4 Congenital Myotonic Dystrophy: a Distinct Disease with a Different Molecular Etiology?

Why do shorter DM1-associated $r(CUG)_{50-\sim3500}$ expansions cause a more severe disease phenotype than the more extended DM2 $r(CCUG)_{75-\sim11\,000}$ expansions and why is CDM linked exclusively to DM1? One hypothesis is that the MBNL proteins have a higher affinity for $r(CUG)_n$ repeats, so free MBNL levels are lower in DM1 cells. Several observations argue against this possibility. First, in vitro RNA binding experiments indicate that MBNL proteins have a higher affinity for r(CCUG), as opposed to r(CUG), repeats and RNA FISH and MBNL immunolocalization experiments indicate that ribonuclear foci are larger in DM2 cells and tissues (Mankodi et al. 2003; Jiang et al. 2004; Kino et al. 2004). Second, the congenital hypotonia and adult-onset muscle-wasting phenotypes are not present in *Mbnl1*$^{\Delta E3/\Delta E3}$ knockout mice, so MBNL loss may not be directly involved in this aspect of CDM and DM1 disease. Another hypothesis is that the more severe DM1 phenotype reflects differences in the expression patterns of *DMPK* and *ZNF9*. However, both genes show overlapping expression patterns during mouse embryogenesis and they are expressed in many of the adult tissues that are affected in DM (Kanadia et al. 2003b; Shimizu et al. 2003; Sarkar et al. 2004b). Indeed, *ZNF9* is generally expressed at a significantly higher level in most tissues. A third hypothesis is that DM2 disease results exclusively from poly r(CCUG) toxicity, while the DM1 phenotype may reflect combinatorial effects of $r(CUG)_n$ toxicity and altered expression of genes at the DM1 locus, particularly the tightly linked *DMWD*, *DMPK* and *SIX5* genes. To examine this last hypothesis, we will first review the evidence for altered expression of genes at the DM1 locus and then end with a discussion of studies assessing cell culture and transgenic mouse models for CDM.

4.1 The DMPK $(CTG)_n$ Expansion Alters the Chromatin Structure and Expression of the DM1 Locus

Early DM disease models invoked gene-specific effects, including *DMPK* haploinsufficiency and chromatin structural effects induced by the $d(CTG)_n$ expansion, to explain DM1 pathogenesis (Otten and Tapscott 1995; Wang et al. 1994). Using normal, DM1 and CDM tissues and derived cell lines, multiple studies reported that DMPK RNA and protein levels were depressed in DM and CDM cells although there have also been a few reports of elevated *DMPK* expression (Carango et al. 1993; Fu et al. 1993; Hofmann-Radvanyi et al. 1993; Novelli et al. 1993; Sabouri et al. 1993; Bhagwati et al. 1996; Hamshere et al. 1997; Laurent et al. 1997; Eriksson et al. 1999, 2000, 2001; Narang et al. 2000; Furling et al. 2001b, 2003; Frisch et al. 2001). The discovery of DMPK ribonuclear foci, and the technical problems associated with RNA extraction from these foci, provided a reasonable explanation for loss of mutant DMPK cytoplasmic mRNA and correspondingly lower DMPK protein levels (Davis et al. 1997; Hamshere et al. 1997). It is noteworthy that *Dmpk*$^{-/-}$, but not *Dmpk*$^{+/-}$, mice develop a late-onset progressive myopathy and muscle weakness, while both homozygous and heterozygous *Dmpk* knockout mice show A-V conduction blocks typically seen in DM (Jansen et al. 1996; Reddy et al. 1996; Berul et al. 1999, 2000).

Although the DM1 expansion only affects the structure of the DMPK transcript, there is evidence that the expression of the downstream gene *SIX5* decreases twofold to fourfold in DM fibroblasts, myoblasts, skeletal muscle and myocardium as well as somatic cell hybrids (Klesert et al. 1997; Thornton et al. 1997; Inukai et al. 2000; Frisch et al. 2001). However, other reports indicate that *SIX5* expression is unaffected in DM cells and tissues (Hamshere et al. 1997; Eriksson et al. 1999). In support of a role for *SIX5* expression changes in DM, heterozygous and homozygous *Six5* knockout mice develop progressive nuclear cataracts, perhaps resulting from increased expression of the *Atp1a1* gene encoding the Na^+/K^+-ATPase α-1 subunit (Klesert et al. 2000; Sarkar et al. 2000); however, the dustlike cataracts characteristic of DM are structurally distinct from these nuclear cataracts. Hypogonadism is another pathophysiological feature of both DM1 and DM2 and testicular atrophy, oligospermia and increased follicle-stimulating hormone levels have been reported in *Six5* knockout mice (Sarkar et al. 2004a). It is unlikely that *SIX5* expression is altered in DM2, although this has not been tested. Nevertheless, there is an intriguing coupling of gene expression at the DM1 locus. The *DMPK* d(CTG) repeats are flanked by CTCF binding sites which form a methylation-sensitive insulator element between the *DMPK* and *SIX5* genes. In CDM, the *DMPK* gene is hypermethylated and CTCF binding is impaired and this might allow interactions between the *SIX5* enhancer and the *DMPK* promoter and increased levels of *DMPK* RNA (Sabourin et al. 1993; Laurent et al. 1997; Steinbach et al. 1998; Filippova et al. 2001).

The *DMWD* gene is ubiquitously expressed in adult tissues, although the highest levels are detectable in synapse-dense regions of the brain and in the testes (Westerlaken et al. 2003). As with *DMPK* and *SIX5*, the effect of the *DMPK* $d(CTG)_n$ expansion on the expression of *DMWD* has been a subject of controversy with either no change or a 20–50% decrease in DM cytoplasmic RNA levels (Alwazzan et al. 1999; Eriksson et al. 2000; Frisch et al. 2001). These observations suggest that the expression profiles of the *DMWD*, *DMPK* and *SIX5* genes should be reevaluated using multiple CDM samples and contemporary optimized techniques for subcellular fractionation and reverse transcription PCR.

4.2 Cell Culture Models for Congenital Myotonic Dystrophy

The presence of neonatal muscle hypotrophy and hypotonia suggests that myogenic differentiation is delayed in CDM and that it might be possible to model this phenomenon in cell culture systems. In support of this possibility, histological analysis of CDM muscle shows myofiber immaturity with elevated numbers of satellite cells. Nuclear RNA foci containing mutant DMPK transcripts are detectable in CDM quadriceps myoblasts which also show reduced proliferative potential in cell culture (Furling et al. 2001a, b). Interestingly, CDM lymphoblast cell doubling is also compromised in culture, suggesting that a reduced in vitro life span is characteristic of all CDM cells (Khajavi et al. 2001). To address the specific step of myogenesis which is impaired by expression of CDM alleles, several groups have studied the effect of expressing mutant *DMPK* minigenes during myogenic differentiation of C2C12 cells (Sabourin et al. 1997; Amack et al. 1999; Bhagwati et al. 1999). An early study reported that C2C12 differentiation was inhibited by transgenes which express full-length wild-type DMPK or just the DMPK 3′-UTR alone (Sabourin et al. 1997). This fusion-inhibitory activity was mapped to a 239-nucleotide fragment immediately upstream of the d(CTG) repeat and loss of fusion correlated with a significant reduction in myogenin levels. Other studies using GFP–DMPK 3′-UTR $d(CTG)_{5 \text{ or } 200}$ reporter constructs have concluded that the d(CTG) repeat expansion is essential to inhibit myoblast fusion, although the same DMPK 3′-UTR upstream (or proximal) region appears to be important for this inhibitory activity (Amack et al. 1999, 2001). Expression of GFP–DMPK 3′-UTR $d(CTG)_{200}$, but not GFP–DMPK 3′-UTR $d(CTG)_5$, led to reduced MyoD expression levels and disruption of myoblast differentiation which was restored following infection with a MyoD-expressing retrovirus (Amack et al. 2002). A potential problem is that the longest repeat used in these studies was $d(CTG)_{200}$ and CDM patients generally possess repeats in excess of 1000 trinucleotide repeat units, so it is unclear if shorter repeat minigenes are replicating the molecular events involved in CDM pathogenesis. One possibility is that the precipitating pathogenic event is the accumulation of mutant DMPK transcripts in the nucleus beyond a threshold level, which leads to a loss of myogenic differentiation. Perhaps the larger d(CTG) repeats associated with CDM are required to attain this threshold in CDM patients, while minigene-induced overexpression of shorter repeats in cell culture replicates this pathogenic pathway. Alternatively, myogenic differentiation is inhibited by overexpression and underexpression of a variety of factors (e.g., TNFα, HuR), so DMPK 3′-UTR $d(CTG)_{200}$ minigene overexpression may cause deleterious effects that block efficient myoblast function but these effects are unrelated to CDM pathogenesis (van der Giessen et al. 2003; Langen et al. 2004).

4.3 Mouse Transgenic and Knockout Models for Congenital Myotonic Dystrophy

Attempts to model CDM in transgenic and knockout mice have failed (reviewed in Wansink and Wieringa 2003). The HSA^{LR} transgenic mouse, which expresses a $d(CTG)_{250}$ in the 3′-UTR of the human skeletal actin gene, develops myotonia but does not show either muscle wasting or neonatal hypotonia, possibly because the *HSA* transgene is expressed later than the endogenous *Dmpk* gene during myogenic differentiation. Additional transgenic lines (*Dmt*-A–*Dmt*-E and DM300) have been generated in which the d(CTG) repeat expansions are driven by a human promoter, so the failure to show neonatal hypotonia might also result from differences in developmental timing between human and mouse myogenesis (Seznec et al. 2001). Recently, Storbeck et al. (2004) produced mice expressing a chimeric transgene composed of a human *DMPK* promoter region, *DMPK* intron 1, a GFP reporter open reading frame and the DMPK 3′-UTR containing either $d(CTG)_{11}$ or $d(CTG)_{91}$ repeats. Both $d(CTG)_{11}$ and $d(CTG)_{91}$ transgenic mice show delayed myogenin expression and muscle development, and type I and II fiber atrophy in 1–3-month-old animals which disappears by 6 months of age. Primary myoblasts obtained from these animals have reduced fusion potential in culture. Although the relevance of these findings to delayed myogenic differentiation in CDM is unclear since these mice were not born hypotonic, the important result is that overexpression of a normal DMPK 3′-UTR with a $d(CTG)_{11}$ repeat has a similar effect on muscle development as a mutant $d(CTG)_{91}$. Interestingly, neither these $d(CTG)_{91}$ transgenic mice nor "humanized" $d(CTG)_{84}$ knockin mice, in which the human *DMPK* 3′-UTR containing 84 repeats is inserted into the mouse *Dmpk* gene, develop myotonia or detectable RNA foci while HSA^{LR} mice with 250 r(CUG) repeats do (Mankodi et al.; van den Broek 2002). This latter observation supports the idea of a threshold effect for Mbnl protein recruitment resulting in postnatal pathogenesis and suggests that it may be possible to model CDM in mice with larger d(CTG) repeats independent of gene context. Alternatively, the generation of mouse knockout models might be appropriate if congenital disease involves loss of both MBNL and unidentified factors that are linked to the DM1 locus or which bind to DMPK RNA and are cosequestered with MBNL proteins in nuclear RNA foci. Most importantly, studies designed to reveal the molecular etiology of CDM will likely provide significant new insights into biochemical pathways crucial for development of the brain and skeletal muscle during embryogenesis.

References

Alwazzan M, Newman E, Hamshere MG, Brook JD (1999) Myotonic dystrophy is associated with a reduced level of RNA from the DMWD allele adjacent to the expanded repeat. Hum Mol Genet 8:1491–1497

Amack JD, Mahadevan MS (2001) The myotonic dystrophy expanded CUG repeat tract is necessary but not sufficient to disrupt C2C12 myoblast differentiation. Hum Mol Genet 10:1879–1887

Amack JD, Paguio AP, Mahadevan MS (1999) Cis and trans effects of the myotonic dystrophy (DM) mutation in a cell culture model. Hum Mol Genet 8:1975–1984

Amack JD, Reagan SR, Mahadevan MS (2002) Mutant DMPK 3′-UTR transcripts disrupt C2C12 myogenic differentiation by compromising MyoD. J Cell Biol 159:419–429

Anant S, Henderson JO, Mukhopadhyay D, Navaratnam N, Kennedy S, Min J, Davidson NO (2001) Novel role for RNA-binding protein CUGBP2 in mammalian RNA editing. J Biol Chem 276:47338–47351

Artero R, Prokop A, Paricio N, Begemann G, Pueyo I, Mlodzik M, Perez-Alonso M, Maylies MK (1998) The muscleblind gene participates in the organization of Z-bands and epidermal attachments of Drosophila muscles and is regulated by Dmef2. Dev Biol 195:131–143

Baldwin BR, Timchenko NA, Zahnow CA (2004) Epidermal growth factor receptor stimulation activates the RNA binding protein CUG-BP1 and increases expression of C/EBPβ-LIP in mammary epithelial cells. Mol Cell Biol 24:3682–3691

Begemann G, Paricio N, Kiss I, Perez-Alonso M, Mlodzik M (1997) Muscleblind, a gene required for photoreceptor differentiation in Drosophila, encodes novel Cys_3His-type zinc-finger-containing proteins. Development 124:4321–4331

Berg J, Jiang H, Thornton CA, Cannon SC (2004) Truncated ClC-1 mRNA in myotonic dystrophy exerts a dominant-negative effect on the Cl current. Neurology 63:2371–2375

Berul CI, Maguire CT, Aronovitz MJ, Greenwood J, Miller C, Gehrmann J, Housman D, Mendelsohn ME, Reddy S (1999) DMPK dosage alterations in atrioventricular conduction abnormalities in a mouse myotonic dystrophy model. J Clin Invest 103:R1–R7

Berul CI, Maguire CT, Gehrmann J, Reddy S (2000) Progressive atrioventricular block in a mouse myotonic dystrophy model. J Interv Card Electrophysiol 4:351–358

Bhagwati S, Ghatpande A, Leung B (1996) Normal levels of DM RNA and myotonin protein kinase in skeletal muscle from adult myotonic dystrophy (DM) patients. Biochim Biophys Acta 1317:155–157

Bhagwati S, Shafiq SA, Xu W (1999) $(CTG)_n$ repeats markedly inhibit differentiation of the C2C12 myoblast cell line: implications for congenital myotonic dystrophy. Biochim Biophys Acta 1453:221–229

Biesiadecki BJ, Elder BD, Yu Z-B, Jin J-P (2002) Cardiac troponin T variants produced by aberrant splicing of multiple exons in animals with high instances of dilated cardiomyopathy. J Biol Chem 277:50275–50285

Brook JD, McCurrach ME, Harley HG, Buckler AJ, Church D, Aburatani H, Hunter K, Stanton VP, Thirion J-P, Hudson T, Sohn R, Zemelman B, Snell RG, Rundle SA, Crow S, Davies J, Shelbourne P, Buxton J, Jones C, Juvonen V, Johnson K, Harper PS, Shaw DJ, Housman DE (1992) Molecular basis of myotonic dystrophy: expansion of a trinucleotide (CTG) repeat at the 3′ end of a transcript encoding a protein kinase family member. Cell 68:799–808

Buj-Bello A, Furling D, Tronchere H, Laporte J, Lerouge T, Butler-Browne GS, Mandel JL (2002) Muscle-specific alternative splicing of myotubularin-related 1 gene is impaired in DM1 muscle cells. Hum Mol Genet 11:2297–2307

Campbell C, Sherlock R, Jacob P, Blayney M (2004) Congenital myotonic dystrophy: assisted ventilation duration and outcome. Pediatrics 113:811–816

Carango P, Nobler JE, Marks HG, Funanage VL (1993) Absence of myotonic dystrophy protein kinase (DMPK) mRNA as a result of a triplet repeat expansion in myotonic dystrophy. Genomics 18:340–348

Caskey CT, Swanson MS, Timchenko LT (1996) Myotonic dystrophy: discussion of molecular mechanism. Cold Spring Harbor Symp Quant Biol 61:607–614

Charlet-B N, Savkur RS, Singh G, Philips AV, Grice EA, Cooper TA (2002) Loss of the muscle-specific chloride channel in type 1 myotonic dystrophy due to misregulated alternative splicing. Mol Cell 10:45–53

Chen M-F, Niggeweg R, Iaizzo PA, Lehmann-Horn F, Jockusch H (1997) Chloride conductance in mouse models is subject to post-transcriptional compensation of the functional Cl^- channel 1 gene dosage. J Physiol 504:75–81

Dansithong W, Paul S, Comai L, Reddy S (2005) MBNL1 is the primary determinant of focus formation and aberrant insulin receptor splicing in DM1. J Biol Chem 280:5773–5780

Davis BM, McCurrach ME, Taneja KL, Singer RH, Housman DE (1997) Expansion of a CUG trinucleotide repeat in the 3′ untranslated region of myotonic dystrophy protein kinase transcripts results in nuclear retention of transcripts. Proc Natl Acad Sci USA 94:7388–7393

Day JW, Ranum LPW (2005) RNA pathogenesis of the myotonic dystrophies. Neuromus Dis 15:5–16

Day JW, Ricker K, Jacobsen JF, Rasmussen LJ, Dick KA, Kress W, Schneider C, Koch MC, Beilman GJ, Harrison AR, Dalton JC, Ranum LPW (2003) Myotonic dystrophy type 2: molecular, diagnostic and clinical spectrum. Neurology 60:657–664

Eriksson M, Ansved T, Edstrom L, Anvret M, Carey N (1999) Simultaneous analysis of expression of the three myotonic dystrophy locus genes in adult skeletal muscle samples: the CTG expansion correlates inversely with DMPK and 59 expression levels, but no DMAHP levels. Hum Mol Genet 8:1053–1060

Eriksson M, Ansved T, Edstrom L, Wells DJ, Watt DJ, Anvret M, Carey N (2000) Independent regulation of the myotonic dystrophy 1 locus genes postnatally and during adult skeletal muscle regeneration. J Biol Chem 275:19964–19969

Eriksson M, Hedberg B, Carey N, Ansved T (2001) Decreased *DMPK* transcript levels in myotonic dystrophy type IIA muscle fibers. Biochem Biophys Res Commun 286:1177–1182

Fardaei M, Larkin K, Brook JD, Hamshere MG (2001) In vivo co-localization of MBNL protein with *DMPK* expanded-repeat transcripts. Nucleic Acids Res 29:2766–2771

Fardaei M, Rogers MT, Thorpe HM, Larkin K, Hamshere MG, Harper PS, Brook JD (2002) Three proteins, MBNL, MBLL and MBXL, co-localize in vivo with nuclear foci of expanded-repeat transcripts in DM1 and DM2 cells. Hum Mol Genet 11:805–814

Faustino NA, Cooper TA (2003) Pre-mRNA splicing and human disease. Genes Dev 17:419–437

Filippova GN, Thienes CP, Penn BH, Cho DH, Hu YJ, Moore JM, Klesert TR, Lobanenkov VV, Tapscott SJ (2001) CTCF-binding sites flank CTG/CAG repeats and form a methylation-sensitive insulator at the *DM1* locus. Nat Genet 28:335–343

Finsterer (2002) Myotonic dystrophy type 2. Eur J Neurol 9:441–447

Frisch R, Singleton KR, Moses PA, Gonzalez IL, Carango P, Marks HG, Funanage VL (2001) Effect of triplet repeat expansion on chromatin structure and expression of *DMPK* and neighboring genes, *SIX5* and *DMWD*, in myotonic dystrophy. Mol Genet Metabol 74:281–291

Fu Y-H, Friedman DL, Richards S, Pearlman JA, Gibbs RA, Pizzuti A, Ashizawa T, Perryman HB, Scarleto G, Fenwick RG, Caskey CT (1993) Decreased expression of myotonin-protein kinase messenger RNA and protein in adult form of myotonic dystrophy. Science 260:235–238

Furling D, Coiffier L, Mouly V, Barbet JP, St Guily JL, Taneja K, Gourdon G, Junien C, Butler-Browne GS (2001a) Defective satellite cells in congenital myotonic dystrophy. Hum Mol Genet 10:2079–2087

Furling D, Lemieux D, Taneja K, Puymirat J (2001b) Decreased levels of myotonic dystrophy protein kinase (DMPK) and delayed differentiation in human myotonic dystrophy myoblasts. Neuromuscul Disord 11:728–735

Furling D, Lam L-T, Agbulut O, Butler-Browne GS, Morris GE (2003) Changes in myotonic dystrophy protein kinase levels and muscle development in congenital myotonic dystrophy. Am J Path 162:1001–1009

Garcia-Blanco MA, Baraniak AP, Lasda EL (2004) Alternative splicing in disease and therapy. Nat Biotech 22:535–546

Good PJ, Chen Q, Warner SJ, Herring DC (2000) A family of human RNA-binding proteins related to the *Drosophila* bruno translational regulator. J Biol Chem 275:28583–28592

Hamshere MG, Newman EE, Alwazzan M, Athwal BS, Brook JD (1997) Transcriptional abnormality in myotonic dystrophy affects *DMPK* but not neighboring genes. Proc Natl Acad Sci USA 94:7394–7399

Harper PS (2001) Myotonic dystrophy. Saunders, London

Ho TH, Bundman D, Armstrong DL, Cooper TA (2005a) Transgenic mice expressing CUG-BP1 reproduce splicing mis-regulation observed in myotonic dystrophy. Hum Mol Genet 14:1539–1547

Ho TH, Charlet-B N, Poulos MG, Singh G, Swanson MS, Cooper TA (2004) Muscleblind proteins regulate alternative splicing. EMBO J 23:3103–3112

Ho TH, Savkur RS, Poulos MG, Mancini MA, Swanson MS, Cooper TA (2005b) Colocalization of muscleblind with RNA foci is separable from mis-regulation of alternative splicing in myotonic dystrophy. J Cell Sci 118:2923–2933

Hofmann-Radvanyi H, Junien C (1993) Myotonic dystrophy: over-expression or/and under expression? A critical review on a controversial point. Neuromuscul Disord 3:497–501

Houseley JM, Wang Z, Brock GJR, Soloway J, Artero R, Perez-Alonso M, O'Dell KMC, Monckton DG (2005) Myotonic dystrophy associated expanded CUG repeat muscleblind positive ribonuclear foci are not toxic to *Drosophila*. Hum Mol Genet 14:873–883

Iakova P, Wang G-L, Timchenko L, Michalak M, Pereira-Smith OM, Smith JR, Timchenko NA (2004) Competition of CUGBP1 and calreticulin for the regulation of p21 translation determines cell fate. EMBO J 23:406–417

Inukai A, Doyu M, Kato T, Liang Y, Kuru S, Yamamoto M, Kobayashi Y, Sobue G (2000) Reduced expression of DMAHP/SIX5 gene in myotonic dystrophy muscle. Muscle Nerve 23:1421–1426

Jansen G, Groenen PJTA, Bächner D, Jap PHK, Coerwinkel M, Oerlemans F, van den Broek W, Gohlsch B, Pette D, Plomp JJ, Molenaar PC, Nederhoff MGJ, van Echteld CJA, Dekker M, Berns A, Hameister H, Wieringa B (1996) Abnormal myotonic dystrophy protein kinase levels produce only mild myopathy in mice. Nat Genet 13:316–324

Jiang H, Mankodi A, Swanson MS, Moxley RT, Thornton CA (2004) Myotonic dystrophy type 1 is associated with nuclear foci of mutant RNA, sequestration of muscleblind proteins and deregulated alternative splicing in neurons. Hum Mol Genet 13:3079–3088

Johnston HM (2003) The floppy weak infant revisited. Brain Dev 25:155–158

Kanadia RN, Johnstone KA, Mankodi A, Lungu C, Thornton CA, Esson D, Timmers AM, Hauswirth WW, Swanson MS (2003a) A muscleblind knockout model for myotonic dystrophy. Science 302:1978–1980

Kanadia RN, Urbinati CR, Cruselle VJ, Luo D, Lee Y-J, Harrison JK, Oh SP, Swanson MS (2003b) Developmental expression of mouse muscleblind genes *Mbnl1*, *Mbnl2* and *Mbnl3*. Gene Expr Patterns 3:459–462

Khajavi M, Tari AM, Patel NB, Tsuji K, Siwak DR, Meistrich ML, Terr NH, Ashizawa T (2001) Mitotic drive of expanded CTG repeats in myotonic dystrophy type 1. Hum Mol Genet 10:855–863

Kim D-H, Langlois M-A, Lee K-B, Riggs AD, Puymirat J, Rossi JJ (2005) HnRNP H inhibits nuclear export of mRNA containing expanded CUG repeats and a distal branch point sequence. Hum Mol Genet 33:3866–3874

Kimura T, Nakamori M, Lueck JD, Pouliquin P, Aoike F, Fujimura H, Dirksen RT, Takahashi MP, Dulhunty AF, Sakoda S (2005) Altered mRNA splicing of the skeletal muscle ryanodine receptor and sarcoplasmic/endoplasmic reticulum Ca^{2+}-ATPase in myotonic dystrophy type 1. Hum Mol Genet 14:2189–2200

Kino Y, Mori D, Oma Y, Takeshita Y, Sasagawa N, Ishiura S (2004) Muscleblind protein, MBNL1/EXP, binds specifically to CHHG repeats. Hum Mol Genet 13:495–507

Klesert TR, Otten AD, Bird TD, Tapscott SJ (1997) Trinucleotide repeat expansion at the myotonic dystrophy locus reduces expression of *DMAHP*. Nat Genet 16:402–406

Klesert TR, Cho DH, Clark JI, Maylie J, Adelman J, Snider L, Yuen EC, Soriano P, Tapscott SJ (2000) Mice deficient in Six5 develop cataracts: implications for myotonic dystrophy. Nat Genet 25:105–109

Ladd AN, Charlet-B N, Cooper TA (2001) The CELF family of RNA binding proteins is implicated in cell-specific and developmentally regulated alternative splicing. Mol Cell Biol 21:1285–1296

Ladd AN, Stenberg MG, Swanson MS, Cooper TA (2005) Dynamic balance between activation and repression regulates pre-mRNA alternative splicing during heart development. Dev Dyn 233:783–793

Landles C, Bates GP (2004) Huntingtin and the molecular pathogenesis of Huntington's disease. EMBO Rep 5:958–963

Langen RC, Van der Velden JL, Schols AM, Kelders MC, Wouters EF, Janssen-Heininger YM (2004) Tumor necrosis factor-alpha inhibits myogenic differentiation through MyoD protein destabilization. FASEB J 18:227–237

Laurent A, Costa JM, Assouline B, Voyer M, Vidaud M (1997) Myotonic dystrophy protein kinase gene expression in skeletal muscle from congenitally affected infants. Ann Genet 40:169–174

Le Ber I, Martinez M, Campion D, Laquerrière A, Bétard C, Bassez G, Girard C, Saugier-Veber P, Raux G, Sergeant N, Magnier P, Maisonobe T, Eymard B, Duyckaerts C, Delacourte A, Frebourg T, Hannequin D (2004) A non-DM1, non-DM2 multisystem myotonic dystrophy disorder with frontotemporal dementia: phenotype and suggestive mapping of the DM3 locus to chromosome 15q21–24. Brain 127:1979–1992

Leppert J, Urbinati CR, Hafner S, Ohlenschlager O, Swanson MS, Gorlach M, Ramachandran R (2004) Identification of NH $\cdots$ H hydrogen bonds by magic angle spinning solid state NMR in a double-stranded RNA associated with myotonic dystrophy. Nucleic Acids Res 32:1177–1183

Liquori CL, Ricker K, Moseley ML, Jacobson JF, Kress W, Naylor SL, Day JW, Ranum LPW (2001) Myotonic dystrophy type 2 caused by a CCTG expansion in intron 1 of *ZNF9*. Science 293:864–867

Machuca-Tzili L, Brook D, Hilton-Jones D (2005) Clinical and molecular aspects of the myotonic dystrophies: a review. Muscle Nerve 32:1–18

Mankodi A, Thornton CA (2002) Myotonic syndromes. Curr Opin Neurol 15:545–552

Mankodi A, Logigian E, Callahan L, McClain C, White R, Henderson D, Krym M, Thornton CA (2000) Myotonic dystrophy in transgenic mice expressing an expanded CUG repeat. Science 289:1769–1772

Mankodi A, Urbinati CR, Yuan Q-P, Moxley RT, Sansone V, Krym M, Henderson D, Schalling M, Swanson MS, Thornton CA (2001) Muscleblind localizes to nuclear foci of aberrant RNA in myotonic dystrophy types 1 and 2. Hum Mol Genet 10: 2165–2170

Mankodi A, Takahashi MP, Jiang H, Beck CL, Bowers WJ, Moxley RT, Cannon SC, Thornton CA (2002) Expanded CUG repeats trigger aberrant splicing of ClC-1 chloride channel pre-mRNA and hyperexcitability of skeletal muscle in myotonic dystrophy. Mol Cell 10:35–44

Mankodi A, Teng-umnuay P, Krym M, Henderson D, Swanson M, Thornton CA (2003) Ribonuclear inclusions in skeletal muscle in myotonic dystrophy types 1 and 2. Ann Neurol 54:760–768

Matlin AJ, Clark F, Smith CWJ (2005) Understanding alternative splicing: towards a cellular code. Nat Rev Mol Cell Biol 6:386–398

Meola G, Moxley RT III (2004) Myotonic dystrophy type 2 and related myotonic disorders. J Neurol 251:1173–1182

Michalowski S, Miller JW, Urbinati CR, Paliouras M, Swanson MS, Griffith J (1999) Visualization of double-stranded RNAs from the myotonic dystrophy protein kinase gene and interactions with CUG-binding protein. Nucleic Acids Res 27:3534–3542

Miller JW, Urbinati CR, Teng-umnuay P, Stenberg MG, Byrne BJ, Thornton CA, Swanson MS (2000) Recruitment of human muscleblind proteins to $(CUG)_n$ expansions associated with myotonic dystrophy. EMBO J 19:4439–4448

Mukhopadhyay D, Houchen CW, Kennedy S, Dieckgraefe BK, Anant S (2003) Coupled mRNA stabilization and translational silencing of cyclooxygenase-2 by a novel RNA binding protein, CUGBP2. Mol Cell 11:113–126

Napierala M, Kryzyzosiak WJ (1997) CUG repeats present in myotonin kinase RNA form metastable slippery hairpins. J Biol Chem 272:31079–31085

Narang MA, Waring JD, Sabourin LA, Korneluk RG (2000) Myotonic dystrophy (DM) protein kinase levels in congenital and adult DM patients. Eur J Hum Genet 8:507–512

Novelli G, Gennareli M, Zelano G, Pizzuti A, Fattorini C, Caskey CT, Dallapiccola B (1993) Failure in detecting mRNA transcripts from the mutated allele in myotonic dystrophy muscle. Biochem Mol Biol Int 29:291–297

Nykamp KR, Swanson MS (2004) Toxic RNA in the nucleus: unstable microsatellite expression in neuromuscular disease. Proc Mol Subcell Biol 35:57–77

Otten AD, Tapscott SJ (1995) Triplet repeat expansion in myotonic dystrophy alters the adjacent chromatin structure. Proc Natl Acad Sci USA 92:5465–5469

Pelargonio G, Dello Russo A, Sanna T, De Martino G, Bellocci F (2002) Myotonic dystrophy and the heart. Heart 88:665–670

Philips AV, Timchenko LT, Cooper TA (1998) Disruption of splicing regulated by a CUG-binding protein in myotonic dystrophy. Science 280:737–741

Pusch M (2002) Myotonia caused by mutations in the muscle chloride channel gene CLCN1. Hum Mutat 19:423–434

Ranum LPW, Day JW (2004) Pathogenic RNA repeats: an expanding role in genetic disease. Trends Genet 20:506–512

Reddy S, Smith DB, Rich MM, Leferovich JM, Reilly P, Davis BM, Tran K, Rayburn H, Bronson R, Cros D, Balice-Gordon RJ, Housman D (1996) Mice lacking the myotonic dystrophy protein kinase develop a late onset progressive myopathy. Nat Genet 13:325–335

Sabouri LA, Mahadevan MS, Narang M, Lee DS, Surh LC, Korneluk RG (1993) Effect of the myotonic dystrophy (DM) mutation on mRNA levels of the DM gene. Nat Genet 4:233–238

Sabourin LA, Tamai K, Narang MA, Korneluk RG (1997) Overexpression of 3′-untranslated region of the myotonic dystrophy kinase cDNA inhibits myoblast differentiation in vitro. J Biol Chem 272:29626–29635

Sarkar PS, Appukuttan B, Han J, Ito Y, Ai C, Tsai W, Chai Y, Stout JT, Reddy S (2000) Heterozygous loss of *Six5* in mice is sufficient to cause ocular cataracts. Nat Genet 25:110–114

Sarkar PS, Paul S, Han J, Reddy S (2004a) *Six5* is required for spermatogenic cell survival and spermiogenesis. Hum Mol Genet 13:1421–1431

Sarkar PS, Han J, Reddy S (2004b) In situ hybridization analysis of Dmpk mRNA in adult mouse tissues. Neuromuscul Dis 14:497–506

Savkur RS, Philips AV, Cooper TA (2001) Aberrant regulation of insulin receptor alternative splicing is associated with insulin resistance in myotonic dystrophy. Nat Genet 29:40–47

Schoser BGH, Kress W, Walter MC, Halliger-Keller B, Lochmuller H, Ricker K (2004a) Homozygosity for CCTG mutation in myotonic dystrophy type 2. Brain 127:1868–1877

Schoser BGH, Ricker K, Schneider-Gold C, Hengstenberg C, Dürre J, Bültmann B, Kress W, Day JW, Ranum LPW (2004b) Sudden cardiac death in myotonic dystrophy type 2. Neurology 63:2402–2404

Sergeant N, Sablonnière B, Schraen-Maschke S, Ghestem A, Maurage C-A, Wattez A, Vermersch P, Delacourte A (2001) Dysregulation of human brain microtubule-associated tau mRNA maturation in myotonic dystrophy type 1. Hum Mol Genet 10:2143–2155

Seznec H, Agbulut O, Sergeant N, Savouret C, Ghestem A, Tabti N, Willer J-C, Ourth L, Duros C, Brisson E, Fouquet C, Butler-Browne G, Delacourte A, Junien C, Gourdon G (2001) Mice transgenic for the human myotonic dystrophy region with expanded CTG repeats display muscular and brain abnormalities. Hum Mol Genet 10:2717–2726

Shimizu K, Chen W, Ashique AM, Moroi R, Li Y-P (2003) Molecular cloning, developmental expression, promoter analysis and functional characterization of the mouse *CNBP* gene. Gene 307:51–62

Sobczak K, de Mezer M, Michlewski G, Krol J, Krzyzosiak WJ (2003) RNA structure of trinucleotide repeats associated with human neurological diseases. Nucleic Acids Res 31:5469–5482

Squillace RM, Chenault DM, Wang EH (2002) Inhibition of muscle differentiation by the novel muscleblind-related protein CHCR. Dev Biol 250:218–230

Steinbach P, Gläser D, Vogel W, Wolf M, Schwemmle S (1998) The DMPK gene of severely affected myotonic dystrophy patients is hypermethylated proximal to the largely expanded CTG repeat. Am J Hum Genet 62:278–285

Storbeck CJ, Drmanic S, Daniel K, Waring JD, Jirik FR, Parry DJ, Ahmed N, Sabourin LA, Ikeda J-E, Korneluk RG (2004) Inhibition of myogenesis in transgenic mice expressing the human DMPK 3′-UTR. Hum Mol Genet 13:589–600

Takahashi N, Sasagawa N, Suzuki K, Ishiura S (2000) The CUG-binding protein binds specifically to UG dinucleotide repeats in a yeast three-hybrid system. Biochem Biophys Res Comm 277:518–523

Taneja KL, McCurrach M, Schalling M, Housman D, Singer RH (1995) Foci of trinucleotide repeat transcripts in nuclei of myotonic dystrophy cells and tissues. J Cell Biol 128:995–1002

Taroni F, DiDonato S (2004) Pathways to motor incoordination: the inherited ataxias. Nat Rev Neurosci 5:641–655

Thornton CA, Wymer JP, Simmons Z, McClain C, Moxley III RT (1997) Expansion of the myotonic dystrophy CTG repeat reduces expression of the flanking DMAHP gene. Nat Genet 16:407–409

Tian B, White RJ, Xia T, Welle S, Turner DH, Mathews MB, Thornton CA (2000) Expanded CUG repeat RNAs form hairpins that activate the double-stranded RNA-dependent protein kinase PKR. RNA 6:79–87

Timchenko NA, Cai Z-J, Welm AL, Reddy S, Ashizawa T, Timchenko LT (2001a) RNA CUG repeats sequester CUGBP1 and alter protein levels and activity of CUGBP1. J Biol Chem 276:7820–7826

Timchenko NA, Iakova P, Cai Z-J, Smith JR, Timchenko LT (2001b) Molecular basis for impaired muscle differentiation in myotonic dystrophy. Mol Cell Biol 21:6927–6938

Timchenko LT, Iakova P, Welm AL, Cai Z-J, Timchenko NA (2002) Calreticulin interacts with C/EBPα and C/EBPβ mRNAs and represses translation of C/EBP proteins. Mol Cell Biol 22:7242–7257

Timchenko LT, Miller JW, Timchenko NA, DeVore DR, Datar KV, Lin L, Roberts R, Caskey CT, Swanson MS (1996) Identification of a (CUG)n triplet repeat RNA-binding protein and its expression in myotonic dystrophy. Nucleic Acids Res 24:4407–4414

Timchenko NA, Wang G-L, Timchenko LT (2005) RNA CUG-binding protein 1 increases translation of a 20-kDa isoform of CCAAT/enhancer-binding protein β by interacting with the α and β subunits of eukaryotic initiation translation factor 2. J Biol Chem 280:20549–20557

Tsai T-F, Jiang Y-H, Bressler J, Armstrong D, Beaudet AL (1999) Paternal deletion from *Snrpn* to *Ube3a* in the mouse causes hypotonia, growth retardation and partial lethality and provides evidence for a gene contributing to Prader-Willi syndrome. Hum Mol Genet 8:1357–1364

van den Broek WJAA, Nelen MR, Wansink DG, Coerwinkel MM, Riele H, Groenen PJTA, Wieringa B (2002) Somatic expansion behavior of the $(CTG)_n$ repeat in myotonic dystrophy knock-in mice is differentially affected by Msh3 and Msh6 mismatch-repair proteins. Hum Mol Genet 11:191–198

van der Giessen K, Di-Marco S, Clair E, Gallouzi IE (2003) RNAi-mediated HuR depletion leads to the inhibition of muscle cell differentiation. J Biol Chem 278:47119–47128

van Engelen BG, Eymard B, Wilcox D (2005) 123rd ENMC international workshop: management and therapy in myotonic dystrophy, 6–8 February 2004, Naarden, The Netherlands. Neuromuscul Disord 15:389–394

Wang J, Pegoraro E, Menegazzo E, Gennarelli M, Hoop RC, Angelini C, Hoffman EP (1995) Myotonic dystrophy: evidence for a possible dominant-negative RNA mutation. Hum Mol Genet 4:599–606

Wang YH, Amirhaeri S, Kang S, Wells RD, Griffith JD (1994) Preferential nucleosome assembly at DNA triplet repeats from the myotonic dystrophy gene. Science 265:669–671

Wansink DG, Wieringa B (2003) Transgenic mouse models for myotonic dystrophy type 1 DM1). Cytogenet Genome Res 100:230–242

Watkins H, McKenna WJ, Thierfelder L, Suk HJ, Anan R, O'Donoghue A, Spirito P, Matsumori A, Moravec CS, Seidman JG, Seidman CE (1995) Mutations in the genes for cardiac troponin T and α-tropomyoson in hypertrophic cardiomyopathy. New Eng J Med 332:1058–1064

Westerlaken JHAM, Van der Zee CEEM, Peters W, Wieringa B (2003) The DMWD protein from the myotonic dystrophy (DM1) gene region is developmentally regulated and is present most prominently in synapse-dense brain areas. Brain Res 971:116–127

Zuker M, Mathews DH, Turner DH (1999) Algorithms and thermodynamics for RNA secondary structure prediction: a practical guide. In: Barciszewski J, Clark BFC (eds) RNA biochemistry and biotechnology. NATO ASI series. Kluwer, Dordrecht, pp 11–43

Nucleic Acids and Molecular Biology, Vol. 19
M. Fry/K. Usdin (Eds.)
Human Nucleotide Expansion Disorders

Spinocerebellar Ataxia Type 8

Katherine A. Dick[1,3] · John W. Day[2,3] · Laura P. W. Ranum[1] (✉)

[1]Department of Genetics, Cell Biology, and Development, University of Minnesota, MMC 206, 420 Delaware St. S.E., Minneapolis, MN 55455, USA
ranum001@umn.edu

[2]Department of Neurology, University of Minnesota, MMC 206, 420 Delaware St. S.E., Minneapolis, MN 55455, USA

[3]Institute of Human Genetics, University of Minnesota, MMC 206, 420 Delaware St. S.E., Minneapolis, MN 55455, USA

1 Introduction: Repeat Expansions and SCA8

Repeat expansion mutations are the underlying genetic mechanism of many neurodegenerative disorders, including Huntington disease, fragile X syndrome (FXS), myotonic dystrophy types 1 and 2 (DM1 and DM2), and nine forms of spinocerebellar ataxia (SCAs) (Warner et al. 1996; Zoghbi and Orr 2000; Ranum and Day 2002). The most common category of pathogenic expansion involves d(CAG) trinucleotide repeats that are translated into extended polyglutamine tracts, which confer pathogenic effects through gain-of-function mechanisms of their corresponding proteins. Repeat length is generally inversely correlated with age of onset and progression, but varies significantly between diseases (Warner et al. 1996; Zoghbi and Orr 2000; Ranum and Day 2002). The SCAs variably affect limb coordination, speech, swallowing, eye movements, and often shorten life span owing to ventilatory problems, aspiration, and frequent pneumonia (Day et al. 2000).

SCA type 8 (*SCA8*) presents as a slowly progressive form of ataxia characterized by dramatic repeat instability and a high degree of reduced penetrance. Among the SCAs, SCA8 was the first example of a dominant SCA not caused by the expansion of a d(CAG) polyglutamine encoding repeat tract (Mosemiller et al. 2003). SCA8 belongs to a class of diseases in which microsatellite repeat expansions are transcribed but do not appear to be translated, which also includes SCA10, SCA12, DM1, DM2, and fragile X associated tremor and ataxia syndrome (FXTAS) (Koob et al. 1999; Matsuura et al. 2000; Hagerman et al. 2001; Holmes et al. 2001; Liquori et al. 2001; Jacquemont et al. 2003). The pathogenic mechanism underlying several of these disorders points to an RNA gain-of-function mechanism.

2
Rapid Cloning of the SCA8 Repeat Expansion

In 1998, we used our RAPID cloning method to identify a previously unknown form of ataxia, that we designated SCA8 and which is caused by a non-coding d(CTG) repeat expansion within a gene of unknown function (Koob et al. 1998, 1999). This novel cloning method eliminated the need for linkage analysis or extensive clinical data, and allowed us to identify the mutation using only a single DNA sample (Koob et al. 1999; Mosemiller et al. 2003). Removing typical biases inherent in standard positional cloning approaches allowed us to identify a mutation characterized by reduced penetrance.

We first performed the repeat expansion detection assay, to identify d(CAG) repeat expansions, on DNA samples of patients with dominant but unknown forms of ataxia (Schalling et al. 1993; Koob et al. 1998, 1999). This screen identified an expansion of 80 repeats in an affected mother and daughter that was previously uncharacterized. We subsequently used RAPID to clone the repeat expansion and to obtain the genomic sequence flanking the repeat (Koob et al. 1998, 1999). Sequence analysis of the original SCA8 clone revealed that the expansion consisted of 80 uninterrupted d(CTG)/d(CAG) repeats preceded by a stretch of 11 d(CTA)/d(TAG) repeats. Although we developed RAPID cloning expecting to identify a polyglutamine-encoding ataxia gene, the only polyglutamine open reading frame (ORF) contained a single methionine followed by a polyglutamine stretch and sequence analysis did not reveal splice donor or acceptor signals that would allow a polyglutamine ORF to extend through the expansion as part of a spliced transcript. Furthermore, no transcripts spanning the repeat in the polyglutamine direction have been detected. These observations made it appear unlikely that the SCA8 expansion could be translated into a polyglutamine tract (Koob et al. 1999).

3
The d(CTG) Repeat Cosegregates with a Novel Form of Ataxia

We used PCR analysis to screen the kindred from which the expansion had been originally cloned, and found that both of the affected individuals and two at-risk family members carried the expansion on one of their two alleles (Koob et al. 1999). We subsequently screened our ataxia family collection (Moseley et al. 1998) and identified probands from 11 additional ataxia kindreds with expanded alleles. From one of these individuals we were able to identify and collect a seven-generation kindred with 92 members (Fig. 1). Mutation analyses showed that all of the affected individuals in the family had an expanded allele and linkage analysis between ataxia and the expansion gave a maximum logarithm of the odds (LOD) score of 6.8 at $\Theta = 0.00$.

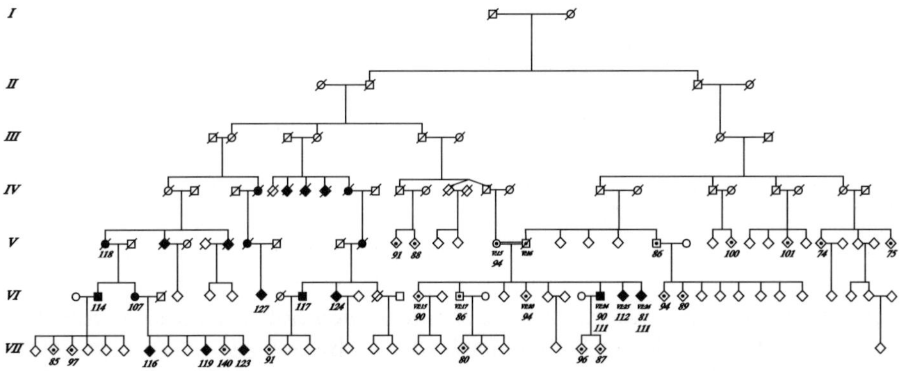

Fig. 1 The large spinocerebellar ataxia type 8 (*SCA8*) kindred (MN-A family). *Filled symbols* indicate individuals with ataxia, *symbols with a dot* indicate individuals who inherited the CTG expansion but are not clinically affected by ataxia. The CTG repeat lengths of expanded alleles are indicated *below the symbols*. Haplotype analyses using five short tandem repeat markers confirm that both branches of the family inherited the expanded repeat from a common founder. Family members homozygous for the SCA8 expansion and their affected heterozygous sibling (individuals *VI*: 24–26) had similar clinical features, with comparable ages of onset and rates disease of progression. (Reproduced from Koob et al. 1999 with permission from © 1999 Nature Publishing Group (http://www.nature.com/ng/index.html))

4
Organization of the SCA8 Gene

SCA8 transcripts are expressed at low levels in the d(CTG) direction with the d(CTG) repeat located at the 3′ end of a highly alternatively spliced transcript (Fig. 2). In humans, low steady-state transcript levels are found throughout the CNS and in testis and kidney, but not in other tissues (Janzen et al. 1999; Koob et al. 1999). The SCA8 d(CTG) repeat tract is conserved in chimpanzees, gorillas, and orangutans, with humans having larger alleles than these other species (Andres et al. 2003, 2004). At the genomic level, the SCA8 gene overlaps the 5′ end of a second gene, *Kelch-like 1* (*KLHL1*), which encodes an actin binding protein that is transcribed in the opposite direction (Koob et al. 1999; Nemes et al. 2000). Although no functional relationship between the two transcripts has been demonstrated, the genomic organization of the two genes and the evolutionary conservation of a much shorter SCA8 mouse gene without the d(CTG) repeat (Benzow and Koob 2002) suggests the possibility that one normal function of the SCA8 transcript may be to regulate *KLHL1* transcripts through an antisense mechanism (Koob et al. 1999; Nemes et al. 2000; Benzow and Koob 2002). Although reverse-transcription PCR shows that SCA8 is transcribed in the d(CTG) orientation and sequence analysis shows a short ORF with 41 amino acids plus the d(CTG) expansion, this ORF appeared unlikely to be translated because of the relatively large number of upstream start and stop codons, lead-

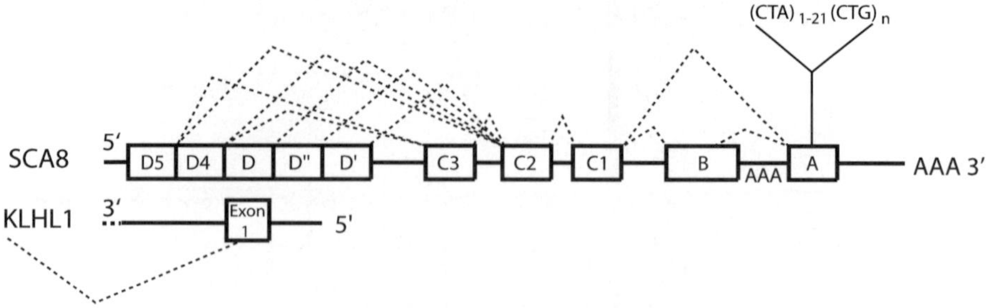

Fig. 2 SCA8 gene organization. Exons are shown as *boxes* and alternate splice forms are indicated by *dashed lines*. The SCA8 repeat tract is located in exon A at the 3′ end of the gene and is transcribed in the CTG orientation. The region of the SCA8 gene that overlaps the 5′ end of KLHL1 is shown. Various alternative splice forms of the SCA8 transcript are indicated. (Reproduced from Mosemiller et al. 2003, with permission from S. Karger AG, Basel)

ing to our hypothesis that SCA8 is mediated by an RNA mechanism similar to DM1 (Mosemiller et al. 2003; Koob et al. 1999).

5 Clinical Features of SCA8

On the basis of clinical evaluations of over 200 patients from 25 separate families, it is apparent that SCA8 presents as a slowly progressive ataxia that largely spares brainstem and cerebral function (Koob et al. 1999; Day et al. 2000; Ikeda et al. 2000a; Juvonen et al. 2000; Silveira et al. 2000; Brusco et al. 2002; Topisirovic et al. 2002; Mosemiller et al. 2003). The disease is characterized by gait and limb ataxia, speech and oculomotor incoordination, dysarthria, and sensory loss. The onset of gait incoordination, commonly one of the initial symptoms, ranged between 13 and 60 years of age within the MN-A family, while the need for mobility aids ranged between 35 and 50 years—generally requiring at least 20 years of disease progression before an aid was needed (Day et al. 2000).

Neurological examinations commonly reveal signs of oculomotor involvement in moderate to severely affected patients (Day et al. 2000; Juvonen et al. 2000; Anderson et al. 2002). Additionally, speech is dysarthric with ataxic and spastic components for all individuals examined (Day et al. 2000). Occasionally, mild athetotic movements of extended fingers and intermittent low-amplitude myoclonic jerks in the fingers and arms are detected. An elicitable Babinski sign is sometimes observed in severely affected individuals, whereas hyperreflexia is a common finding (Day et al. 2000; Juvonen et al. 2000). Impaired vibratory perception, indicative of mild sensory loss, was an intermittent clinical finding (Day et al. 2000).

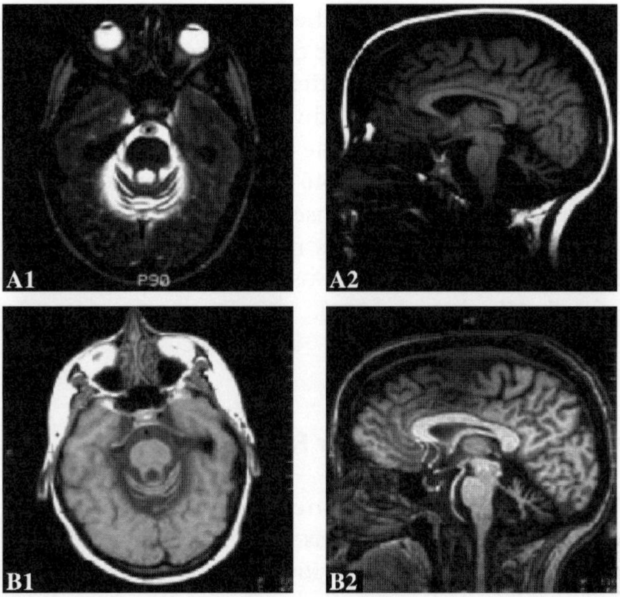

Fig. 3 Serial MRI scans of an affected individual. Horizontal (**a1**, **b1**) and sagittal (**a2**, **b2**) MRI scans from an affected individual at ages 26 (**a**) and 35 (**b**) years. The earlier image is 9 years after onset (17 years). There is marked cerebellar atrophy, minimal brainstem atrophy, and no evidence of cerebral involvement. There is very little change over the 9-year period between scans, which is consistent with the slow progression of the disease. (Reproduced from Day et al. 2000 with permission from Lippincott Williams & Wilkins (http://ww.com))

Atrophy of the cerebellar hemispheres and vermis is apparent on MRI analysis of affected SCA8 individuals (Day et al. 2000; Ikeda et al. 2000a; Topisirovic et al. 2002), with brainstem involvement appearing minimal. A typical SCA8 patient was tracked over a 9-year period with MRI; scans revealed little change, characteristic of the slowly progressive course of the disease (Day et al. 2000) (Fig. 3). The imaging also showed that the cerebral hemispheres, white matter, and basal ganglia were spared. In contrast, Zeman et al. (2004) reported a patient having had two MRI scans separated by 4 years—the initial scan was determined to be normal, while the second scan showed clear cerebellar atrophy.

6 Disease Penetrance Affected by d(CTG) Repeat Length: the MN-A Family

In the MN-A family, 17 individuals ranging in age from 14 to 74 years carried an expansion but were not clinically affected at the time of examination (Koob et al. 1999). These asymptomatic carriers had a mean age of 43

(±17 years), which is comparable to the mean age of examination of affected individuals. The expansion size for the carrier group was significantly smaller ($p < 10^{-8}$) than the size found in the affected individuals (mean 90 and 116 repeats, respectively) and all but one individual with an expansion greater than 107 repeats were clinically affected. The one exception was a 42-year-old individual who carried an expansion of 140 repeats. Since SCA8 is an adult-onset disorder with a documented age of onset as old as 65 years of age, an asymptomatic status for this individual was not unanticipated. These data clearly demonstrate that disease penetrance was affected by the d(CTG) repeat length in the MN-A family (Koob et al. 1999).

7 Reduced Penetrance of SCA8 in Other Families

In the MN-A family, SCA8 is transmitted in an autosomal dominant pattern with reduced penetrance of alleles less than 110 combined repeats. In other families, SCA8 shows a complex inheritance pattern in which only a subset of expansion carriers from a given family is affected (Koob et al. 1999; Day et al. 2000; Ikeda et al. 2000b; Cellini et al. 2001; Topisirovic et al. 2002). Representative SCA8 pedigrees are shown in Fig. 4 (Ikeda et al. 2004). Family A appears to transmit ataxia in a dominant pattern with affected individuals in multiple generations. Family B appears recessive with multiple affected individuals in a single generation, while the affected individual in family C presents as a sporadic case with no other affected family members. In contrast to the relatively large number of affected patients in the MN-A family ($n = 13$), 25 of the remaining 36 ataxia families had only a single affected individual, nine families had two affected individuals, and only two families had three affected

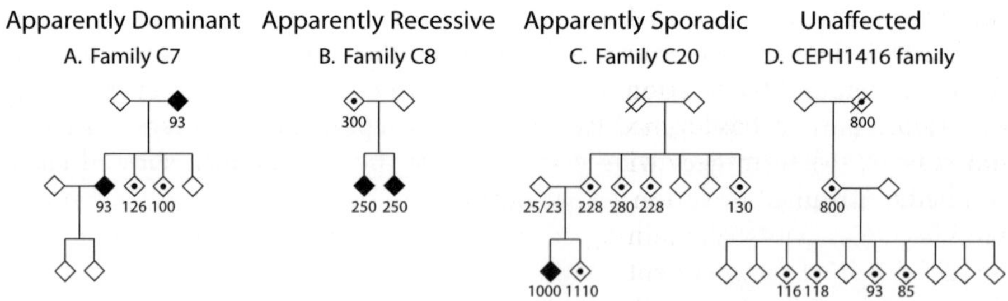

Fig. 4 SCA8 pedigrees with varying degrees of disease penetrance. *Black symbols* are for individuals affected by ataxia, and unaffected expansion carriers are indicated by *symbols with a dot inside* them. A *diagonal line* through a symbol denotes an individual who is deceased. The size of the expanded SCA8 allele is shown *below* the individuals. (Reproduced with permission from University of Chicago Press and Ikeda et al. (2004) Am J Hum Genet 75:3–16. © 2004 by The American Society of Human Genetics. All rights reserved)

individuals. Although only a subset of the expansion carriers in the MN-A family developed ataxia (13/35), the penetrance of disease was significantly higher in the MN-A pedigree than in the 36 smaller ataxia families we have studied as well as families reported by other groups (Koob et al. 1999; Day et al. 2000; Ikeda et al. 2000b, 2004; Juvonen et al. 2000; Cellini et al. 2001; Topisirovic et al. 2002). Of note, MRI analysis of a 71-year-old patient, who was clinically unaffected, showed mild cerebellar atrophy (Ikeda et al. 2000b), indicating that asymptomatic individuals may still show signs of cerebellar atrophy in imaging studies.

In summary, the tight correlation between repeat size and pathogenesis found in the MN-A family is not found in other ataxia families that have been reported (Ikeda et al. 2004). Among the additional SCA8 families examined, repeat sizes among affected and unaffected expansion carriers overlap and often exceed the pathogenic threshold found in the MN-A family. These data demonstrate that SCA8 expansions found among ataxia patients vary dramatically in size and that the presence of an SCA8 expansion cannot be used to predict whether or not an asymptomatic individual will develop ataxia (Ranum et al. 1999; Moseley et al. 2000; Worth et al. 2000; Ikeda et al. 2004).

8
SCA8 Expansions on Control Chromosomes

Surprisingly, SCA8 expansions have also been found in control samples we and others have screened (Vincent et al. 2000; Worth et al. 2000; Ikeda et al. 2004). Out of 2626 unrelated control chromosomes analyzed in Minnesota and Canada, we identified ten SCA8 alleles (0.4%) larger than 74 combined d(CTA)/d(CTG) repeats, which is the smallest expansion found in an ataxia patient (Ikeda et al. 2004). One of the control expansions was from a CEPH grandmother (family 1416) (Fig. 4). Medical histories indicate that neither this woman nor her son (54 years old, 800 repeats) are affected by ataxia. All six of the SCA8 expansion carriers in this family were asymptomatic at the time of clinical evaluation, although the expansion-positive individuals in generation III were children when they were clinically evaluated and thus it is not yet clear whether they will be at higher risk of developing ataxia.

In our original collection of probands from genetically undefined ataxia families, expansions containing more than 74 combined repeats occurred on 12/292 (4%) independent chromosomes. Although this frequency is significantly higher than in the general population (10/2626 chromosomes, $p = 4 \times 10^{-25}$), the relative frequency of alleles with more than 74 combined repeats in the general population (0.4%) is higher than that for all forms of ataxia (approximately 1/10 000). Taken together, these data suggest that the d(CTG) repeat can cause ataxia but that environmental or genetic modifiers, including repeat length, affect disease penetrance (Ikeda et al. 2004).

9
SCA8 Expansions Cosegregate with Ataxia in Small Families

In the MN-A family, previous studies have shown that the cosegregation of the SCA8 expansion and ataxia is highly significant (LOD 6.8, $\Theta = 0.00$) (Koob et al. 1999). To distinguish between the possibility that the SCA8 expansions are found by chance in the 36 additional smaller ataxia families versus the possibility that the expansions predispose carriers to ataxia, we examined the incidence of cosegregation of the expansion with ataxia in family members other than the probands (Ikeda et al. 2004). For example, if the SCA8 expansions do not predispose patients to ataxia but are merely found by chance in these 36 families, then we would expect that the frequency of SCA8 expansions in additional affected first-degree relatives would be 50%. In contrast, we found that 12 of the 13 affected first-degree relatives available for analysis also inherited the SCA8 expansion, indicating that the expansion cosegregates with ataxia in these small families ($p = 0.0038$). Linkage analysis was performed on the remaining small families with multiple affected individuals. Although the highest LOD score for a single family was only 0.34 at a recombination fraction of 0.00, the LOD scores were consistently positive and when combined exceeded the threshold level of 2.0, considered significant for testing linkage to a single specific locus (Ott 1991). The only exception was found in a family in which two sisters were affected with a form of ataxia clinically distinct from SCA8 by being a markedly more severe disease with rapid disease progression, pronounced choreiform movements, a severe sensory neuronopathy, and neuromyotonic discharges seen by electromyography. The cosegregation of the SCA8 expansion among additional affected relatives in the group of small ataxia families further indicates that the SCA8 expansion directly predisposes individuals to developing ataxia (Ikeda et al. 2004).

10
Haplotype Analysis of SCA8 Expansion Chromosomes

To better understand the origin of the SCA8 expansion and the reduced penetrance of the disease, haplotype analysis was performed on a panel of 37 SCA8 families from the USA, Canada, Japan, and Mexico, 13 SCA8 expansion-positive samples sent to Athena Diagnostics for ataxia testing, seven control samples with expansions, and 14 expansion carriers with psychiatric diseases (Ikeda et al. 2004). A total of 17 polymorphic short tandem repeat markers were analyzed, including 13 newly developed markers that span an approximately 1 Mb region flanking the SCA8 d(CTG) repeat.

Two ancestrally related haplotypes (A and A′) were observed in the Caucasian population, which included SCA8 and psychiatric patients, and controls—indicating a common origin for the pathogenic and nonpathogenic

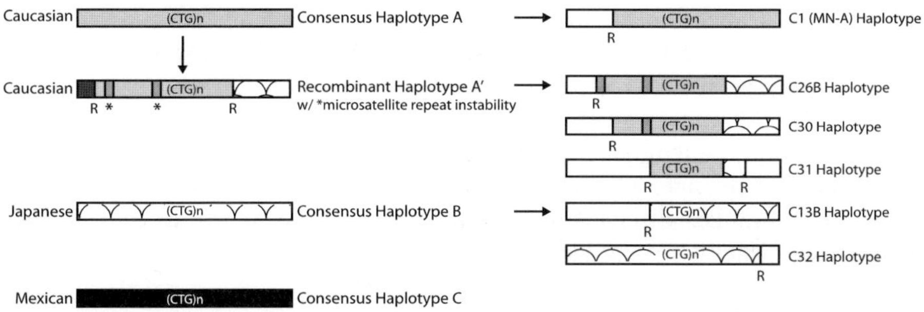

Fig. 5 Proposed summary of the ancestral origins of the SCA8 expansion haplotypes based on the analysis of 37 SCA8 families. The current haplotypes are likely to have arisen from a small number of ancestral recombination and microsatellite instability events as illustrated. *R* indicates a recombination event and the *asterisk* symbolizes an area with microsatellite repeat instability. (Reproduced with permission from University of Chicago Press and Ikeda et al. (2004) Am J Hum Genet 75:3–16. © 2004 by The American Society of Human Genetics. All rights reserved)

expansions within the Caucasian population (Ikeda et al. 2004) (Fig. 5). Two other distinct haplotypes were identified for the Japanese and Mexican ataxia families (B and C, respectively) (Fig. 5). These results indicate that independently arising SCA8 expansions are found in ataxia families with various ethnic backgrounds, further supporting the direct role of the d(CTG) expansion in disease pathogenesis.

11 Factors that May Influence SCA8 Disease Penetrance

Possible *cis* modifiers that could affect penetrance of the SCA8 expansion include d(CTG) repeat length, sequence interruptions within the repeat tract, and the size of the d(CTA) tract preceding the d(CTG) repeat, all of which show remarkable variation independent of haplotype (Ikeda et al. 2000a, b, 2004; Moseley et al. 1998, 2000, 2002).

11.1 The d(CTA) Repeat Tract

A polymorphic but stably transmitted d(CTA) repeat tract containing from one to 21 repeats precedes the d(CTG) expansion, with the overall configuration d(CTA)$_n$d(CTG)$_{exp}$ (Koob et al. 1999; Moseley et al. 2000; Stevanin et al. 2000; Mosemiller et al. 2003). In most studies, a simple PCR assay that detects the overall size of the combined repeats has been used to amplify the SCA8 expansions, with the respective lengths of the d(CTA) and d(CTG) repeat tracts not being determined. Although the SCA8 expansion in the MN-

A and other Caucasian SCA8 expansion families descends from a common founder mutation, a notable molecular difference in the repeat tract of the MN-A family versus that of other families with lower disease penetrance is that the d(CTA) tract is much smaller in the MN-A family (Ikeda et al. 2004), suggesting that the length of the d(CTA) repeat may contribute to differences in disease penetrance. The size variability of both the d(CTA) and d(CTG) repeat tracts makes direct comparisons between repeat length and disease penetrance difficult among families.

11.2 Interruptions Within the d(CTG) Expansion

An unusual feature of the SCA8 expansions is that the expanded alleles often have triplet interruptions within the repeat tract, with one or more d(CCG), d(CTA), d(CTC), d(CCA), or d(CTT) motifs found within the d(CTG) expansion (Moseley et al. 2000; Mosemiller et al. 2003). These interruptions, which are generally clustered at the 5′ end of the expansion, often duplicate during transmission—resulting in offspring with alleles that vary from the affected parent both in repeat tract length and sequence configuration. In general, most normal d(CTG) repeat tracts do not have sequence interruptions, although Sobrido et al. (2001) described a normal allele with 23 combined repeats, in which the d(CTG) tract had a d(CAG) interruption. Although both interrupted and pure d(CTG) repeat tracts are found in SCA8 ataxia families, the high frequency of interruptions in the MN-A family suggests that the d(CCG) interruptions in this family may play a role in the relatively high disease penetrance (Moseley et al. 2000).

11.3 Repeat Instability During Transmission

In addition to changes in the sequence of the SCA8 expansion, the SCA8 expansion alleles also show dramatic intergenerational changes in repeat length (Koob et al. 1999; Mosemiller et al. 2003). The changes in SCA8 expansion size are generally larger than in the other dominantly inherited SCAs, but are typically not as large as for DM1 (Tsilfidis et al. 1992; Chung et al. 1993; Maciel et al. 1995; Maruyama et al. 1995; Cancel et al. 1997; David et al. 1997; Jodice et al. 1997; Zhuchenko et al. 1997; Koob et al. 1999). As a general rule, paternal transmissions result in a contraction of the repeat tract (– 86 to + 7), while maternal transmissions result in expansions (– 11 to + 900), with extreme examples of large maternally transmitted increases in repeat length including + 250, + 375, + 600, and + 900 (Koob et al. 1999; Corral et al. 2005). A histogram depicting the intergenerational changes in the repeat length, which distinguishes between maternal and paternal transmission, is shown (Fig. 6). The maternal bias for repeat tract expansion has not been observed for other

SCAs, but is reminiscent of transmission tendencies for two other noncoding expansion disorders—FXS and DM1 (Groenen and Wieringa 1998; Koob et al. 1999; Jin and Warren 2000; Mosemiller et al. 2003).

In the MN-A family the maternal expansion and paternal deletion biases affect disease penetrance, with 90% of the transmissions that resulted in ataxia being maternally transmitted and the remaining 10% involving the transmission of expanded alleles from both parents (Fig. 1) (Koob et al. 1999; Mosemiller et al. 2003). In contrast, 16 of the 19 asymptomatic individuals who carried repeat expansions received the SCA8 expansion from their father. This maternal penetrance bias observed in the MN-A family is consistent with a higher frequency of female transmissions resulting in expansions above the pathogenic threshold of approximately 110 combined repeats, while paternal transmissions tend to result in alleles in which the repeat tract has contracted below the pathogenic threshold (Koob et al. 1999; Day et al. 2000). However, this maternal penetrance bias seen in the MN-A family is not evident in many of the SCA8 families examined (Juvonen et al. 2000).

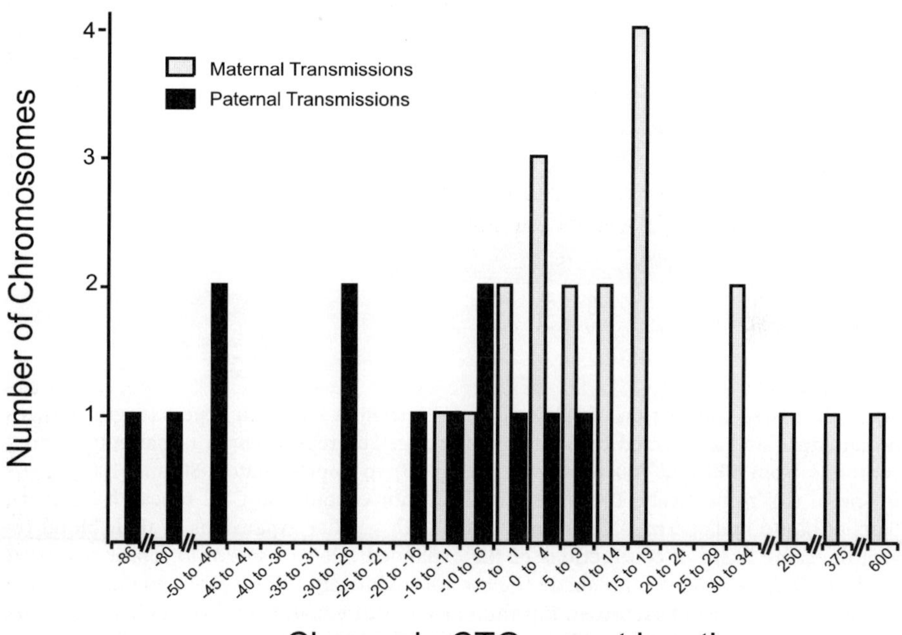

Fig. 6 Intergenerational variation in repeat number for maternal and paternal transmissions. Repeat variation is shown as a decrease or an increase of CTG repeat units. Maternal and paternal transmissions are represented by *gray bars* and *black bars*, respectively. (Reproduced from Koob et al. 1999 with permission from © 1999 Nature Publishing Group (http://www.nature.com/ng/index.html))

11.4 En Masse d(CTG) Repeat Contractions in Sperm

To further investigate the SCA8 repeat instability and the paternal bias towards d(CTG) repeat contraction, we examined sperm samples from men who had expansions ranging in size from 80 to 800 repeats in the blood (Moseley et al. 2000; Mosemiller et al. 2003). Southern blot analysis on sperm DNA from two unrelated individuals showed that each expanded allele underwent a massive contraction—into a size range less often associated with ataxia (from 500 to approximately 90 and from 800 to approximately 110)

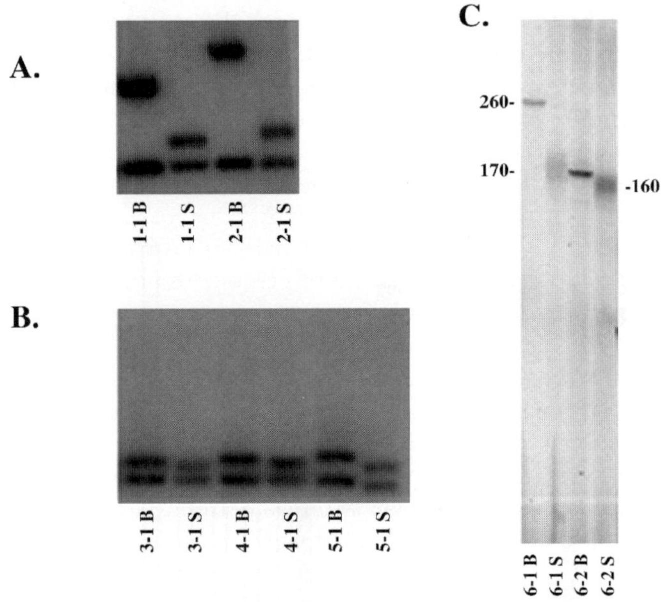

Fig. 7 En masse contraction of SCA8 alleles in sperm. **a** Dramatic repeat length changes in patients 1 and 2 detected by Southern blotting. The repeat length of patients 1 and 2 contracts from 500 and 800 repeats in blood (*B*) to approximately 80 and 100 repeats in sperm (*S*), respectively. The probe used did not contain the CTG repeat. **b** Southern blots of blood and sperm DNA from patients with smaller expansions in their blood reveal the same trend of contractions of the expanded allele in sperm to repeat sizes that are less often associated with ataxia (below approximately 100 repeats). Again, the equal intensities of the bands representing the normal and expanded alleles indicate that repeat contractions occurred in all or nearly all of the sperm with expanded alleles. **c** PCR analysis of SCA8 contractions in two patients from a family with paternal disease transmission. Although contraction of repeats in sperm is again observed, the resulting alleles remain within a more penetrant size range (more than 100 CTGs). (Reproduced from Moseley et al. SCA8 CTG repeat: en masse concentrations in sperm and intergenerational sequence changes may play a role in reduced penetrance. Human Molecular Genetics (2000) 9(14):2125–2130 with permission from Oxford University)

(Fig. 7a). A similar trend was also observed for individuals with smaller somatic expansions; the expanded allele contracted in sperm to a non-pathogenic range, generally below approximately 100 repeats (Fig. 7b). The equal intensities of the bands representing the normal and expanded alleles indicate that all or nearly all of the expanded allele in the sperm contracted. The tendency for the SCA8 expanded allele to contract in sperm most likely contributes to the reduced penetrance and maternal bias observed in some SCA8 families (Moseley et al. 2000; Silveira et al. 2000).

12
Molecular Parallels with Myotonic Dystrophy

Similar to SCA8, DM1 is also caused by a d(CTG) repeat expansion that is transcribed but not translated (Tapscott 2000). In 1992, the DM1 mutation was identified as a d(CTG) expansion in the 3′ untranslated region of the *DMPK* gene; however, the molecular mechanism underlying the pathogenesis of this expansion was not clear. In 2001, we identified the mutation that causes DM2, which is characterized by the same multisystemic features as DM1, but results from a noncoding r(CCUG) expansion in intron 1 of the *zinc finger protein 9* (*ZNF9*) gene (Liquori et al. 2001). The *DMPK* and *ZNF9* genes and the surrounding regions bear no obvious similarity, but the fact that both mutations involve similar expanded repeat motifs that are transcribed but untranslated pointed to an RNA gain-of-function mechanism. The identification of the DM2 mutation and experiments demonstrating ribonuclear foci formation and downstream alternative splicing of other genes have established that the clinical features common to DM1 and DM2 are caused by an RNA gain-of-function mechanism. Molecular parallels between SCA8, DM1, and DM2 mutations, along with the known toxic properties of transcripts containing expanded r(CUG) repeats, suggest the possibility that a similar mechanism may play a role in SCA8 pathogenesis (Mosemiller et al. 2003). The SCA8 gene is almost exclusively expressed in the CNS but the *DMPK* and *ZNF9* genes are broadly expressed, consistent with the differing clinical features of the diseases (Koob et al. 1999; Liquori et al. 2001).

13
Modeling SCA8 Pathogenesis in the Fly

Mutsuddi et al. (2004) have developed a *Drosophila* model of SCA8. These investigators have shown that expression of SCA8 transcripts with both the normal and the expanded repeat tracts in the *Drosophila* retina induces a late-onset, progressive neurodegeneration. Using this neurodegenerative phenotype as a sensitized background for a genetic modifier screen, this group performed a targeted screen of a panel of available mutants in RNA binding proteins to look for dominant modifiers of the SCA8 phenotype. Three enhancer mutations in *muscleblind*, *split ends*, and *staufen* and one suppressor mutation in *CG3249*, which encodes a putative protein kinase A anchor protein (PKAAP) with a K-homology-type RNA binding motif, were identified. All four of these genes encode neuronally expressed RNA binding proteins that are conserved in *Drosophila* and humans. Although expression of both normal and expanded repeat tracts causes neurodegeneration in this model, the interaction between muscleblind and SCA8 varies in relation to d(CTG) repeat size. These experiments suggest that the SCA8 expansion can alter interactions with RNA binding proteins, which could in turn play a role in disease pathogenesis (Mutsuddi et al. 2004).

14 Conclusion

The data summarized in this review describe various known pieces of a complex puzzle—detailing the genetic features of the SCA8 expansion and variations in disease penetrance in the MN-A family and in other populations. SCA8 differs in numerous respects from many of the other dominant SCAs identified to date, most notably in the reduced penetrance of the d(CTG) expansion. Because the SCA8 mutation was isolated from a single ataxia patient using our RAPID cloning method (Koob et al. 1998, 1999) instead of the positional cloning approaches used to identify the SCA1, SCA2, SCA3, and SCA6 mutations, all of which depend on the collection and characterization of large families, it is not surprising that the genetic characteristics and disease penetrance do not follow the pattern of previously defined SCAs. To date, not all comparisons have taken into account the unprecedented genetic complexity associated with SCA8; often the genetics are oversimplified and the focus is only on the combined repeat size of the expansion rather than on examining the d(CTA), d(CTG), and interruptions to the d(CTG) tract. While additional molecular information from humans will help further define the sequence variation at the SCA8 locus, additional experiments in cell culture and animal models will be needed to understand the molecular mechanisms of the disease and the biology underlying the reduced penetrance.

References

Anderson JH, Yavuz MC, Kazar BM, Christova P, Gomez CM (2002) The vestibulo-ocular reflex and velocity storage in spinocerebellar ataxia 8. Arch Ital Biol 140:323–329

Andres AM, Soldevila M, Saitou N, Volpini V, Calafell F, Bertranpetit J (2003) Understanding the dynamics of spinocerebellar ataxia 8 (SCA8) locus through a comparative genetic approach in humans and apes. Neurosci Lett 336:143–146

Andres AM, Soldevila M, Lao O, Volpini V, Saitou N, Jacobs HT, Hayasaka I, Calafell F, Bertranpetit J (2004) Comparative genetics of functional trinucleotide tandem repeats in humans and apes. J Mol Evol 59:329–339

Benzow KA, Koob MD (2002) The KLHL1-antisense transcript (KLHL1AS) is evolutionarily conserved. Mamm Genome 13:134–141

Brusco A, Cagnoli C, Franco A, Dragone E, Nardacchione A, Grosso E, Mortara P, Mutani R, Migone N, Orsi L (2002) Analysis of SCA8 and SCA12 loci in 134 Italian ataxic patients negative for SCA1–3, 6 and 7 CAG expansions. J Neurol 249:923–929

Cancel G, Durr A, Didierjean O, Imbert G, Burk K, Lezin A, Belal S, Benomar A, Abadabendib M, Vial C, Guimaraes J, Chneiweiss H, Stevanin G, Yvert G, Abbas N, Saudou F, Lebre A, Yahyaoui M, Hentati F, Vernant J, Klockgether T, Mandel J, Agrid Y, Brice A (1997) Molecular and clinical correlations in spinocerebellar ataxia 2—a study of 32 families. Hum Mol Genet 6:709–715

Cellini E, Nacmias B, Forleo P, Piacentini S, Guarnieri BM, Serio A, Calabro A, Renzi D, Sorbi S (2001) Genetic and clinical analysis of spinocerebellar ataxia type 8 repeat expansion in Italy. Arch Neurol 58:1856–1859

Chung M-Y, Ranum LPW, Duvick LA, Servadio A, Zoghbi HY, Orr HT (1993) Evidence for a mechanism predisposing to intergenerational CAG repeat instability in spinocerebellar ataxia type 1. Nat Genet 5:254–258

Corral J, Genis D, Banchs I, San Nicolas H, Armstrong J, Volpini V (2005) Giant SCA8 alleles in nine children whose mother has two moderately large ones. Ann Neurol 57:549–553

David G et al. (1997) Cloning of the SCA7 gene reveals a highly unstable CAG repeat expansion. Nat Genet 17:65–70

Day JW, Schut LJ, Moseley ML, Durand AC, Ranum LPW (2000) Spinocerebellar ataxia type 8: clinical features in a large family. Neurology 55:649–657

Groenen P, Wieringa B (1998) Expanding complexity in myotonic dystrophy. Bioessays 20:901–912

Hagerman RJ, Leehey M, Heinrichs W, Tassone F, Wilson R, Hills J, Grigsby J, Gage B, Hagerman PJ (2001) Intention tremor, parkinsonism, and generalized brain atrophy in male carriers of fragile X. Neurology 57:127–130

Holmes SE, Hearn EO, Ross CA, Margolis RL (2001) SCA12: an unusual mutation leads to an unusual spinocerebellar ataxia. Brain Res Bull 56:397–403

Ikeda Y, Shizuka M, Watanabe M, Okamoto K, Shoji M (2000a) Molecular and clinical analyses of spinocerebellar ataxia type 8 in Japan. Neurology 54:950–955

Ikeda Y, Shizuka-Ikeda M, Watanabe M, Schmitt M, Okamoto K, Shoji M (2000b) Asymptomatic CTG expansion at the SCA8 locus is associated with cerebellar atrophy on MRI. J Neurol Sci 182:76–79

Ikeda Y, Dalton JC, Moseley ML, Gardner KL, Bird TD, Ashizawa T, Seltzer WK, Pandolfo M, Milunsky A, Potter NT, Shoji M, Vincent JB, Day JW, Ranum LP (2004) Spinocerebellar ataxia type 8: molecular genetic comparisons and haplotype analysis of 37 families with ataxia. Am J Hum Genet 75:3–16

Jacquemont S, Hagerman RJ, Leehey M, Grigsby J, Zhang L, Brunberg JA, Greco C, Des Portes V, Jardini T, Levine R, Berry-Kravis E, Brown WT, Schaeffer S, Kissel J, Tassone F, Hagerman PJ (2003) Fragile X premutation tremor/ataxia syndrome: molecular, clinical, and neuroimaging correlates. Am J Hum Genet 72:869–878

Janzen MA, Moseley ML, Benzow KA, Day JW, Koob MD, Ranum LPW (1999) Limited expression of SCA8 is consistent with cerebellar pathogenesis and toxic gain of function RNA model. Am J Hum Genet 65:A267

Jin P, Warren ST (2000) Understanding the molecular basis of fragile X syndrome. Hum Mol Genet 9:901–908

Jodice C, Mantuano E, Veneziano L, Trettel F, Sabbadini G, Calandriello L, Francia A, Spadaro M, Pierelli F, Salvi F, Ophoff R, Frants R, Frontali M (1997) Episodic ataxia type 2 (EA2) and spinocerebellar ataxia type 6 (SCA6) due to CAG repeat expansion in the CACNA1A gene on chromosome 19p. Hum Mol Genet 6:1973–1978

Juvonen V, Hietala M, Paivarinta M, Rantamaki M, Hakamies L, Kaakkola S, Vierimaa O, Penttinen M, Savontaus ML (2000) Clinical and genetic findings in Finnish ataxia patients with the spinocerebellar ataxia 8 repeat expansion. Ann Neurol 48:354–361

Koob MD, Benzow KA, Bird TD, Day JW, Moseley ML, Ranum LPW (1998) Rapid cloning of expanded trinucleotide repeat sequences from genomic DNA. Nat Genet 18:72–75

Koob MD, Moseley ML, Schut LJ, Benzow KA, Bird TD, Day JW, Ranum LPW (1999) An untranslated CTG expansion causes a novel form of spinocerebellar ataxia (SCA8). Nat Genet 21:379–384

Liquori C, Ricker K, Moseley ML, Jacobsen JF, Kress W, Naylor S, Day JW, Ranum LPW (2001) Myotonic dystrophy type 2 caused by a CCTG expansion in intron 1 of ZNF9. Science 293:864–867

Maciel P, Gaspar C, DeStefano AL, Silveira I, Coutinho P, Radvany J, Dawson DM, Sudarsky L, Guimaraes J, Loureiro JEL, Nezarati MM, Corwin LI, Lopes-Cendes I, Rooke K, Rosenburg R, MacLeod D, Farrer LA, Sequeiros J, Roulean GA (1995) Correlation between CAG repeat length and clinical features in Machado-Joseph disease. Am J Hum Genet 57:54–61

Maruyama H, Nakamura S, Matsuyama Z, Sakai T, Doyu M, Sobue G, Seto M, Tsujihata M, Oh-i T, Nishio T, Sunohara N, Takahashi R, Hayashi M, Nishino I, Ohtake T, Oda T, Nishimura M, Saida T, Matsumoto H, Baba M, Kawaguchi Y, Kakizuka A, Kawakimi H (1995) Molecular features of the CAG repeats and clinical manifestation of Machado-Joseph disease. Hum Mol Genet 4:807–812

Matsuura T, Yamagata T, Burgess DL, Rasmussen A, Grewal RP, Watase K, Khajavi M, McCall AE, Davis CF, Zu L, Achari M, Pulst SM, Alonso E, Noebels JL, Nelson DL, Zoghbi HY, Ashizawa T (2000) Large expansion of the ATTCT pentanucleotide repeat in spinocerebellar ataxia type 10. Nat Genet 26:191–194

Moseley ML, Benzow KA, Schut LJ, Bird TD, Gomez CM, Barkhaus PE, Blindauer KA, Labuda M, Pandolfo M, Koob MD, Ranum LPW (1998) Incidence of dominant spinocerebellar and Friedreich triplet repeats among 361 ataxia families. Neurology 51:1666–1671

Moseley ML, Schut LJ, Bird TD, Koob MD, Day JW, Ranum LPW (2000) SCA8 CTG repeat: en masse contractions in sperm and intergenerational sequence changes may play a role in reduced penetrance. Hum Mol Genet 9:2125–2130

Mosemiller AK, Dalton JC, Day JW, Ranum LPW (2003) Molecular genetics of spinocerebellar ataxia type 8 (SCA8). Cytogenet Genome Res 100:175–183

Mutsuddi M, Marshall C, Benzow K, Koob M, Rebay I (2004) The spinocerebellar ataxia 8 noncoding RNA causes neurodegeneration and associates with staufen in Drosophila. Curr Biol 14:302–308

Nemes JP, Benzow KA, Moseley ML, Ranum LPW, Koob MD (2000) The SCA8 transcript is an antisense RNA to a brain-specific transcript encoding a novel actin-binding protein (KLHL1). Hum Mol Genet 9:1543–1551, correction/addition Hum Mol Genet 9:2777

Ott J (1991) Analysis of human genetic linkage. The Johns Hopkins University Press, Baltimore

Ranum LP, Day JW (2002) Dominantly inherited, non-coding microsatellite expansion disorders. Curr Opin Genet Dev 12:266–271
Ranum LPWM, Leppert ML, Guan den Eng MF, La Spada AR, Koob MD, Day J (1999) Massive CTG expansions and deletions reduce penetrance of spinocerebellar ataxia type 8. Am J Hum Genet 65:A:466
Schalling M, Hudson T, Buetow K, Housman D (1993) Direct detection of novel expanded trinucleotide repeats in the human genome. Nat Genet 4:135–139
Silveira I, Alonso I, Guimaraes L, Mendonca P, Santos C, Maciel P, Fidalgo De Matos JM, Costa M, Barbot C, Tuna A, Barros J, Jardim L, Coutinho P, Sequeiros J (2000) High germinal instability of the $(CTG)_n$ at the SCA8 locus of both expanded and normal alleles. Am J Hum Genet 66:830–840
Sobrido MJ, Cholfin JA, Perlman S, Pulst SM, Geschwind DH (2001) SCA8 repeat expansions in ataxia: a controversial association. Neurology 57:1310–1312
Stevanin G, Herman A, Durr A, Jodice C, Frontali M, Agid Y, Brice A (2000) Are (CTG)n expansions at the SCA8 locus rare polymorphisms? Nat Genet 24:213, discussion 215
Tapscott SJ (2000) Deconstructing myotonic dystrophy. Science 289:1701–1702
Topisirovic I, Dragasevic N, Savic D, Ristic A, Keckarevic M, Keckarevic D, Culjkovic B, Petrovic I, Romac S, Kostic VS (2002) Genetic and clinical analysis of spinocerebellar ataxia type 8 repeat expansion in Yugoslavia. Clin Genet 62:321–324
Tsilfidis C, MacKenzie AE, Mettler G, Barcelo J, Korneluk RG (1992) Correlation between CTG trinucleotide repeat length and frequency of severe congenital myotonic dystrophy. Nat Genet 1:192–195
Vincent JB, Neves-Pereira ML, Paterson AD, Yamamoto E, Parikh SV, Macciardi F, Gurling HM, Potkin SG, Pato CN, Macedo A, Kovacs M, Davies M, Lieberman JA, Meltzer HY, Petronis A, Kennedy JL (2000) An unstable trinucleotide-repeat region on chromosome 13 implicated in spinocerebellar ataxia: a common expansion locus. Am J Hum Genet 66:819–829
Warner JP, Barron LH, Goudie D, Kelly K, Dow D, Fitzpatrick DR, Brock DJ (1996) A general method for the detection of large CAG repeat expansions by fluorescent PCR. J Med Genet 33:1022–1026
Worth PF, Houlden H, Giunti P, Davis MB, Wood NW (2000) Large, expanded repeats in SCA8 are not confined to patients with cerebellar ataxia. Nat Genet 24:214–215
Zeman A, Stone J, Porteous M, Burns E, Barron L, Warner J (2004) Spinocerebellar ataxia type 8 in Scotland: genetic and clinical features in seven unrelated cases and a review of published reports. J Neurol Neurosurg Psychiatry 75:459–465
Zhuchenko O, Bailey J, Bonnen P, Ashizawa T, Stockton DW, Amos C, Dobyns WB, Subramony SH, Zoghbi HY, Lee CC (1997) Autosomal dominant cerebellar ataxia (SCA6) associated with small polyglutamine expansions in the alpha-1A-voltage-dependent calcium channel. Nat Genet 15:62–69
Zoghbi HY, Orr HT (2000) Glutamine repeats and neurodegeneration. Annu Rev Neurosci 23:217–247

Nucleic Acids and Molecular Biology, Vol. 19
M. Fry/K. Usdin (Eds.)
Human Nucleotide Expansion Disorders

Recent Progress in Spinocerebellar Ataxia Type 10

Xi Lin · Tetsuo Ashizawa (✉)

Department of Neurology, University of Texas Medical Branch,
301 University Boulevard, JSA 9.128, Galveston, TX 77555, USA
teashiza@utmb.edu

1 Introduction

Hereditary ataxias are a treasure trove for neurogeneticists. Numerous genetic loci and mutations, the jewels and diamonds in the eyes of geneticists, have been discovered in recent years. The reward is phenomenal and carries far-reaching significance in many fields of biomedical science, ranging from molecular biology and genetics, protein chemistry, to pathophysiology of neurodegeneration. To date, 24 autosomal dominant ataxias (ADCA) have been genetically defined, including spinocerebellar ataxias (SCA) 1–8, 10–19, 21–23 and 25–27, dentatorubral-pallidoluysian atrophy (DRPLA), and ataxia caused by mutations in fibroblast growth factor 14 (FGF14) (Schols et al. 2004). Many of these are caused by abnormal expansion of d(CAG)/d(CTG) trinucleotide repeats in the respective genes. Different pathogenic mechanisms have been uncovered, the most important of which is centered on the toxicity of polyglutamine expansions encoded by d(CAG) repeats.

SCA10 is unique and caused by a novel microsatellite expansion primarily composed of d(ATTCT) pentanucleotide repeats. The repeats are located in intron 9 of a novel gene on chromosome 22q13.3, previously known as *E46L* now designated as *ATXN10*. The molecular mechanisms of how such an expansion leads to the typical disease phenotypes in SCA10 remain unknown, and pose a challenge, as well as opportunities, for neurogenetic research. Important questions include how d(ATTCT) expansion arises, what cellular and molecular processes are preferentially affected and why the cerebellum is particularly susceptible to this mutation. Answers to these questions will help us gain insight into some of the fundamental genetic processes, such as the control of genetic stability and DNA replication. They are also likely to advance our understanding of the molecular and physiological properties of the cerebellum.

2 Clinical Presentations

Clinically, most ADCAs are poorly differentiated. Genetics has played an important role in dissecting this heterogeneous entity. Research development of many dominantly inherited spinocerebellar ataxias unfolds in similar ways. Large families are invaluable for the initial gene mapping and detection of mutations. Subsequent identification of similar mutations in unrelated families and characterization of a full spectrum of signs and symptoms further establishes a distinct entity and refines the genotype–phenotype relationship.

Studies of two large Mexican-American families were instrumental in the initial recognition of SCA10, which has a relatively pure cerebellar ataxia typically accompanied by seizures (Grewal et al. 1998; Matsuura et al. 1999; Zu et al. 1999). Linkage to chromosome 22q13.3 and identification of a d(ATTCT) pentanucleotide repeat expansion were primarily based on data gathered from these two families (Matsuura et al. 2000). Subsequent reports on four unrelated families from Mexico further revealed more diverse phenotypes; associated polyneuropathy, pyramidal signs, cognitive and neuropsychiatric impairment were often associated with the classic pictures of cerebellar ataxia plus seizures (Rasmussen et al. 2001). Recently, five Brazilian families were diagnosed to have SCA10, and all the 28 patients have cerebellar ataxia without seizures, representing one end of the whole spectrum of SCA10 phenotypes (Teive et al. 2004).

2.1 Cerebellar Signs and Symptoms

All SCA10 patients have progressive cerebellar dysfunctions (Grewal et al. 1998, 2002; Lin and Ashizawa 2003; Matsuura et al. 1999, 2000; Rasmussen et al. 2001). The function of the cerebellum is to continuously adjust, coordinate and refine complex motor movements involving multiple muscles in a smooth and integrated pattern. Fluidity and accuracy of movement is lost with cerebellar dysfunction. Typical cerebellar signs include limb and gait ataxia, dysarthria and ocular disturbances. The first sign of disease is usually manifested as unbalanced gait and stance with variable degrees of limb ataxia, which is characterized by jerky or uncoordinated movements unexplained by motor weakness or sensory loss. On neurological examination, patients show wide-based ataxic gait with impaired tandem-walk, dysdiadokinesia, intention tremor and dysmetria. Ataxia is followed by speech difficulties and dysarthria; the speech is slow and slurred, or explosive with frequent hesitations and inappropriate pauses; hence called "scanning" speech. Ocular abnormalities common in SCA10 patients include ocular dyskinesia, most commonly presented as intrusions of hypometric saccade during pursuit, which may progress to overt ocular flutter with brief conjugate oscillations of the eyes during attempted fixation or movement (ocular dysmetria). Gaze-evoked nystagmus may be seen in some patients with SCA10. Irregularity in pursuit eye movements is often seen in early stages of the disease. Neuroimaging studies with MRI or CT showed specific cerebellar atrophy involving the vermis and both hemispheres, while other brain structures, such as the cerebral cortex and brain stem, were only minimally involved.

2.2 Seizures

Epileptic seizures are variably associated with SCA10. In the currently known SCA10 families with over 100 affected members (two Mexican-American families, four Mexican families and five Brazilian families), seizures are present only in families of Mexican origin, and the prevalence between families is also significantly different, ranging from 25 to 80% (Grewal et al. 1998, 2002; Lin and Ashizawa 2003; Matsuura et al. 1999, 2000; Rasmussen et al. 2001). The epilepsy usually presents as generalized motor seizure and/or complex partial seizures a few years after the start of cerebellar ataxia. Anti-epileptic drugs, such as carbamazepine, phenytoin and valproic acid, are effective for most cases. The interictal EEG is abnormal in many SCA10 patients with epilepsy and in some without seizures. The most common findings were diffuse cortical dysfunction with slow, fused and disorganized activities; focal cortical irritability or slow activity was also observed in some individuals. Two patients developed status epilepticus, and one died as a direct result. It is likely that the epileptic activities reflect the pathogenic effects of d(ATTCT) expansion outside the cerebellum, particularly the cerebral cortex.

2.3 Other Extracerebellar Signs and Symptoms

Some SCA10 patients of Mexican origin have additional phenotypes beyond cerebellar degeneration and epileptic seizures (Grewal et al. 1998, 2002; Lin and Ashizawa 2003; Matsuura et al. 1999, 2000; Rasmussen et al. 2001). More extra-cerebellar signs and non-neuronal involvement have been observed in some families. Variable degrees of pyramidal signs, including hyperreflexia, leg spasticity and Babinski's sign, were reported. Affected individuals often complain of mild sensory loss in distal lower extremities, and nerve conduction studies confirmed the presence of polyneuropathy. Some patients have low intelligence quotient (IQ), and brief neuropsychiatric evaluation by MMPI demonstrated depressive, aggressive and/or irritable traits. Again, these extra-cerebellar phenotypes further suggested that other neural tissues have different susceptibilities to the d(ATTCT) expansion in the *SCA10* gene. Interestingly, one family also showed hepatic, cardiac and hematological abnormalities in the affected members. But it is unclear whether these were a part of the SCA10 syndrome or due to another disease which happened to co-segregate with SCA10. Investigation of more SCA10 families/patients and better understanding of the molecular mechanisms of pathogenesis will be important to answer this question.

3 Genetic Features

3.1 d(ATTCT) Pentanucleotide Repeat Expansion

By positional cloning, we identified an expansion of a d(ATTCT) pentanucleotide repeat, which is located in intron 9 of the *SCA10* gene in chromosome 22q13.3, as the disease-causing mutation for this disease (Matsuura et al. 2000). The *SCA10* gene consists of 12 exons spanning 172.8 kb of genomic DNA with an open reading frame of 1428 bp, encoding 475 amino acids. The number of d(ATTCT) repeat units is polymorphic and ranges from 10 to 29 in the normal population, whereas the number of expanded allele ranges from 800 (4 kb) to 4500 (22.5 kb). Thus, the SCA10 repeat expansion is one of the largest microsatellite repeat expansions known to exist in the human genome. Several lines of evidence support the notion that the expansion of d(ATTCT) repeats is the pathogenic mutation for SCA10. First, the d(ATTCT) repeat expansion co-segregates with the SCA10 phenotype in the affected families, and was found to be absent in over 1000 normal chromosomes. Most importantly, the size of the expanded repeat shows an inverse correlation with the age of onset, arguing against the possibility of another closely linked mutation as the direct cause of the disease. Moreover, *SCA10* is strongly expressed in the brain; disruption of its expression might preferentially affect the CNS, such as the cerebellum and cerebral cortex.

Since the mutation in SCA10 was identified, many families with autosomal dominant cerebellar ataxia from different ethnic background have been screened (Fujigasaki et al. 2002). All the currently known SCA10 families originate in Latin America, including six Mexican families and five Brazilian families. In collaborations with Astrid Rasmussen's group at the National Institute of Neurology and Neurosurgery in Mexico City, Mexico, and Hélio Teive's group at the University of Prana, Curitiba, Brazil, we have recently identified additional five Mexican and two Brazilian families (unpublished data). Thus, SCA10 is the second most common SCA in these populations, secondary to SCA2 in Mexico and Machado–Joseph disease in southern Brazil. In contrast, other ethnic populations, including those of their European ancestors, have no identifiable SCA10 families. The genealogical histories and physical characteristics of these patients suggested an admixture of Native American with Spanish/Portuguese ancestry in all SCA10 families.

SCA10 mutation might have arisen in the New World as a result of a founder-effect. To elucidate the basis for this disorder, haplotype analysis of short tandem repeat polymorphisms and single nucleotide polymorphisms in regions of linkage disequilibrium around the *SCA10* gene should be informative. However, it is still theoretically possible that the same mutation is prevalent across different ethnic populations. The mutation might be manifested as dominant ataxia only in certain genetic backgrounds. Other phenotypes, such as seizure or psychiatric disorders, might be predominant. Screening for a d(ATTCT) repeat expansion in other hereditary neurological conditions could be valuable.

3.2 Anticipation, Repeat Instability and Genotype–Phenotype Correlation

Anticipation is a clinical phenomenon, describing progressively earlier age of onset and/or more severe phenotypes in successive generations. While it can have non-biological components such ascertainment biases and phenotype-copying among family members, anticipation frequently reflects the dynamic nature of certain genetic mutations, which can change from generation to generation. It is a characteristic finding in a number of neurological diseases which are caused by trinucleotide repeat expansions, such as Fragile X syndrome, Huntington's disease, myotonic dystrophy type 1 (DM1) and several SCAs. The underlying mechanism is instability of repeat expansion during DNA replication; the affected offspring tend to have longer repeats than the parents, resulting in a more severe disease with earlier manifestation.

Multiple factors are potentially involved in determining how faithfully the expanded repeats are replicated during meiosis and mitosis. These include the motif and configuration of different repeat units, the length of the repeats, *cis* elements of surrounding DNA sequences and chromosomal structures, *trans*-acting factors involved in DNA repair and recombination, and the gender of the transmitting parent (Lin and Ashizawa 2003). Analysis of the stability of expanded d(ATTCT) repeats in SCA10 revealed a rather complex picture (Matsuura et al. 2004). The repeats are highly unstable during paternal transmission, whereas maternal transmission is relatively stable. Sperm DNA has tremendous heterogeneity in the size of the expanded allele, indicating high degree of instability in male germ line. Mosaicism of the repeat size is also present in somatic tissues. It is of note that distinct patterns of intergenerational changes in the repeat size were observed in different families, indicating family-dependent factors, such as those involved in repeat stability as above (Grewal et al. 1998, 2002; Lin and Ashizawa 2003; Matsuura et al. 1999, 2000; Rasmussen et al. 2001). Moreover, different tissues also seem to have different degrees of heterogeneity, suggesting a potential role of neuron-specific instability in the pathogenesis (Matsuura et al. 2004). It would be of great interest to see if d(ATTCT) repeats are longer in the CNS, particularly in the cerebellum, the principal target of SCA10 pathogenesis. On the other hand, this inter-tissue variability raises the issue of DNA sampling, which is routinely done using peripheral blood leukocytes (PBLs). Caution should be used in diagnosis and counseling when the repeat length derived from PBL DNA is used to correlate with pathology in the CNS and age of onset. Family-dependent intertissue variability might provide an explanation for the paradox in one SCA10 family: anticipation with apparent repeat contraction, which is based on the repeat length in PBL.

4 Molecular Studies

4.1 DNA Structure

Normal *SCA10* allele has 10–29 d(ATTCT) repeats, while the disease allele expands enormously, ranging from 800 to 4500 repeats (Matsuura et al. 2000). d(ATTCT) repeats are unique in its high content of A-T. Structurally, it lacks symmetric elements to form hairpins, intramolecular triplexes or quadruplexes. High A-T content is a feature of DNA sequences that form unpaired structures when under torsional stress, called DNA unwinding elements (DUE) (Potaman et al. 2003). DUE have been known to function as replication origins in both prokaryotes and eukaryotes (Berberich et al. 1995; Miller et al. 1999). Indeed, when cloned into plasmids under superhelical torsion, d(ATTCT) repeats form an unpaired DNA structure, which can be demonstrated by two-dimensional gel electrophoresis and atomic force microscopy (Potaman et al. 2003). Furthermore, the unpaired DNA structure formed by d(ATTCT) repeats can support complete plasmid replication in a HeLa cell extract, indicating that expanded d(ATTCT) repeats can function as an aberrant replication origin (Potaman et al. 2003). These observations may have important implications for repeat instability and pathogenic mechanisms.

In the chromosomal environment, d(ATTCT) repeats would be subject to variable torsional stress, as determined by the pattern of nuclear matrix attachment and tracking of DNA and RNA polymerases (i.e. replication and transcription status of the locus) (Kramer et al. 1999; Kramer and Sinden 1997; Ljungman and Hanawalt 1992). Expansion of d(ATTCT) repeats might facilitate DNA unwinding and formation of unpaired structures, resulting in aberrant replication initiation. With large expansion of the repeats, DNA unwinding might reoccur at the still-extending, nascent double-stranded d(ATTCT) repeats, and replication could re-initiate at the new unpaired structure. This anomalous replication may lead to both repeat expansion and deletion by different mechanisms, such as strand breakage and recombination, resulting in heterogeneity of the expanded allele (Potaman et al. 2003). DNA replication in eukaryotes initiates from multiple replication origins. There is a tight regulation of when and where to start DNA replication from a multitude of origins. Different development stages and differentiation status can have different temporal and spatial patterns of active replication origins, since DNA replication initiates neither from all the potential origins nor at the same time. This tight control is more relaxed and less selective during gametogenesis and early embryogenesis (Martinez-Salas et al. 1988; Simon et al. 1999; Wiekowski et al. 1997). Therefore, expanded d(ATTCT) repeat-mediated anomalous replication may be more likely to occur during sperm formation, resulting in enhanced heterogeneity of SCA10 expanded alleles in male germ line.

Different tissues have distinct rate of cell division, and *SCA10* gene is differentially expressed in various tissues. These properties reflect activities of DNA and RNA polymerases, which may be important in determining the torsional stress in *SCA10* locus. Thus the frequency of d(ATTCT) repeat-mediated aberrant replication may be variable in different tissues, resulting in inter-tissue variability in d(ATTCT) repeat expansion. It is likely that in the cerebellum, where *SCA10* is abundantly expressed by potent transcription, tracking of RNA polymerase along *SCA10* DNA strands creates stronger torsional stress, facilitating the d(ATTCT) repeat expansion to unwind and form an unpaired structure. Subsequent aberrant DNA replication might lead to further expansion of the repeats in cerebellar neurons. Alternatively, cell cycle proteins can be recruited, resulting in aberrant cell cycle entry and neuronal apoptosis/degeneration in the cerebellum.

4.2 Protein Function

SCA10 encodes a protein of 475 amino acids with an apparent molecular weight of 53 000 (Marz et al. 2004; Matsuura et al. 2000). It is highly homologous to a mouse protein, E46, the function of which is unknown. Thus the protein encoded by *SCA10* is known as ataxin-10 or E46L (mouse E46-like protein). Analysis of the amino acid sequence of the human E46L protein did not reveal any transmembrane domain, nuclear localization signal or other known functional motifs. However, in the carboxyl terminus, E46L contains two armadillo repeat domains that have been found in membrane-associated proteins such as β-catenin; thus E46L belongs to the family of armadillo repeat proteins. Related genes have also been found in flies, worms, plants and yeast, suggesting that *SCA10* and its homologues across different species are evolutionarily conserved. E46L mRNA is abundantly expressed throughout the brain. Outside the CNS, it is also highly expressed in the testis, heart, kidney and skeletal muscles, with a medium level in the liver, and low level in PBL.

E46L is a cytoplasmic protein, which forms homotrimeric complexes via a "tip-to-tip" contact with the concave sides of the molecules facing each other (Marz et al. 2004). Immunostaining of mouse and human brain sections with an antibody against rat ataxin-10 revealed a predominantly cytoplasmic and perinuclear localization. "Knockdown" of *SCA10* in primary cerebellar and cortical neurons in culture by small interfering RNAs (siRNAs) caused increased apoptosis (Marz et al. 2004). It is of note that cerebellar neurons were significantly more sensitive to reduced level of ataxin-10. It would be interesting to see (1) if ataxin-10 in SCA10 patients is reduced, (2) if expanded d(ATTCT) repeats cause reduced ataxin-10 expression and (3) if reduced expression in animal models results in cerebellar degeneration. Answers to these questions will be important for establishing a loss-of-function mechanism in SCA10 pathogenesis. If such a mechanism can be established, a therapeutic approach can be designed to supplement E46L expression.

To understand E46L function, we have identified several E46L-interacting proteins by yeast two-hybrid screening (unpublished). One of these molecules is a subunit of heterotrimeric GTP-binding protein (G-protein). Preliminary studies suggest that E46L potentiates heterotrimeric G-protein signaling, leading to enhanced neurite formation. It would be interesting to investigate if E46L is required for G-protein signaling and if reduced expression of E46L causes dendritic degeneration.

4.3 RNA Gain of Function

A RNA gain of function has been proposed as a pathogenic mechanism for several neurodegenerative disorders caused by repeat expansions (Liquori et al. 2001; Mankodi et al. 2002; Miller et al. 2000; Ranum and Day 2004; Savkur et al. 2001; Taneja et al. 1995; Timchenko et al. 1996, also reviewed in this volume). In myotonic dystrophy type 1 (DM1), a large d(CTG) repeat expansion in the 3′ untranslated region (UTR) of the *DMPK* gene results in a transcript containing expanded r(CUG) repeats. The r(CUG) repeats at the 3′ UTR retain the DMPK messenger RNA (mRNA) in the nucleus, forming distinct loci with reduced presence in the cytoplasm. The mutant transcript interacts with a number of RNA binding proteins, such as CUG-BP1, ETR3, MBLL, MBNL and MBXL. Some of these proteins are known to be important in precursor mRNA processing, such as splicing and nuclear export. Accumulation of mutant DMPK transcript in the nucleus might disrupt the normal functions of these RNA-binding proteins, affecting the splicing efficiency and expression of other genes, such as cardiac troponin T, insulin receptor and chloride channel. These contribute to cardiomyopathy, myotonia and insulin resistance, phenotypes related to the multisystemic involvement in DM1. A parallel mechanism was indicated in DM2; a large expansion of d(CCTG) repeats in the first intron of the *zinc finger 9* (*ZNF9*) gene results in accumulation of r(CCUG) repeat RNA. Like in DM1, the mutant RNA transcript containing expanded r(CCUG) repeats accumulates in nuclear foci, sequesters a similar array of RNA binding proteins, and results in the multisystemic feature of pathogenesis. Thus, RNA containing an expanded repeat transcribed from both intronic and extronic regions could play a major role in the pathogenesis of neurodegenerative diseases caused by repeat expansions.

In SCA10, d(ATTCT) repeat expansion is located in intron 9 of the *sca10* gene. Our preliminary studies suggest that both normal and mutant alleles are expressed at similar level and RNA with expanded r(AUUCU) repeats is present at detectable levels. When plasmids expressing a pure d(ATTCT) repeat tract were transfected into cells in culture, RNA foci were observed by in situ hybridization; thus overexpression of d(ATTCT) repeats can lead to accumulation of r(AUUCU) RNA and formation of RNA foci. It is very likely that RNA gain of function is an integral part of SCA10 pathogenesis. A *trans*-dominant gain of function by a large expansion of the r(AUUCU) repeat in the *E46L* RNA might recruit RNA binding proteins and perturb their normal cellular functions. An essential next step in establishing a potential gain-of-function mechanism for SCA10 is the identification of proteins that bind r(AUUCU) repeats.

5 Concluding Remarks

As more SCA10 families were identified, the complete phenotypic spectrum is emerging. Central to this spectrum is cerebellar ataxia, with the extra-cerebellar involvement being variable and dependent on different genetic backgrounds. Perhaps the most exciting and important finding in the studies of SCA10 is the novel mutation of d(ATTCT) repeat expansion which raises several important questions for geneticists. Investigation of how d(ATTCT) repeat expansion arises and maintains the stability in SCA10 families will provide insight into the fundamentals of molecular genetics, such as DNA structures, replication and stability. Equally exciting questions still remain to be answered as to why the cerebellum is particularly susceptible and what the pathogenic mechanism is. Analysis of the primary disease tissue and establishing a mouse model will surely pay off.

Acknowledgement This work is supported by a grant from NIH to TA (NS41547).

References

Berberich S, Trivedi A, Daniel DC, Johnson EM, Leffak M (1995) In vitro replication of plasmids containing human c-myc DNA. J Mol Biol 245:92–109

Fujigasaki H, Tardieu S, Camuzat A, Stevanin G, LeGuern E, Matsuura T, Ashizawa T, Durr A, Brice A (2002) Spinocerebellar ataxia type 10 in the French population. Ann Neurol 51:408–409

Grewal RP, Achari M, Matsuura T, Durazo A, Tayag E, Zu L, Pulst SM, Ashizawa T (2002) Clinical features and ATTCT repeat expansion in spinocerebellar ataxia type 10. Arch Neurol 59:1285–1290

Grewal RP, Tayag E, Figueroa KP, Zu L, Durazo A, Nunez C, Pulst SM (1998) Clinical and genetic analysis of a distinct autosomal dominant spinocerebellar ataxia. Neurology 51:1423–1426

Kramer PR, Sinden RR (1997) Measurement of unrestrained negative supercoiling and topological domain size in living human cells. Biochemistry 36:3151–3158

Kramer PR, Fragoso G, Pennie W, Htun H, Hager GL, Sinden RR (1999) Transcriptional state of the mouse mammary tumor virus promoter can affect topological domain size in vivo. J Biol Chem 274:28590–28597

Lin X, Ashizawa T (2003) SCA10 and ATTCT repeat expansion: clinical features and molecular aspects. Cytogenet Genome Res 100:184–188

Liquori CL, Ricker K, Moseley ML, Jacobsen JF, Kress W, Naylor SL, Day JW, Ranum LP (2001) Myotonic dystrophy type 2 caused by a CCTG expansion in intron 1 of ZNF9. Science 293:864–867

Ljungman M, Hanawalt PC (1992) Localized torsional tension in the DNA of human cells. Proc Natl Acad Sci USA 89:6055–6059

Mankodi A, Takahashi MP, Jiang H, Beck CL, Bowers WJ, Moxley RT, Cannon SC, Thornton CA (2002) Expanded CUG repeats trigger aberrant splicing of ClC-1 chloride channel pre-mRNA and hyperexcitability of skeletal muscle in myotonic dystrophy. Mol Cell 10:35–44

Martinez-Salas E, Cupo DY, DePamphilis ML (1988) The need for enhancers is acquired upon formation of a diploid nucleus during early mouse development. Genes Dev 2:1115–1126

Marz P, Probst A, Lang S, Schwager M, Rose-John S, Otten U, Ozbek S (2004) Ataxin-10, the spinocerebellar ataxia type 10 neurodegenerative disorder protein, is essential for survival of cerebellar neurons. J Biol Chem 279:35542–35550

Matsuura T, Achari M, Khajavi M, Bachinski LL, Zoghbi HY, Ashizawa T (1999) Mapping of the gene for a novel spinocerebellar ataxia with pure cerebellar signs and epilepsy. Ann Neurol 45:407–411

Matsuura T, Yamagata T, Burgess DL, Rasmussen A, Grewal RP, Watase K, Khajavi M, McCall AE, Davis CF, Zu L, Achari M, Pulst SM, Alonso E, Noebels JL, Nelson DL, Zoghbi HY, Ashizawa T (2000) Large expansion of the ATTCT pentanucleotide repeat in spinocerebellar ataxia type 10. Nat Genet 26:191–194

Matsuura T, Fang P, Lin X, Khajavi M, Tsuji K, Rasmussen A, Grewal RP, Achari M, Alonso ME, Pulst SM, Zoghbi HY, Nelson DL, Roa BB, Ashizawa T (2004) Somatic and germline instability of the ATTCT repeat in spinocerebellar ataxia type 10. Am J Hum Genet 74:1216–1224

Miller CA, Umek RM, Kowalski D (1999) The inefficient replication origin from yeast ribosomal DNA is naturally impaired in the ARS consensus sequence and in DNA unwinding. Nucleic Acids Res 27:3921–3930

Miller JW, Urbinati CR, Teng-Umnuay P, Stenberg MG, Byrne BJ, Thornton CA, Swanson MS (2000) Recruitment of human muscleblind proteins to $(CUG)_n$ expansions associated with myotonic dystrophy. EMBO J 19:4439–4448

Potaman VN, Bissler JJ, Hashem VI, Oussatcheva EA, Lu L, Shlyakhtenko LS, Lyubchenko YL, Matsuura T, Ashizawa T, Leffak M, Benham CJ, Sinden RR (2003) Unpaired structures in SCA10 $(ATTCT)_n$.$(AGAAT)_n$ repeats. J Mol Biol 326:1095–1111

Ranum LP, Day JW (2004) Pathogenic RNA repeats: an expanding role in genetic disease. Trends Genet 20:506–512

Rasmussen A, Matsuura T, Ruano L, Yescas P, Ochoa A, Ashizawa T, Alonso E (2001) Clinical and genetic analysis of four Mexican families with spinocerebellar ataxia type 10. Ann Neurol 50:234–239

Savkur RS, Philips AV, Cooper TA (2001) Aberrant regulation of insulin receptor alternative splicing is associated with insulin resistance in myotonic dystrophy. Nat Genet 29:40–47

Schols L, Bauer P, Schmidt T, Schulte T, Riess O (2004) Autosomal dominant cerebellar ataxias: clinical features, genetics, and pathogenesis. Lancet Neurol 3:291–304

Simon I, Tenzen T, Reubinoff BE, Hillman D, McCarrey JR, Cedar H (1999) Asynchronous replication of imprinted genes is established in the gametes and maintained during development. Nature 401:929–932

Taneja KL, McCurrach M, Schalling M, Housman D, Singer RH (1995) Foci of trinucleotide repeat transcripts in nuclei of myotonic dystrophy cells and tissues. J Cell Biol 128:995–1002

Teive HA, Roa BB, Raskin S, Fang P, Arruda WO, Neto YC, Gao R, Werneck LC, Ashizawa T (2004) Clinical phenotype of Brazilian families with spinocerebellar ataxia 10. Neurology 63:1509–1512

Timchenko LT, Miller JW, Timchenko NA, DeVore DR, Datar KV, Lin L, Roberts R, Caskey CT, Swanson MS (1996) Identification of a $(CUG)_n$ triplet repeat RNA-binding protein and its expression in myotonic dystrophy. Nucleic Acids Res 24:4407–4414

Wiekowski M, Miranda M, Nothias JY, DePamphilis ML (1997) Changes in histone synthesis and modification at the beginning of mouse development correlate with the establishment of chromatin mediated repression of transcription. J Cell Sci 110:1147–1158

Zu L, Figueroa KP, Grewal R, Pulst SM (1999) Mapping of a new autosomal dominant spinocerebellar ataxia to chromosome 22. Am J Hum Genet 64:594–599

Part III
Disorders Associated with Coding Repeats

Nucleic Acids and Molecular Biology, Vol. 19
M. Fry/K. Usdin (Eds.)
Human Nucleotide Expansion Disorders

Polyglutamine Diseases

Meyer J. Friedman · Shi-Hua Li · Xiao-Jiang Li (✉)

Department of Human Genetics, Emory University School of Medicine, 615 Michael St., Atlanta, GA 30322, USA
xiaoli@genetics.emory.edu

1 Introduction

Dynamic mutations in the coding region of various genes can result in human pathology, including the polyglutamine (polyQ) and polyalanine diseases. The molecular pathogenesis of the former group, comprising spinal and bulbar muscular atrophy (SBMA), dentatorubral-pallidoluysian atrophy (DRPLA), Huntington's disease (HD), and spinocerebellar ataxia (SCA) 1, 2, 3, 6, 7, and 17, has received considerable attention for more than a decade (Zoghbi and Orr 2000; Li and Li 2004a,b). Trinucleotide repeat expansion was first linked to the SBMA locus in 1991 (La Spada et al. 1991), and, 2 years later, cloning of the HD locus by a large collaboration of researchers revealed the same type of mutational event (Huntington's Disease Collaborative Research Group 1993). Although the proteins encoded by the nine polyQ disease genes do not share sequence homology outside of the polyQ domain, there is some overlap in the clinical phenotypes induced by repeat expansion. Moreover, the tendency of the expanded proteins to self-associate results in the formation of neuronal aggregates, which constitute a histopathological hallmark of the polyQ diseases. However, each polyQ disease displays selective and distinct neurodegeneration, which is not necessarily a feature of other trinucleotide repeat diseases that arise from noncoding mutations but may also affect neuronal function. Different regions of the brain or cell types within a particular structure are affected in each polyQ disease (Fig. 1). Several mechanisms have been invoked to explain the molecular pathogenesis of the polyQ diseases, including defects in both nuclear and cytoplasmic functions that are crucial to cell viability. Elucidation of the underlying molecular mechanisms will prove valuable in the development of therapeutic strategies to counter polyQ-induced neuropathology.

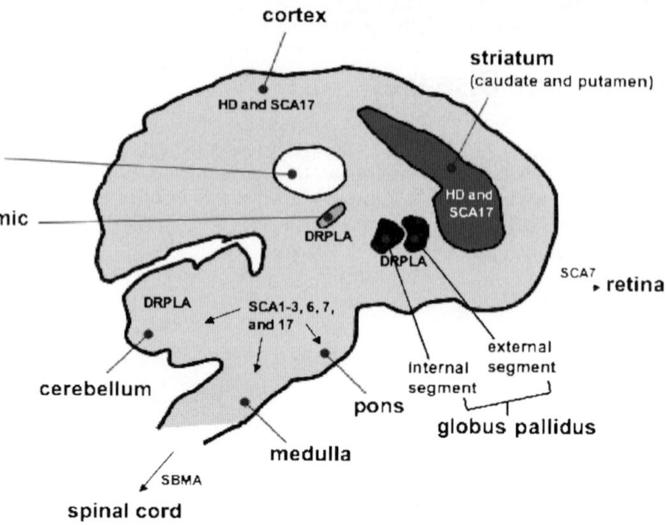

Fig. 1 A cartoon of a sagittal brain section indicating regions that are prominently affected in the various polyglutamine (*polyQ*) diseases. Although degeneration in the cerebellum and brainstem is observed in all of the spinocerebellar ataxias (*SCAs*), the members of this subgroup can be distinguished by characteristic neuropathology within these structures. Notably, retinal degeneration occurs only in SCA7. (Modified from Rubinsztein et al. 2002)

2
Genetics of PolyQ Diseases

The genetic basis of polyQ disease is expansion of a d(CAG) trinucleotide repeat in the coding region of one of at least nine different genes (Table 1). The d(CAG) repeat is translated into a polyQ domain that resides in different regions of the disease proteins. Although these repeats are largely conserved in the great apes, namely chimpanzees, gorillas, and orangutans, expansion and the resulting pathology have only been observed in humans. At both expanding and nonexpanding loci, the mean d(CAG) repeat tract size is not markedly different between humans and their closest primate relatives. However, human alleles of disease genes demonstrate the greatest within-species variance in repeat length. This high variance is characteristic of expanding alleles alone, suggesting that the probability of expansion and repeat size variability for a given gene may be correlated (Andres et al. 2004). Heterogeneity in the nucleotide composition of the repeat, due either to a mixture of d(CAG)/d(CAA) codons or to the presence of non-glutamine codons, appears to be relevant to expansion. Notably, loss of these interruptions by deletion or chromosomal rearrangement has been associated with intergenerational expansion (Maltecca 2003). Moreover, polyQ disease proteins often harbor several homopeptide sequences outside of the glutamine tract. In hu-

Table 1 Polyglutamine (*polyQ*) disease proteins

Diseases	Normal repeats	Expanded repeats	Protein	M_r (kDa)	Subcellular distribution of normal protein in the brain [a]	Proposed or established function
HD	6–35	36–150	Huntingtin	348	Cytoplasmic	A scaffold protein involved in trafficking (Li and Li 2004a, b)
DRPLA	3–36	49–88	Atrophin-1	124	Cytoplasmic	Transcriptional corepressor (Wood et al. 2000)
SCA1	6–39	40–83	Ataxin-1	87	Nuclear [c]	RNA binding protein (Yue et al. 2001)
SCA2	14–32	33–77	Ataxin-2	90	Cytoplasmic	RNA metabolism (Ralser et al. 2005)
SCA3/MJD	12–40	55–86	Ataxin-3	42	Cytoplasmic	Polyubiquitin binding protein (Chai et al. 2004)
SCA6	4–18	21–30	$\alpha_{1A}P/Q$ Calcium channel	160–250	Membrane	Principal subunit of voltage-gated Ca^{2+} channel
SCA7	7–17	38–200	Ataxin-7	95	Cytoplasmic	transcriptional coactivator (Helmlinger et al. 2004)
SBMA [b]	9–36	38–62	Androgen receptor	99	Nuclear	Nuclear hormone receptor
SCA17	25–42	47–55	TBP	41	Nuclear	General transcription factor

M_r molecular weight, *HD* Huntington's disease, *DRPLA* dentatorubral-palidoluysian atrophy, *SCA* spinocerebellar ataxia, *SBMA* spinal and bulbar muscular atrophy, *MJD* Machado–Joseph disease, *TBP* TATA-box binding protein

[a] Predominant localization indicated, but some of the cytoplasmic proteins can translocate to the nucleus

[b] Also referred to as Kennedy's disease

[c] Mainly cytoplasmic in peripheral tissues

man huntingtin (htt), for example, the polyQ domain is immediately followed by two polyproline tracts. The non-glutamine homopeptides are generally encoded by more than a single codon, presumably making them impervious to expansion (Karlin and Burge, 1996).

With the exception of Kennedy's disease, or SBMA, all of the polyQ diseases are inherited as autosomal dominant traits. The causative mutation in Kennedy's disease is a polyQ expansion in the androgen receptor, which is encoded by a gene that resides on the long arm of the X chromosome. Because female carriers of a pathogenic allele are largely asymptomatic, as a consequence of relatively low levels of circulating androgens (Katsuno et al. 2002), an X-linked recessive pattern of inheritance is observed for this disease (Everett and Wood 2004). Despite this discrepancy, and consistent with the dominant transmission of the other polyQ diseases, there is tremendous evidence that expansion of a polyQ tract has a gain-of-function effect. The fact that the disease phenotypes can be largely recapitulated in transgenic but not in knockout mice corroborates this idea. However, since the normal functions of only three of the polyQ disease proteins (androgen receptor, TATA-box binding protein, (TBP), and the SCA6 protein) have been well characterized (Table 1), the potential contribution of partial loss of function to molecular pathogenesis remains unclear (Evert et al. 2003). Notably, combined loss-of-function and gain-of-function effects have been reported for the polyQ-expanded androgen receptor (Lieberman et al. 2002).

The clinical profiles of families affected by polyQ disease are primarily determined by d(CAG) repeat length, but various genetic and environmental factors are also influential. In general, the length of the d(CAG) repeat is positively correlated with disease severity and negatively correlated with age of onset; however, the strength of these correlations can be undermined by clinical heterogeneity between and even within families. In SCA17, for example, reduced penetrance has been associated with pathogenic repeat lengths near the disease threshold (Zuhlke et al. 2003a,b). In both human patients and transgenic mice, the dosage of the polyQ-expanded protein is significant. Although individuals homozygous for pathogenic d(CAG)-repeat alleles are rare, there is evidence of incomplete dominance for some of the polyQ diseases. In SCA3 (Lang et al. 1994), DRPLA (Sato et al. 1995), and SCA6 (Kato et al. 2000), homozygosity results in an earlier age of onset and, although there can be considerable variability (Lerer et al. 1996), a more severe manifestation of the disease. HD was originally considered an example of complete dominance (Wexler et al. 1987), as the age of onset is not significantly different between homozygotes and heterozygotes; however, it is now clear that disease progression is markedly enhanced in the former genotype (Squitieri et al. 2003). The same phenomenon is observed in SCA17 (Toyoshima et al. 2004; Zuhlke et al. 2003a,b), thus, the molecular mechanisms responsible for the age of onset and disease progression may be separable (Squitieri et al. 2003).

The d(CAG) repeats in disease genes are vulnerable to both meiotic and mitotic instability, allowing for somatic mosaicism and genetic anticipation, respectively. The extent of both types of instability varies among the different disease genes. Anticipation, defined by a decrease in the age of onset and an increase in disease severity between generations, is characteristic of trinucleotide repeat diseases in general. In the polyQ diseases, pathogenic alleles tend to be more unstable, or prone to expansion, in paternal rather than maternal transmission (Everett and Wood 2004). Examination of individual sperm from a small number of male patients with different polyQ diseases has suggested a direct relationship between the variance in the change in repeat size and genetic anticipation. The highest variance was observed in sperm from men with DRPLA, in which marked anticipation is typical, whereas sperm from men with SBMA, a disease characterized by very limited intergenerational instability, displayed a small variance in repeat size (Leeflang et al. 1995; Zhang et al. 1995; Takiyama et al. 1997, 1999). Moreover, for at least two diseases, namely, Machado–Joseph disease and DRPLA, there is molecular evidence of biased meiotic segregation in favor of the expanded mutant allele in the sperm of affected men. Thus, meiotic drive might explain the non-Mendelian transmission of the disease trait evident in some polyQ disease pedigrees (Ikeuchi et al. 1996; Takiyama et al. 1997).

As the correlation between the age of onset and d(CAG) repeat length cannot account for all of the phenotypic variability observed in the different polyQ diseases, the existence of various genetic and environmental modifiers has been proposed. In HD, 69% of the variance in the age of onset can be explained by d(CAG) repeat size. A candidate loci approach, based on knowledge of pathways relevant to HD pathology, identified a specific allele of the gene encoding the GluR6 kainate receptor, which has been renamed GRIK2, as a genetic modifier of the age of onset (Rubinsztein et al. 1997). A similar directed approach, in which variations in genes encoding either htt-interacting proteins or apoptotic proteins were examined, revealed single nucleotide polymorphisms in the genes for the transcription factor p53 and human caspase activated DNase that could account for some of the remaining variance in the age of onset (Chattopadhyay et al. 2004). Notably, a whole genome scan for genetic modifiers has indicated strong linkage at a couple of locations, including 4p16, where the HD locus happens to reside (J.L. Li et al. 2003).

3 Neuropathology of PolyQ Diseases

Most of the polyQ disease proteins are widely expressed within and outside of the brain. However, expansion of the polyQ tract results in an essentially neuronal-specific phenotype in patients. Moreover, each of the polyQ diseases is distinguished by a unique profile of selective neurodegeneration that can be evinced radiographically or by postmortem analysis (Fig. 1). Although the pathological relevance remains controversial, the polyQ disease brain is characterized by the presence of aggregates or inclusions. These structures, the subcellular localization of which depends on the polyQ disease protein, generally have not been observed outside the central nervous system. In SBMA (Li et al. 1998) and SCA7 (Jonasson et al. 2002), however, there is evidence of nuclear inclusions in certain peripheral tissues. Nuclear aggregation in neuronal tissue is prominent in all of the polyQ diseases except SCA2 (Huynh et al. 1999) and SCA6 (Ishikawa et al. 1999, 2001). Cytoplasmic aggregates are present in some of the polyQ diseases, including SCA2 (Huynh et al. 2000), SCA6 (Ishikawa et al. 1999), HD (DiFiglia et al. 1997; Gutekunst et al. 1999), DRPLA (Hayashi et al. 1998a,b), and SMBA (Adachi et al. 2005). In HD, cytoplasmic aggregates, which primarily localize in neuronal processes such as axons and dendrites, have been extensively characterized (DiFiglia et al. 1997; Gutekunst et al. 1999).

Mouse models of polyQ diseases have proven valuable for study of nuclear accumulation of mutant polyQ proteins in neuronal tissues. In knockin and transgenic mouse models, it is often possible to arrange these distinct labeling patterns in a histochemical time course (Michalik and Broeckhoven 2003). Diffuse nuclear staining, which increases in intensity with age and probably represents the presence of abundant microaggregates, is the initial histological event. Multiple puncta eventually become discernable within the diffuse immunoreactive signal. Ultimately, loss of the diffuse staining pattern is coincident with the emergence of, in most cases, a single neuronal intranuclear inclusion (NII) (Yvert et al. 2000; Schilling et al. 2001). The duration of each phase within this time course is directly related to the length of the polyQ tract in the transgene-encoded protein. Comparison of two different lines of SCA17 transgenic mice of the same genetic background, which express TBP with a polyQ tract of either 71 or 105 residues under a prion protein promoter, demonstrates this relationship. At 2–2.5 months of age, NII are detected prominently in the cerebellum of 105Q female mice, whereas immunoreactive cerebellar neurons in identically aged 71Q female mice are characterized by diffuse albeit intense nuclear staining with occasional puncta (M. Friedman and X.J.Li, unpublished data). Intriguingly, a conditional mouse model of HD has provided evidence that polyQ-mediated neuropathology may be reversible. In this model, aggregate formation and pathology in striatal neurons are contingent on the continued production of a mutant htt fragment (Yamamoto et al. 2000).

Although nuclear accumulation of mutant protein has received considerable attention as a neuropathological feature of the polyQ diseases, the relevance of this phenomenon to neurodegeneration is not entirely clear. Neurons containing nuclear aggregates are not necessarily prevalent in the brain regions that selectively degenerate in a given polyQ disease. Conversely, nuclear aggregates can be abundant in mildly affected or even unaffected areas of the brain. In HD, for example, neuronal death is most prominent in the caudate and putamen, but intranuclear inclusions are sparse in the striatum of patients. Aggregates abound in the lesser-affected HD cortex, however (Gutekunst et al. 1999). Moreover, within the striatum, aggregates are rarely observed in medium spiny neurons, which are selectively degraded in HD, but are prevalent in spared interneurons (Kuemmerle et al. 1999). Neuropathological evaluation of postmortem brains from patients with SCA17 as well as other polyQ diseases has revealed similar discrepancies (Fujigasaki et al. 2001; Adachi et al. 2005; Yamada et al. 2001). Interestingly, in most transgenic mouse models, despite the rapid appearance of aggregates due to overexpression of a particular polyQ disease protein or a fragment thereof, neurodegeneration is absent or not obvious (Clark et al. 1997; Abel et al. 2001; Schilling et al. 1999a,b; Mangiarini et al. 1996; Ordway et al. 1997). The short life span of mice may limit the extent of neurodegeneration, which can precede symptom manifestation in patients with polyQ disease (Albin et al. 1992) but generally becomes pronounced in the late stages of pathology.

4 PolyQ-Dependent Misfolding and Aggregation

The presence of an expanded polyQ tract invariably results in protein misfolding, but protein context modulates both repeat threshold and the kinetics of aggregation. For example, the most common TBP allele in Caucasians contains 38 polyQ-encoding d(CAG) repeats, while a polyQ stretch of this length would be conducive to aggregation as well as pathogenic in five of the eight remaining polyQ disease proteins (Reid et al. 2003). PolyQ aggregation has been investigated intensively in vitro by use of synthetic polyQ peptides and recombinant polyQ proteins as well as cellular models of polyQ disease (Perutz et al. 1994; Scherzinger et al. 1997; Hackam et al. 1999; Poirier et al. 2005). Max Perutz, the esteemed structural biologist, provided some of the earliest theoretical insight as well as empirical data regarding the nature of polyQ interactions. On the basis of molecular modeling, he suggested that polyQ domains might self-associate as antiparallel β-strands that are connected by an elaborate array of hydrogen bonds involving both main chain and side chain amide groups. Thus, by analogy to leucine zippers that link α-helices, Perutz envisioned a polar zipper structure for polyQ aggregates (Perutz et al. 1994).

PolyQ aggregates resemble amyloids in appearance in electron micrographs and display some of the same histochemical and kinetics properties in vitro (Scherzinger et al. 1997, 1999; Chen et al. 2001, 2002a,b). Aggregation of synthetic polyQ peptides occurs by nucleation-dependent polymerization. Specifically, an initial nucleation event, which may actually involve the misfolding of a single polyQ monomer rather than the formation of an unstable oligomer, is followed by the rapid addition of polyQ monomer in the elongation phase (Chen et al. 2002b). Whereas fast elongation ensures that a single aggregate forms in vitro, multiple nucleation events may occur in the neurons of patients or mouse models. Also, the elongation process is markedly protracted in the context of the cellular environment. In neuronal nuclei, individual aggregates, after slowly polymerizing as separate entities, may eventually fuse to form a single, large inclusion. Notably, this progression would be consistent with the histochemical time course of nuclear polyQ accumulation detailed already (Michalik and Broeckhoven 2003). Furthermore, as the rate of nucleus formation is directly related to the length of the synthetic polyQ tract, it has been suggested that polyQ aggregation kinetics may underlie the correlation between repeat length and the age of onset in the polyQ diseases (Chen et al. 2002b).

In a cellular model of HD, the cytoplasmic and nuclear environments are not differentially conducive to aggregate formation. Moreover, the subcellular localization of aggregates, which can be manipulated by the attachment of either a nuclear localization signal or a nuclear export signal to mutant htt fragments, does not modulate the toxicity of expanded polyQ in cultured cells (Hackam et al. 1999). Nevertheless, there are some differences in the nature of nuclear and cytoplasmic aggregates. The latter are generally smaller in size, at least when present in neuronal processes (Li et al. 1999). Additionally, whereas NII in all of the polyQ diseases colocalize with ubiquitin, perikaryal aggregates in SCA2 (Huynh et al. 2000) and SCA6 (Ishikawa et al. 1999) neurons and neuropil htt aggregates lack this decoration (Gutekunst et al. 1999; Li et al. 1999). The presence of ubiquitin as well as proteasomal subunits in NII probably indicates the involvement of the ubiquitin–proteasome system (UPS) in aggregate clearance (Everett and Wood 2004). However, markedly reduced nuclear aggregation is observed in SCA1 transgenic mice that lack the E6-AP ubiquitin ligase (Cummings et al. 1999). This paradoxical finding suggests that ubiquination may actually stabilize aggregates in some fashion.

Whereas polyQ expansion in nuclear proteins can be sufficient to produce NII, the same is not true in large, cytoplasmic polyQ proteins, like htt and atrophin-1. Rather, proteolytic processing may be a prerequisite for nuclear accumulation as well as intraceullular aggregate formation by the latter. Loss of htt and atrophin-1 carboxy-terminal antigenicity in NII is consistent with the occurrence of potentially extensive processing (Schilling et al. 1999a,b; Gutekunst et al. 1999), and cleavage sites for various proteases have been identified in vitro (Kim et al. 2001; Gafni and Ellerby 2002; Wellington et al. 2002; Lunkes et al. 2002; Zhou et al. 2003; Nucifora et al. 2003; Gafni et al. 2004). Perikaryal and neuropil htt aggregates also consist of polyQ-containing fragments (DiFiglia et al. 1997; Gutekunst et al. 1999). Similarly, in vitro experiments (Wellington 1998) and immunhistochemical examination of NII in brains of patients have revealed evidence of proteolytic cleavage of other polyQ-expanded proteins, including ataxin-3 (Goti et al. 2004), ataxin-7 (Garden et al. 2002), and androgen receptor (Li et al. 1998). It is possible that conformational differences between soluble and aggregated proteins may contribute to some of the discrepancies in immunoreactivity. Nevertheless, comparison of transgenic mice expressing full-length mutant htt (YAC 46, 72, and 128) and *N*-terminal htt fragments (N171 or R6/2 lines) demonstrates that truncated polyQ proteins with an expanded polyQ tract are not only sufficient to induce aggregation and neuropathology but may actually be more toxic than their unprocessed counterparts (Mangiarini et al. 1996; Schilling et al. 1999a,b; Hodgson et al. 1999; Slow et al. 2003).

In addition to various components of the UPS, polyQ aggregates are also immunoreactive for a number of molecular chaperones, including, most notably, Hsp40 and Hsp70 family members. The former, which are considered cochaperones, recognize and deliver misfolded proteins to the latter. Hsp70 chaperones have an intrinsic ATPase activity that facilitates refolding; however, recalcitrant proteins are ubiquitinated and targeted to the proteasome for degradation. In sum, the colocalization data, which have been collected from cellular (Wyttenbach et al. 2000; Chai et al. 1999a,b; Stenoien et al. 1999; Jana et al. 2001) and mouse models (Jana et al. 2001; Hay et al. 2004; Adachi et al. 2003; Cummings et al. 1998) as well as brain tissue of patients (Chai et al. 1999a,b; Cummings et al. 1998), indicate that polyQ aggregates trigger the normal cellular response to misfolded protein (Fig. 2). Screens for genetic modifiers of polyQ-induced neurodegeneration have substantiated the involvement of the protein folding machinery (Fernandez-Funez et al. 2000; Kazemi-Esfarjani and Benzer 2000). Although biochemical purification of polyQ aggregates is indicative of sequestration (Suhr et al. 2001), live cell imaging has demonstrated that the interaction between Hsp70 and these structures can be dynamic (Kim et al. 2002).

The refolding and clearance of misfolded polyQ proteins by chaperones and the UPS, respectively, may impact the subcellular distribution of mutant polyQ. Biochemical analysis of brains from HD repeat knockin mice, in which a 150 d(CAG) repeat is present in the endogenous mouse htt (Hdh) gene, indicates that a collection of truncated htt fragments accumulate in neuronal nuclei in association with an age-dependent decrease in proteasomal function (Zhou et al. 2003). Most of these *N*-terminal htt fragments are smaller than the size threshold for passive diffusion through the nuclear pore complex, and recent evidence suggests that their entry into the nucleus occurs by a Ran GTPase-independent process. PolyQ expansion decreases the interaction of *N*-terminal htt with a component of the nuclear export machinery, which can explain the accumulation and concomitant aggregation of mutant htt fragments in the nucleus (Cornett et al. 2005). It is unclear if this mechanism of nuclear accumulation applies to other polyQ disease proteins, particularly those that normally localize to the nucleus.

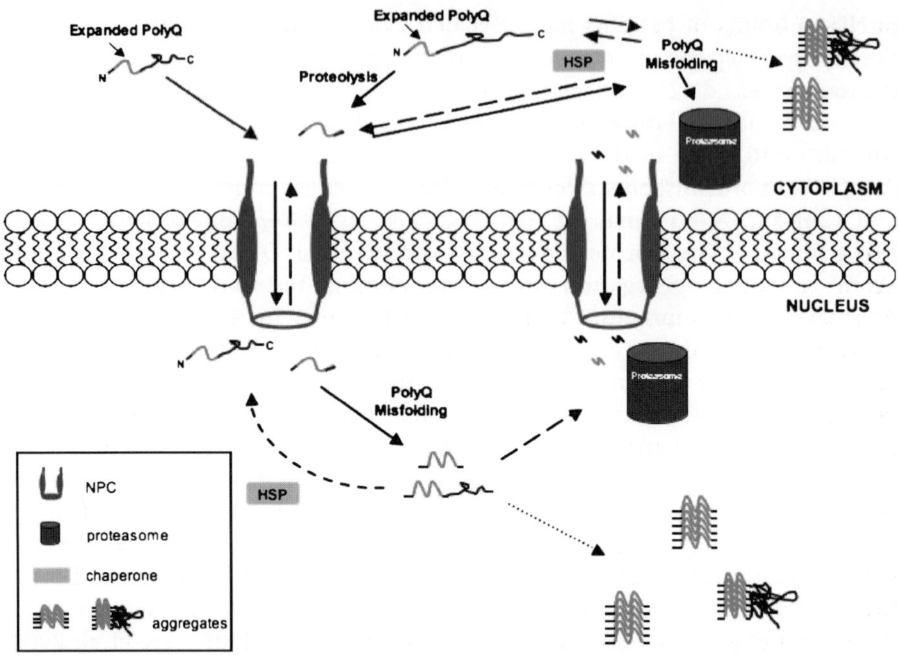

Fig. 2 Intracellular aggregation of expanded polyQ proteins. PolyQ aggregates can form in both the cytoplasm and the nucleus, depending on the polyQ disease protein. The initial step in aggregate formation is polyQ-mediated protein misfolding (*black lines*). Molecular chaperones recognize misfolded polyQ proteins and attempt to reintroduce the proper conformation. Chaperone substrates that cannot be refolded are targeted to the proteasome for degradation (*light dashed line*). Importantly, misfolded polyQ proteins that are refractory to refolding can aggregate (*dark dotted line*) if not degraded. Moreover, age-dependent decline in proteasome function would result in increased aggregation over time. Proteolytic processing often precedes polyQ-mediated aggregate formation in both the nucleus and the cytoplasm and may be a prerequisite for the nuclear accumulation of mutant atrophin-1 and mutant huntingtin (*htt*). Nuclear entry (*solid line*) can be facilitated by classic nuclear localization signals or, at least in the case of htt fragments, may occur by a Ran GTPase-independent process (see text for details). *NPC* nuclear pore complex, *HSP* heat shock protein

5
Pathogenesis of the PolyQ Diseases

A variety of mechanisms have been proposed to explain the molecular pathogenesis of the polyQ diseases, including, most prominently, transcriptional dysregulation and disruption of intracellular trafficking. Contributions from mitochondrial dysfunction, proteasomal impairment, and excitotoxicity are probably important in HD pathogenesis and could be relevant to polyQ-mediated pathology in general. It is now clear that disruption of critical nuclear or cytoplasmic functions can be cytotoxic, and the involvement of either may be contingent on the normal subcellular localization of the polyQ disease protein.

5.1 PolyQ Diseases as "Transcriptionopathies"

When localized in the nucleus, polyQ-expanded proteins aberrantly interact with a variety of transcription factors, many of which contain a polyQ or glutamine-rich domain (Table 2). Certain transcription pathways, namely those involving the cyclic AMP response element (CRE)-binding protein (CREB) and specificity protein-1 (Sp1) have been implicated in the pathogenesis of multiple polyQ diseases. Interestingly, the cofactor TBP-associated factor 4 (TAF4; formerly $TAF_{II}130$), which was independently identified in a yeast two-hybrid screen for nuclear proteins that interact with polyQ tracts (Shimohata et al. 2000), mediates transcriptional activation by both CREB and Sp1 (Fig. 3, top). TAF4 is a component of the general transcription factor TFIID, a multi-subunit complex that comprises TBP and at least 12 TAFs (Muller and Tora 2004). Different glutamine-rich subdomains in TAF4 facilitate its interaction with Sp1 and CREB (Saluja et al. 1998). Although CRE-mediated transcription is constitutive at a subset of promoters (Conkright et al. 2003), recruitment of the cofactor CBP (or the related protein p300), which is contingent on the phosphorylation CREB at a single serine residue, is generally a prerequisite for transcriptional activation (Johannessen et al. 2004).

Various members of the CREB and Sp1 transcription pathways have been reported to interact with soluble and/or aggregated, mutant polyQ proteins (Table 2). As these two possibilities are not mutually exclusive, both forms of polyQ protein could contribute to transcriptional deregulation. Moreover, the sequestration of a given transcription factor in NII would have the same consequence as a soluble interaction of increased affinity. In either case, the transcription factor would be effectively titrated from its cognate promoter binding site (Schaffar et al. 2004; S.H. Li et al. 2002; Dunah et al. 2002) (Fig. 3, bottom). Consistently, reporter assays carried out in cellular models of certain polyQ diseases indicate that expanded polyQ antagonizes both CRE-mediated (Shimohata et al. 2000; Nucifora et al. 2001) and Sp1-dependent transcription (S.H. Li et al. 2002; Dunah et al. 2002). Overexpression of either TAF4 (Shimohata et al. 2000) or CBP (Nucifora et al. 2001) can rescue CRE-mediated transcription, while overexpression of both Sp1 and TAF4 is required to attenuate the effects of mutant htt on Sp1-dependent reporter activity (Dunah et al. 2002). Downregulation of CRE-mediated transcription has been corroborated by expression profiling in both cellular (Wyttenbach et al. 2001) and mouse (Luthi-Carter et al. 2000) models of HD. Unexpectedly, upregulation of the same transcriptional pathway was observed upon

Table 2 A survey of transcription factors that bind polyQ disease proteins

Transcription factor or cofactor	Interacting polyQ disease protein	Colocalizes with polyQ aggregates[a]	Binds soluble polyQ protein(s)[b]	References
CA150	htt	√	√	Holbert et al. (2001)
CBP	htt, atrophin-1, AR, ataxin-1, ataxin-3, TBP, ataxin-7	√	√	Nucifora et al. (2001), Steffan et al. (2000), McCampbell et al. (2000), Chai et al. (2001), Stenoien et al. (2002), Swope et al. (1996), La Spada et al. (2001)
CREB	Atrophin-1, ataxin-3		√	Shimohata et al. (2000)
Crx	Ataxin-7		√	Chen et al. (2004)
CtBP*	htt		√	Kegel et al. (2002)
ETO/MTG8*	Atrophin-1		√	Wood et al. (2000)
HYP-B	htt		√	Faber et al. (1998)
mSin3a*	htt	√		Boutell et al. (1999)
NCoR*	htt		√	Boutell et al. (1999)
NF-κB	htt, TBP		√	Takano and Gusella (2002), Schmitz et al. (1995)
p53	htt, TBP	√	√	Steffan et al. (2000), Truant et al. (1993)
p300	Ataxin-3, TBP		√	Li et al. (2002), Swope et al. (1996)
P/CAF	Ataxin-3, htt		√	Steffan et al. (2001), Li et al. (2002b)
REST/NRSF*	htt, TBP		√	Zuccato et al. (2003), Murai et al. (2004)
SMRT*	Ataxin-1	√	√	Tsai et al. (2004)
Sp1	htt, TBP, atrophin-1, ataxin-3	√	√	Shimohata et al. (2000), Dunah et al. (2002), S.H. Li et al. (2002), Emili A (1994)

Table 2 (continued)

Transcription factor or cofactor	Interacting polyQ disease protein	Colocalizes with polyQ aggregates[a]	Binds soluble polyQ protein(s)[b]	References
TAF4[c]	htt, atrophin-1, ataxin-2, ataxin-3	√	√	Shimohata et al. (2000), Dunah et al. (2002)
TAF10[d]	Ataxin-7	√		Yvert et al. 2001
TBP	htt, atrophin-1, ataxin-3	√	√	Huang et al. (1998), Schaffar et al. (2004), Shimohata et al. (2000)

TBP interacts with various transcription factors in the context of transcriptional initiation by all three nuclear RNA polymerases. It is presently unclear which of these myriad interactions may be relevant to polyQ-mediated pathogenesis (in SCA17) and most have been excluded from this table. An *asterisk* denotes repressor or corepressor activity. *Bold font* indicates the presence of a polyQ or glutamine-rich domain in the transcription factor.

AR androgen receptor, *CA150* coactivator 150, *CBP* cyclic AMP response element binding protein, *CtBP* C-terminal binding protein, *Crx* cone–rod homeobox containing gene, *htt* huntingtin, *HYP-B* htt–yeast partner, *mSin3a* mammalian Sin3 protein-A, *MTG8* myeloid translocation gene on 8q22, *NCoR* nuclear receptor corepressor; *NF-κB* nuclear factor-κB, *P/CAF* p300/CBP-associated factor, *REST/NRSF* repressor element-1 transcription factor/neuron restrictive silencer factor, *SMRT* silencing mediator of retinoid and thyroid hormone receptors, *Sp1* specificity protein-1, *TAF4* TBP-associated factor 4, *TAF10* TBP-associated factor 10

[a] As determined by double immunolabeling and microscopy or biochemical analysis of aggregate content

[b] As determined by yeast two-hybrid, in vitro binding, or coimmunoprecipitation

[c] A component of TFIID as well as other transcriptional complexes; formerly known as $TAF_{II}130$

[d] A component of TFIID as well as other transcriptional complexes; formerly known as $TAF_{II}30$

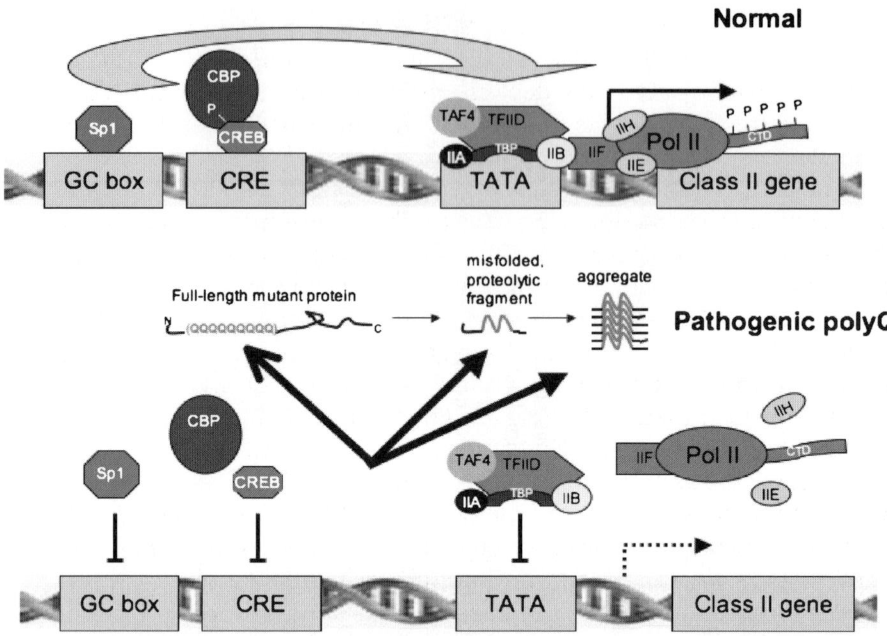

Fig. 3 Transcriptional dysregulation in the presence of nuclear, polyQ-expanded proteins. Under normal circumstances (*top*), transcriptional activators bind upstream promoter elements and interact (*curved arrow*) with various components of the RNA polymerase II (*Pol II*) preinitiation complex (*PIC*). These interactions, which are important for PIC recruitment to class II promoters, are often facilitated by polyQ or glutamine-rich domains that are present in many activator proteins and some general transcription factors. However, in the polyQ diseases (*bottom*), both the soluble and the aggregated versions of mutant polyQ can act as a sink for the same transcriptional activators, effectively titrating the latter from their cognate DNA binding sites (*arrows*). It should be noted that many changes in gene expression are disease-specific and may also involve aberrant interactions with transcriptional repressors or altered chromatin acetylation. Moreover, targeting of expanded polyQ to the core promoter, as occurs in SCA17 but in none of the other polyQ diseases, may have a unique transcriptional impact. *IIA*, *IIB*, *IIE*, *IIF*, and *IIH* represent TFIIA, TFIIB, TFIIE, TFIIF, and TFIIH, respectively. *CRE* cyclic AMP response element, *CREB* CRE binding protein, *CTD* carboxy-terminal domain of Pol II, P — phosphorylation, *Sp1* specificity protein-1, *TAF4* TATA-box binding protein associated factor 4

introduction of a CRE-regulated reporter transgene into an HD mouse model (R6/2) (Obrietan and Hoyt 2004). The basis of this discrepancy is currently unclear.

Microarray experiments, utilizing brain messenger RNAs (mRNAs) from various polyQ mouse models, have revealed some overlap in the expression changes induced by the different polyQ disease proteins (Sugars and Rubinsztein 2003). Moreover, a comparison of cerebellar mRNAs derived from a DRPLA (At-65Q) and an HD (N171-82Q) mouse model demonstrated that the transcriptional impact of expanded polyQ is partly context independent (Luthi-Carter et al. 2002). Despite a difference in the length of the polyQ tract, the two models have an identical genetic background and the respective transgenes are driven by the same promoter. These data support the idea that particular transcription pathways may be disrupted in polyQ disease.

In addition to changes in activated transcription, there is some evidence that basal gene expression may also be dysregulated by expanded polyQ. Aberrant interactions involving TBP, which is required for transcription by all three nuclear RNA polymerases, have been documented in several of the polyQ diseases (Table 2). TBP has been reported to colocalize in HD (Schaffar et al. 2004), DRPLA, and SCA3 aggregates (Shimohata et al. 2000), and the transcription factor also preferentially coimmunoprecipitates with soluble, mutant htt (Schaffar et al. 2004). Functional deactivation of TBP in the presence of mutant htt has been demonstrated in vitro (Schaffar et al. 2004); however, it is unlikely that a polyQ expansion in TBP, which is causative for SCA17, abrogates its function. Heterozygous TBP knockout mice are phenotypically normal, but nullizygous embryos do not develop beyond the blastocyst stage (Martianov et al. 2002). Consistently, below a certain pathogenic repeat threshold, polyQ-expanded TBP upregulates a CRE-regulated reporter gene in a cellular model of SCA17 (Reid et al. 2003).

Although the relevance of general transcriptional repression to polyQ-mediated pathogenesis has not been firmly established, it is clear that histone acetylation is disrupted in the presence of mutant polyQ (Bodai et al. 2003). Three histone acetyltransferases (HATs), including CBP, p300, and p300/CBP-associated factor (P/CAF), interact directly with ataxin-3 (F. Li et al. 2002). htt exon 1 protein also binds CBP and P/CAF (Steffan et al. 2001). CBP physically interacts with the soluble and aggregated forms of various polyQ disease proteins (Table 2). However, since p300 contains only a short polyQ tract and P/CAF lacks this domain entirely, these interactions are not contingent on the association of independent polyQ domains. Rather, mutant htt has been reported to impair the HAT activity of CBP and P/CAF by binding to their acetyltransferase domains (Steffan et al. 2001). Hypoacetylation of histones H3 and H4, which has been documented in multiple polyQ disease models (McCampbell et al. 2001; Steffan et al. 2001), could potentially have widespread effects on transcription. It is well established that acetylation of particular histone residues is associated with euchromatin and active genes. Either local or global changes in acetylation could significantly impact the expression of genes that are important for cell function and viability. Interestingly, acetylation of two lysine residues on histone H3, namely, K9 and K14, may be necessary for the recruitment of TFIID to promoters (Agalioti et al. 2002). Accordingly, polyQ-induced histone deacetylation might antagonize this crucial step in preinitiation complex assembly on certain RNA polymerase II promoters. Alternatively, it is noteworthy that HATs also modulate the function of various non-histone proteins, like p53, by domain-specific acetylation (Guand and Roeder 1997).

Although the outcome is not surprising, it is not entirely clear how transcriptional dysregulation leads to neuronal cell death. Interestingly, mice that lack CREB in the postnatal forebrain, as a result of conditional disruption, display neurodegeneration in the hippocampus and striatum (Mantamadiotis et al. 2002). Nevertheless, changes in gene expression can be detected in early symptomatic (Luthi-Carter et al. 2000) as well as presymptomatic (Lin et al. 2000; Serra et al. 2004) transgenic polyQ mouse models, long before any evidence of neurodegeneration has emerged. Thus, in many cases, neuronal dysfunction and not neuronal death may be responsible for the polyQ-induced phenotype (Hientz and Zoghbi 2000).

5.2
Disruption of Cytoplasmic Activities in PolyQ Disease

As nuclear localization of mutant polyQ protein is limited at best in SCA2 (Huynh et al. 1999) and SCA6 (Ishikawa et al. 2001), any influence of expanded polyQ on transcription or other nuclear activities in these two diseases is probably indirect. In the polyQ diseases that are characterized by soluble or aggregated, mutant protein in the cytoplasm, it is possible that polyQ-mediated changes outside the nucleus are most relevant to pathology. In particular, both disruption of axonal transport (Gunawardena et al. 2003) and potentially related defects in synaptic function (Li et al. 2000; Usdin et al. 1999) have received considerable attention recently.

Similar to the situation in the nucleus, both soluble and aggregated polyQ may contribute to cytoplasmic problems. Neuropil aggregates have been most carefully examined in HD (Li et al. 2000), but, at least in cultured cells, expression of several polyQ disease proteins can induce their formation (Piccioni et al. 2002; Gunawardena et al. 2003). These cytoplasmic structures, which can localize to axons (Li et al. 1999), dendrites, or dendritic spines (Gutekunst et al. 1999), may be an early maker of HD pathology, distinguishing them from perikaryal and nuclear aggregates (Gutekunst et al. 1999; Sapp et al. 1999). In HD repeat knockin mice, neuropil aggregates form progressively in the lateral globus pallidus (LGP) and substantia nigra pars reticulata (SNr). These htt aggregates, distinct from the larger NII that abound in the nuclei of striatal neurons, reside in the axons of medium spiny neurons that project to both of these regions. The presence of neuropil aggregates, which actually can be very large relative to the size of an axon terminal, has been associated with axonal degeneration in electron micrographs. Consistent with the possibility of axonal occlusion, synaptic vesicles are less abundant in terminals that harbor aggregates (Li et al. 2001). The neuropathology in this HD mouse line recapitulates the selective neurodegeneration that occurs early in HD, as loss of striatal projection neurons that target the LGP and SNr is evident in presymptomatic patients (Albin et al. 1990, 1992). Neuritic degeneration is also observed in striatal neurons transfected with mutant htt (Li et al. 2000, 2001), and this event precedes nuclear fragmentation that is indicative of apoptosis (Li et al. 2001).

In vitro experiments indicate that polyQ-expanded htt in axonal terminals can undermine normal synaptic function. Mutant htt causes deficiencies in both glutamate uptake by synaptic vesicles (Li et al. 2000) and glutamate release into the synaptic cleft during high-frequency stimulation. The latter phenomenon, which results in reduced long-term potentiation (Usdin et al. 1999), may contribute to cognitive impairments characteristic of HD patients. Moreover, defects in synaptic transmission should ultimately impact neuronal viability (J.Y. Li et al. 2003). Interestingly, the synaptic disturbances might be explained by the direct binding of htt fragments to synaptic vesicles, since the strength of the interaction is enhanced by polyQ expansion (Li et al. 2000). As various proteins involved in synaptic activities interact with normal htt (Li and Li 2004a), it is unclear if other polyQ diseases may also be characterized as synaptopathies.

The presence of expanded polyQ in neuronal process also dramatically affects fast axonal transport (FAT) (Morfini et al. 2005; Gunawardena and Goldstein 2005). Several ideas have been proposed to explain the mechanism by which mutant polyQ proteins block the movement of microtubule-based molecular motors in axons (Fig. 4). Physical obstruction by neuropil aggregates (Lee et al. 2004; Gunawardena et al. 2003), which increase in size with disease progression (Gutekunst et al. 1999), is an attractive possibility. In late-stage HD, neuropil aggregates in excess of 30 nm^2 are not uncommon, but even the smaller aggregates that predominate in early HD could at least partially occlude a typical nonmotor axon. Alternatively, titration of motor proteins away from microtubules by aggregated (Trushina et al. 2004) or soluble, mutant polyQ could undermine axonal trafficking; however, these explanations are not supported by in vitro experiments carried out in squid axoplasm with purified polyQ proteins (Morfini et al. 2005). Both polyQ-expanded androgen receptor and mutant *N*-terminal htt, but not the normal versions of these proteins, antagonize FAT in this context. Inhibition occurs upon addition of either mutant protein to the isolated axoplasm at a concentration 100-fold lower than endogenous levels of trafficking motors and in the absence of any detectable aggregation (Szebenyi et al. 2003; Morfini et al. 2005).

It is now clear that normal htt, unlike the other polyQ disease proteins, has a role in axonal transport (Gunawardena et al. 2003; Gauthier et al. 2004; Trushina et al. 2004) (Table 2); thus, loss of normal htt function as a result of polyQ expansion might contribute to FAT inhibition in HD. There is at present no evidence of diminished htt function in the presence of other polyQ-expanded proteins. It is also conceivable that disruption of certain intracellular signaling pathways could undermine FAT, but specific pathways

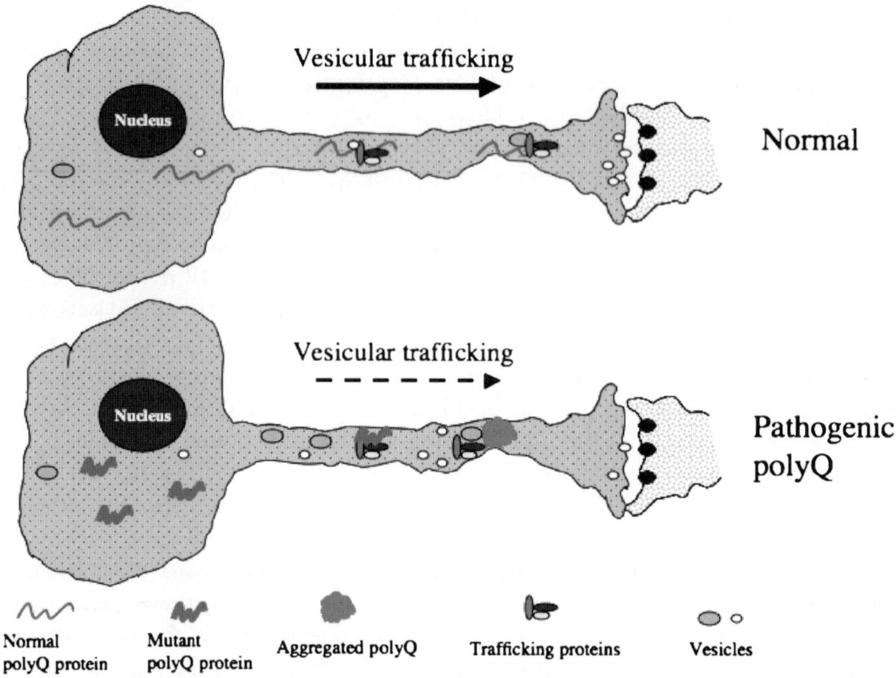

Fig. 4 Disruption of microtubule-based transport by cytoplasmic, polyQ-expanded proteins. Both soluble and aggregated versions of mutant polyQ can sequester trafficking proteins. Also, aggregated polyQ may sterically hinder transport in narrow-diameter axons. As low concentrations of mutant polyQ are capable of antagonizing transport in vitro, other possibilities may also account for observed disturbances in axonal trafficking

or kinases have yet to be directly implicated in this polyQ-mediated effect (Morfini et al. 2005).

5.3 Other Contributors to Pathogenesis

Several alternative mechanisms have been invoked to explain the molecular pathogenesis of HD, including mitochondrial defects and glial dysfunction. While the contribution of the former to HD is more firmly established than that of the latter, the relevance of either possibility to polyQ disease in general is unclear. Altered mitochondrial function, which is a feature of several neurodegenerative diseases, can result in cell death by more than one pathway (Grunewald and Beal 1999). Mitochondria in the HD brain display both increased oxidative stress (Browne et al. 1999) and defective calcium homeostasis (Panov et al. 2002). Moreover, deficiencies in oxidative phosphorylation presumably exacerbate the production of reactive oxygen species. Dramatic but selective reduction in the activity of certain mitochondrial enzymes, including complex II/III of the electron transport chain and the tricarboxylic acid cycle component aconitase, has been demonstrated in HD postmortem brains (Tabrizi et al. 1999; Browne et al. 1997). Notably, administration of 3-nitropropionic acid or malonate, which are inhibitors of complex II, can recapitulate HD striatal pathology in wild-type mice (Ludolph et al. 1992; Beal et al. 1993).

Glia, which serve several important functions, including the release of cytokines and trophic factors as well as the removal of extracellular, excitotoxic neurotransmitters, have been implicated in various neurodegenerative conditions (Hirsch et al. 2003; Mrak and Griffin 2005). Mutant aggregates have been observed in glia in DRPLA and HD patient brains (Hayashi et al. 1998a,b; Singhrao et al. 1998). Additionally, expression of expanded-polyQ proteins in glia alone results in aggregation and can impact both behavior and viability in fly models (Kretzschmar et al. 2004; Lievens et al. 2005). Nevertheless, it is currently unclear to what extent cytokine-induced inflammatory responses or disruption of trophic support may be involved in polyQ-dependent pathology. Moreover, decreased expression of glutamate transporters in glial cells, or specifically astrocytes (Lievens et al. 2001, 2005; Behrens et al. 2002), could contribute to *N*-methyl-D-aspartate receptor mediated excitotoxicty. Given the overabundance of glia relative to neurons and the possible localization of many of the polyQ proteins in the former, glial dysfunction may be proof that polyQ pathogenesis is not necessarily a neuron autonomous event.

6 Potential Therapeutic Strategies

There is currently no effective treatment for polyQ-mediated pathology; however, numerous potential therapeutic targets have emerged from the study of polyQ-dependent pathogenesis. Small molecule screening efforts, carried out in various polyQ model systems, also have identified several interesting candidates (Smith et al. 2001; Tanaka et al. 2004; Zhang et al. 2005; Pollitt et al. 2003). In general, two major strategies, namely, attenuation of polyQ aggregation and inhibition of histone deacetylation, have received considerable attention by investigators in recent years.

Among the multitude of agents capable of modulating polyQ aggregation (Sanchez et al. 2003; Smith et al. 2001; Tanaka et al. 2004; Zhang et al. 2005; Pollitt et al. 2003), molecular chaperones have been the most intensively investigated. There are numerous reports of reduced aggregation in polyQ cell models that overexpress Hsp70 and/or its cochaperone Hsp40 (Cummings et al. 1998; Stenoien et al. 1999; Jana et al. 2000; Kobayashi et al. 2000), although variability between cell types has been observed (Wyttenbach et al. 2000). Genetic manipulation of chaperone levels also can attenuate polyQ aggregation and toxicity in both *Caenorhabditis elegans* (Satyal et al. 2000) and *Drosophila melanogaster* (Warrick et al. 1999; Chan et al. 2000; Kazemi-Esfarjani and Benzer 2000; Fernandez-Funez et al. 2000); however, chaperone overexpression in mouse models of different polyQ diseases has not been similarly efficacious. Overexpression of Hsp70 in R6/2 HD mice, a well-characterized transgenic line that expresses htt exon 1 with 150 d(CAG) repeats and displays rapid and pervasive neuronal aggregation (Davies et al. 1997; Li et al. 2001), had little or no effect on both neuropathology and phenotype (Hansson et al. 2003; Hay et al. 2004). Similarly, simultaneous overexpression of Hsp40 and Hsp70 in SCA7 transgenic mice did not prevent NII formation or neuronal cell death (Helmlinger et al. 2004). In a mouse model of SBMA, however, introduction of an Hsp70 transgene substantially improved various phenotypic parameters, including survival rate. Amelioration in the double transgenic mice coincided with a reduction in both aggregated and soluble, mutant androgen receptor in muscle and spinal cord tissue (Adachi et al. 2003). Additionally, mild improvements in neuropathology and motor function without any change in nuclear aggregation have been reported for SCA1 mice overexpressing Hsp70 (Cummings et al. 2001). Thus, the benefit of elevated levels of molecular chaperones in vivo may depend on the polyQ disease protein. With regard to therapy, activation of the heat shock response by certain drugs may be a practical alternative to genetic manipulation, but this strategy has only been tested in organotypic slice culture (Hay et al. 2004).

The potential for histone deacetylase (HDAC) inhibitors in the treatment of polyQ disease appears promising (Bodai et al. 2003), although limited in vivo data are available at present. HDACs were first implicated in polyQ pathology in a screen for genetic modifiers of SCA1 neurodegeneration. Loss-of-function mutations in two HDAC proteins, namely, Rdp3 and the cofactor Sin3A, suppressed a mutant ataxin-1-induced rough eye phenotype (Fernandez-Funez et al. 2000). Subsequently, administration of the HDAC inhibitors butyrate and suberoylanilide hydroxamic acid (SAHA) was shown to rescue polyQ-dependent neurodegeneration in the fly eye with efficacy comparable to that of Sin3A heterozygosity (i.e., a 50% reduction in Sin3A dose) (Steffan et al. 2001). Similarly, sodium butyrate can ameliorate various disease phenotypes in a mouse model of SBMA (Minamiyama et al. 2004), while SAHA improves motor function in R6/2 HD mice (Hockly et al. 2003). Neither compound has any effect on nuclear localization or aggregation of the respective polyQ disease proteins. Notably, the efficacy of these drugs is tempered by their toxicity, which is considerable outside of a limited dose window (Minamiyama et al. 2004; Hockly et al. 2003).

Paradoxically, a recent report indicates that the antioxidant resveratrol, a component of red wine and an activator of sirtuin deacetylases, can attenuate polyQ-mediated neuronal death. Resveratrol treatment was effective in both a *C. elegans* HD model, which expressed an *N*-terminal htt fragment with 128 glutamines, and striatal neurons cultured from HD repeat knockin mice (Parker et al. 2005). It is unclear how this finding can be reconciled with the available data on HDAC inhibitors.

7 Concluding Remarks

Despite a common genetic basis and the well-defined neuropathology of the polyQ diseases, the pathogenesis of each disorder remains to be fully elucidated. All of the polyQ diseases display late-onset neurological symptoms and neuropathology; however, each disease is characterized by distinct and selective neurodegeneration despite the widespread expression of most polyQ disease proteins. The controversial role of the polyQ aggregates notwithstanding, investigation of these structures has been informative with regard to possible pathological pathways mediated by polyQ proteins. It is now apparent that the subcellular localization of polyQ disease proteins and the protein context of expanded polyQ tracts are critically relevant to neuropathology. Mutant polyQ proteins in the nucleus can impact gene transcription by abnormally binding certain transcription factors. Large inclusions may exacerbate polyQ toxicity by sequestering the same transcription factors. Alternatively, polyQ inclusions may recruit soluble, polyQ-expanded protein and thereby reduce the availability of toxic polyQ. When present in axons and nerve terminals, mutant polyQ proteins can disrupt intracellular transport and synaptic transmission.

PolyQ inclusions reflect abundant protein misfolding and impaired intracellular clearance of toxic proteins, pathological scenarios that are not necessarily associated with neurodegeneration but that are likely to mediate neuronal dysfunction. As a result of differences in the subcellular localization and function of polyQ disease proteins, multiple pathways may be involved in polyQ-mediated pathology. This could explain why most therapeutic approaches, each of which may be effective against one of these pathways, have limited efficacy in animal models of polyQ disease. Thus, although much progress has been made in the study of the molecular pathogenesis of the polyQ diseases, additional insight will prove critical in the development of targeted therapies.

Acknowledgements The work from author's laboratory was supported by grants from the NIH (AG19206, NS36232, NS 41669) and High Q foundation.

References

Abel A, Walcott J, Woods J, Duda J, Merry DE (2001) Expression of expanded repeat androgen receptor produces neurologic disease in transgenic mice. Hum Mol Genet 10:107–116

Adachi H, Katsuno M, Minamiyama M, Sang C, Pagoulatos G, Angelidis C, Kusakabe M, Yoshiki A, Kobayashi Y, Doyu M, Sobue G (2003) Heat shock protein 70 chaperone overexpression ameliorates phenotypes of the spinal and bulbar muscular atrophy transgenic mouse model by reducing nuclear-localized mutant androgen receptor protein. J Neurosci 23:2203–2211

Adachi H, Katsuno M, Minamiyama M, Waza M, Sang C, Nakagomi Y, Kobayashi Y, Tanaka F, Doyu M, Inukai A, Yoshida M, Hashizume Y, Sobue G (2005) Widespread nuclear and cytoplasmic accumulation of mutant androgen receptor in SBMA patients. Brain 128:659–670

Agalioti T, Chen G, Thanos D (2002) Deciphering the transcriptional histone acetylation code for a human gene. Cell 111:381–392

Albin RL, Young AB, Penney JB, Handelin B, Balfour R, Anderson KD, Markel DS, Tourtellotte WW, Reiner A (1990) Abnormalities of striatal projection neurons and N-methyl-D-aspartate receptors in presymptomatic Huntington's disease. N Engl J Med 322:1293–1298

Albin RL, Reiner A, Anderson KD, Dure LSt, Handelin B, Balfour R, Whetsell WO Jr, Penney JB, Young AB (1992) Preferential loss of striato-external pallidal projection neurons in presymptomatic Huntington's disease. Ann Neurol 31:425–430

Andres AM, Soldevila M, Lao O, Volpini V, Saitou N, Jacobs HT, Hayasaka I, Calafell F, Bertranpetit J (2004) Comparative genetics of functional trinucleotide tandem repeats in humans and apes. J Mol Evol 59:329–339

Beal MF, Brouillet E, Jenkins B, Henshaw R, Rosen B, Hyman BT (1993) Age-dependent striatal excitotoxic lesions produced by the endogenous mitochondrial inhibitor malonate. J Neurochem 61:1147–1150

Behrens PF, Franz P, Woodman B, Lindenberg KS, Landwehrmeyer GB (2002) Impaired glutamate transport and glutamate-glutamine cycling: downstream effects of the Huntington mutation. Brain 125:1908–1922

Bodai L, Pallos J, Thompson LM, Marsh JL (2003) Altered protein acetylation in polyglutamine diseases. Curr Med Chem 10:2577–2587

Boutell JM, Thomas P, Neal JW, Weston VJ, Duce J, Harper PS, Jones AL (1999) Aberrant interactions of transcriptional repressor proteins with the Huntington's disease gene product, huntingtin. Hum Mol Genet 8:1647–1655

Browne SE, Bowling AC, MacGarvey U, Baik MJ, Berger SC, Muqit MM, Bird ED, Beal MF (1997) Oxidative damage and metabolic dysfunction in Huntington's disease: selective vulnerability of the basal ganglia. Ann Neurol 41:646–653

Browne SE, Ferrante RJ, Beal MF (1999) Oxidative stress in Huntington's disease. Brain Pathol 9:147–163

Chai Y, Koppenhafer SL, Bonini NM, Paulson HL (1999a) Analysis of the role of heat shock protein (Hsp) molecular chaperones in polyglutamine disease. J Neurosci 19:10338–10347

Chai Y, Koppenhafer SL, Shoesmith SJ, Perez MK, Paulson HL (1999b) Evidence for proteasome involvement in polyglutamine disease: localization to nuclear inclusions in SCA3/MJD and suppression of polyglutamine aggregation in vitro. Hum Mol Genet 8:673–682

Chai Y, Wu L, Griffin JD, Paulson HL (2001) The role of protein composition in specifying nuclear inclusion formation in polyglutamine disease. J Biol Chem 276:44889–44897

Chai Y, Berke SS, Cohen RE, Paulson HL (2004) Poly-ubiquitin binding by the polyglutamine disease protein ataxin-3 links its normal function to protein surveillance pathways. J Biol Chem 279:3605–3611

Chan HY, Warrick JM, Gray-Board GL, Paulson HL, Bonini NM (2000) Mechanisms of chaperone suppression of polyglutamine disease: selectivity, synergy and modulation of protein solubility in Drosophila. Hum Mol Genet 9:2811–2820

Chattopadhyay B, Baksi K, Mukhopadhyay S, Bhattacharyya NP (2005) Modulation of age at onset of Huntington disease patients by variations in TP53 and human caspase activated DNase (hCAD) genes. Neurosci Lett 374:81–86

Chen S, Berthelier V, Yang W, Wetzel R (2001) Polyglutamine aggregation behavior in vitro supports a recruitment mechanism of cytotoxicity. J Mol Biol 311:173–182

Chen S, Berthelier V, Hamilton JB, O'Nuallain B, Wetzel R (2002a) Amyloid-like features of polyglutamine aggregates and their assembly kinetics. Biochemistry 41:7391–7399

Chen S, Ferrone FA, Wetzel R (2002b) Huntington's disease age-of-onset linked to polyglutamine aggregation nucleation. Proc Natl Acad Sci USA 99:11884–11889

Chen S, Peng GH, Wang X, Smith AC, Grote SK, Sopher BL, La Spada AR (2004) Interference of Crx-dependent transcription by ataxin-7 involves interaction between the glutamine regions and requires the ataxin-7 carboxy-terminal region for nuclear localization. Hum Mol Genet 13:53–67

Clark HB, Burright EN, Yunis WS, Larson S, Wilcox C, Hartman B, Matilla A, Zoghbi HY, Orr HT (1997) Purkinje cell expression of a mutant allele of SCA1 in transgenic mice leads to disparate effects on motor behaviors, followed by a progressive cerebellar dysfunction and histological alterations. J Neurosci 17:7385–7395

Conkright MD, Canettieri G, Screaton R, Guzman E, Miraglia L, Hogenesch JB, Montminy M (2003) TORCs: transducers of regulated CREB activity. Mol Cell 12:413–423

Cornett J, Cao F, Wang CE, Ross CA, Bates GP, Li SH, Li XJ (2005) Polyglutamine expansion of huntingtin impairs its nuclear export. Nat Genet 37:198–204

Cummings CJ, Mancini MA, Antalffy B, DeFranco DB, Orr HT, Zoghbi HY (1998) Chaperone suppression of aggregation and altered subcellular proteasome localization imply protein misfolding in SCA1. Nat Genet 19:148–154

Cummings CJ, Reinstein E, Sun Y, Antalffy B, Jiang Y, Ciechanover A, Orr HT, Beaudet al Zoghbi HY (1999) Mutation of the E6-AP ubiquitin ligase reduces nuclear inclusion frequency while accelerating polyglutamine-induced pathology in SCA1 mice. Neuron 24:879–892

Cummings CJ, Sun Y, Opal P, Antalffy B, Mestril R, Orr HT, Dillmann WH, Zoghbi HY (2001) Over-expression of inducible HSP70 chaperone suppresses neuropathology and improves motor function in SCA1 mice. Hum Mol Genet 10:1511–1518

Davies SW, Turmaine M, Cozens BA, DiFiglia M, Sharp AH, Ross CA, Scherzinger E, Wanker EE, Mangiarini L, Bates GP (1997) Formation of neuronal intranuclear inclusions underlies the neurological dysfunction in mice transgenic for the HD mutation. Cell 90:537–548

DiFiglia M, Sapp E, Chase KO, Davies SW, Bates GP, Vonsattel JP, Aronin N (1997) Aggregation of huntingtin in neuronal intranuclear inclusions and dystrophic neurites in brain. Science 277:1990–1993

Dunah AW, Jeong H, Griffin A, Kim YM, Standaert DG, Hersch SM, Mouradian MM, Young AB, Tanese N, Krainc D (2002) Sp1 and TAFII130 transcriptional activity disrupted in early Huntington's disease. Science 296:2238–2243

Emili A, Greenblatt J, Ingles CJ (1994) Species-specific interaction of the glutamine-rich activation domains of Sp1 with the TATA box-binding protein. Mol Cell Biol 14:1582–1593

Everett CM, Wood NW (2004) Trinucleotide repeats and neurodegenerative disease. Brain 127:2385–2405

Evert BO, Vogt IR, Vieira-Saecker AM, Ozimek L, de Vos RA, Brunt ER, Klockgether T, Wullner U (2003) Gene expression profiling in ataxin-3 expressing cell lines reveals distinct effects of normal and mutant ataxin-3. J Neuropathol Exp Neurol 62:1006–1018

Faber PW, Barnes GT, Srinidhi J, Chen J, Gusella JF, MacDonald ME (1998) Huntingtin interacts with a family of WW domain proteins. Hum Mol Genet 7:1463–1474

Fernandez-Funez P, Nino-Rosales ML, de Gouyon B, She WC, Luchak JM, Martinez P, Turiegano E, Benito J, Capovilla M, Skinner PJ, McCall A, Canal I, Orr HT, Zoghbi HY, Botas J (2000) Identification of genes that modify ataxin-1-induced neurodegeneration. Nature 408:101–106

Fujigasaki H, Martin JJ, De Deyn PP, Camuzat A, Deffond D, Stevanin G, Dermaut B, Van Broeckhoven C, Durr A, Brice A (2001) CAG repeat expansion in the TATA box-binding protein gene causes autosomal dominant cerebellar ataxia. Brain 124:1939–1947

Gafni J, Ellerby LM (2002) Calpain activation in Huntington's disease. J Neurosci 22:4842–4849

Gafni J, Hermel E, Young JE, Wellington CL, Hayden MR, Ellerby LM (2004) Inhibition of calpain cleavage of huntingtin reduces toxicity: accumulation of calpain/caspase fragments in the nucleus. J Biol Chem 279:20211–20220

Garden GA, Libby RT, Fu YH, Kinoshita Y, Huang J, Possin DE, Smith AC, Martinez RA, Fine GC, Grote SK, Ware CB, Einum DD, Morrison RS, Ptacek LJ, Sopher BL, La Spada AR (2002) Polyglutamine-expanded ataxin-7 promotes non-cell-autonomous purkinje cell degeneration and displays proteolytic cleavage in ataxic transgenic mice. J Neurosci 22:4897–4905

Gauthier LR, Charrin BC, Borrell-Pages M, Dompierre JP, Rangone H, Cordelieres FP, De Mey J, MacDonald ME, Lessmann V, Humbert S, Saudou F (2004) Huntingtin controls neurotrophic support and survival of neurons by enhancing BDNF vesicular transport along microtubules. Cell 118:127–138

Goti D, Katzen SM, Mez J, Kurtis N, Kiluk J, Ben-Haiem L, Jenkins NA, Copeland NG, Kakizuka A, Sharp AH, Ross CA, Mouton PR, Colomer V (2004) A mutant ataxin-3 putative-cleavage fragment in brains of Machado-Joseph disease patients and transgenic mice is cytotoxic above a critical concentration. J Neurosci 24:10266–10279

Grunewald T, Beal MF (1999) Bioenergetics in Huntington's disease. Ann NY Acad Sci 893:203–213

Gu W, Roeder RG (1997) Activation of p53 sequence-specific DNA binding by acetylation of the p53 C-terminal domain. Cell 90:595–606

Gunawardena S, Goldstein LS (2005) Polyglutamine diseases and transport problems: deadly traffic jams on neuronal highways. Arch Neurol 62:46–51

Gunawardena S, Her LS, Brusch RG, Laymon RA, Niesman IR, Gordesky-Gold B, Sintasath L, Bonini NM, Goldstein LS (2003) Disruption of axonal transport by loss of huntingtin or expression of pathogenic polyQ proteins in Drosophila. Neuron 40:25–40

Gutekunst CA, Li SH, Yi H, Mulroy JS, Kuemmerle S, Jones R, Rye D, Ferrante RJ, Hersch SM, Li XJ (1999) Nuclear and neuropil aggregates in Huntington's disease: relationship to neuropathology. J Neurosci 19:2522–2534

Hackam AS, Hodgson JG, Singaraja R, Zhang T, Gan L, Gutekunst CA, Hersch SM, Hayden MR (1999) Evidence for both the nucleus and cytoplasm as subcellular sites of pathogenesis in Huntington's disease in cell culture and in transgenic mice expressing mutant huntingtin. Philos Trans R Soc Lond Ser B Biol Sci 354:1047–1055

Hansson O, Nylandsted J, Castilho RF, Leist M, Jaattela M, Brundin P (2003) Overexpression of heat shock protein 70 in R6/2 Huntington's disease mice has only modest effects on disease progression. Brain Res 970:47–57

Hay DG, Sathasivam K, Tobaben S, Stahl B, Marber M, Mestril R, Mahal A, Smith DL, Woodman B, Bates GP (2004) Progressive decrease in chaperone protein levels in a mouse model of Huntington's disease and induction of stress proteins as a therapeutic approach. Hum Mol Genet 13:1389–1405

Hayashi Y, Kakita A, Yamada M, Egawa S, Oyanagi S, Naito H, Tsuji S, Takahashi H (1998a) Hereditary dentatorubral-pallidoluysian atrophy: ubiquitinated filamentous inclusions in the cerebellar dentate nucleus neurons. Acta Neuropathol 95:479–482

Hayashi Y, Kakita A, Yamada M, Koide R, Igarashi S, Takano H, Ikeuchi T, Wakabayashi K, Egawa S, Tsuji S, Takahashi H (1998b) Hereditary dentatorubral-pallidoluysian atrophy: detection of widespread ubiquitinated neuronal and glial intranuclear inclusions in the brain. Acta Neuropathol 96:547–552

Heintz N, Zoghbi HY (2000) Insights from mouse models into the molecular basis of neurodegeneration. Annu Rev Physiol 62:779–802

Helmlinger D, Bonnet J, Mandel JL, Trottier Y, Devys D (2004) Hsp70 and Hsp40 chaperones do not modulate retinal phenotype in SCA7 mice. J Biol Chem 279:55969–55977

Helmlinger D, Hardy S, Sasorith S, Klein F, Robert F, Weber C, Miguet L, Potier N, Van-Dorsselaer A, Wurtz JM, Mandel JL, Tora L, Devys D (2004) Ataxin-7 is a subunit of GCN5 histone acetyltransferase-containing complexes. Hum Mol Genet 13:1257–1265

Hirsch EC, Breidert T, Rousselet E, Hunot S, Hartmann A, Michel PP (2003) The role of glial reaction and inflammation in Parkinson's disease. Ann NY Acad Sci 991:214–228

Hockly E, Richon VM, Woodman B, Smith DL, Zhou X, Rosa E, Sathasivam K, Ghazi-Noori S, Mahal A, Lowden PA, Steffan JS, Marsh JL, Thompson LM, Lewis CM, Marks PA, Bates GP (2003) Suberoylanilide hydroxamic acid, a histone deacetylase inhibitor, ameliorates motor deficits in a mouse model of Huntington's disease. Proc Natl Acad Sci USA 100:2041–2046

Hodgson JG, Agopyan N, Gutekunst CA, Leavitt BR, LePiane F, Singaraja R, Smith DJ, Bissada N, McCutcheon K, Nasir J, Jamot L, Li XJ, Stevens ME, Rosemond E, Roder JC, Phillips AG, Rubin EM, Hersch SM, Hayden MR (1999) A YAC mouse model for Huntington's disease with full-length mutant huntingtin, cytoplasmic toxicity, and selective striatal neurodegeneration. Neuron 23:181–192

Holbert S, Denghien I, Kiechle T, Rosenblatt A, Wellington C, Hayden MR, Margolis RL, Ross CA, Dausset J, Ferrante RJ, Neri C (2001) The Gln-Ala repeat transcriptional activator CA150 interacts with huntingtin: neuropathologic and genetic evidence for a role in Huntington's disease pathogenesis. Proc Natl Acad Sci USA 98:1811–1816

Huang CC, Faber PW, Persichetti F, Mittal V, Vonsattel JP, MacDonald ME, Gusella JF (1998) Amyloid formation by mutant huntingtin: threshold, progressivity and recruitment of normal polyglutamine proteins. Somat Cell Mol Genet 24:217–233

Huntington's Disease Collaborative Research Group (1993) A novel gene containing a trinucleotide repeat that is expanded and unstable on Huntington's disease chromosomes. Cell 72:971–983

Huynh DP, Del Bigio MR, Ho DH, Pulst SM (1999) Expression of ataxin-2 in brains from normal individuals and patients with Alzheimer's disease and spinocerebellar ataxia 2. Ann Neurol 45:232–241

Huynh DP, Figueroa K, Hoang N, Pulst SM (2000) Nuclear localization or inclusion body formation of ataxin-2 are not necessary for SCA2 pathogenesis in mouse or human. Nat Genet 26:44–50

Ikeuchi T, Igarashi S, Takiyama Y, Onodera O, Oyake M, Takano H, Koide R, Tanaka H, Tsuji S (1996) Non-Mendelian transmission in dentatorubral-pallidoluysian atrophy and Machado-Joseph disease: the mutant allele is preferentially transmitted in male meiosis. Am J Hum Genet 58:730–733

Ishikawa K, Fujigasaki H, Saegusa H, Ohwada K, Fujita T, Iwamoto H, Komatsuzaki Y, Toru S, Toriyama H, Watanabe M, Ohkoshi N, Shoji S, Kanazawa I, Tanabe T, Mizusawa H (1999) Abundant expression and cytoplasmic aggregations of [alpha]1A voltage-dependent calcium channel protein associated with neurodegeneration in spinocerebellar ataxia type 6. Hum Mol Genet 8:1185–1193

Ishikawa K, Owada K, Ishida K, Fujigasaki H, Shun Li M, Tsunemi T, Ohkoshi N, Toru S, Mizutani T, Hayashi M, Arai N, Hasegawa K, Kawanami T, Kato T, Makifuchi T, Shoji S, Tanabe T, Mizusawa H (2001) Cytoplasmic and nuclear polyglutamine aggregates in SCA6 Purkinje cells. Neurology 56:1753–1756

Jana NR, Tanaka M, Wang G, Nukina N (2000) Polyglutamine length-dependent interaction of Hsp40 and Hsp70 family chaperones with truncated N-terminal huntingtin: their role in suppression of aggregation and cellular toxicity. Hum Mol Genet 9:2009–2018

Jana NR, Zemskov EA, Wang G, Nukina N (2001) Altered proteasomal function due to the expression of polyglutamine-expanded truncated N-terminal huntingtin induces apoptosis by caspase activation through mitochondrial cytochrome c release. Hum Mol Genet 10:1049–1059

Johannessen M, Delghandi MP, Moens U (2004) What turns CREB on? Cell Signalling 16:1211–1227

Jonasson J, Strom AL, Hart P, Brannstrom T, Forsgren L, Holmberg M (2002) Expression of ataxin-7 in CNS and non-CNS tissue of normal and SCA7 individuals. Acta Neuropathol 104:29–37

Karlin S, Burge C (1996) Trinucleotide repeats and long homopeptides in genes and proteins associated with nervous system disease and development. Proc Natl Acad Sci USA 93:1560–1565

Kato T, Tanaka F, Yamamoto M, Yosida E, Indo T, Watanabe H, Yoshiwara T, Doyu M, Sobue G (2000) Sisters homozygous for the spinocerebellar ataxia type 6 (SCA6)/CACNA1A gene associated with different clinical phenotypes. Clin Genet 58:69–73

Katsuno M, Adachi H, Kume A, Li M, Nakagomi Y, Niwa H, Sang C, Kobayashi Y, Doyu M, Sobue G (2002) Testosterone reduction prevents phenotypic expression in a transgenic mouse model of spinal and bulbar muscular atrophy. Neuron 35:843–854

Kazemi-Esfarjani P, Benzer S (2000) Genetic suppression of polyglutamine toxicity in Drosophila. Science 287:1837–1840

Kegel KB, Meloni AR, Yi Y, Kim YJ, Doyle E, Cuiffo BG, Sapp E, Wang Y, Qin ZH, Chen JD, Nevins JR, Aronin N, DiFiglia M (2002) Huntingtin is present in the nucleus, interacts with the transcriptional corepressor C-terminal binding protein, and represses transcription. J Biol Chem 277:7466–7476

Kim S, Nollen EA, Kitagawa K, Bindokas VP, Morimoto RI (2002) Polyglutamine protein aggregates are dynamic. Nat Cell Biol 4:826–831

Kim YJ, Yi Y, Sapp E, Wang Y, Cuiffo B, Kegel KB, Qin ZH, Aronin N, DiFiglia M (2001) Caspase 3-cleaved N-terminal fragments of wild-type and mutant huntingtin are present in normal and Huntington's disease brains, associate with membranes, and undergo calpain-dependent proteolysis. Proc Natl Acad Sci USA 98:12784–12789

Kobayashi Y, Kume A, Li M, Doyu M, Hata M, Ohtsuka K, Sobue G (2000) Chaperones Hsp70 and Hsp40 suppress aggregate formation and apoptosis in cultured neuronal cells expressing truncated androgen receptor protein with expanded polyglutamine tract. J Biol Chem 275:8772–8778

Kretzschmar D, Tschape J, Bettencourt Da Cruz A, Asan E, Poeck B, Strauss R, Pflugfelder GO (2005) Glial and neuronal expression of polyglutamine proteins induce behavioral changes and aggregate formation in Drosophila. Glia 49:59–72

Kuemmerle S, Gutekunst CA, Klein AM, Li XJ, Li SH, Beal MF, Hersch SM, Ferrante RJ (1999) Huntington aggregates may not predict neuronal death in Huntington's disease. Ann Neurol 46:842–849

La Spada AR, Fu YH, Sopher BL, Libby RT, Wang X, Li LY, Einum DD, Huang J, Possin DE, Smith AC, Martinez RA, Koszdin KL, Treuting PM, Ware CB, Hurley JB, Ptacek LJ, Chen S (2001) Polyglutamine-expanded ataxin-7 antagonizes CRX function and induces cone-rod dystrophy in a mouse model of SCA7. Neuron 31:913–927

La Spada AR, Taylor JP (2003) Polyglutamines placed into context. Neuron 38:681–684

La Spada AR, Wilson EM, Lubahn DB, Harding AE, Fischbeck KH (1991) Androgen receptor gene mutations in X-linked spinal and bulbar muscular atrophy. Nature 352:77–79

Lang AE, Rogaeva EA, Tsuda T, Hutterer J, St George-Hyslop P (1994) Homozygous inheritance of the Machado-Joseph disease gene. Ann Neurol 36:443–447

Lee WC, Yoshihara M, Littleton JT (2004) Cytoplasmic aggregates trap polyglutamine-containing proteins and block axonal transport in a Drosophila model of Huntington's disease. Proc Natl Acad Sci USA 101:3224–3229

Leeflang EP, Zhang L, Tavare S, Hubert R, Srinidhi J, MacDonald ME, Myers RH, de Young M, Wexler NS, Gusella JF et al (1995) Single sperm analysis of the trinucleotide repeats in the Huntington's disease gene: quantification of the mutation frequency spectrum. Hum Mol Genet 4:1519–1526

Lerer I, Merims D, Abeliovich D, Zlotogora J, Gadoth N (1996) Machado-Joseph disease: correlation between the clinical features, the CAG repeat length and homozygosity for the mutation. Eur J Hum Genet 4:3–7

Li F, Macfarlan T, Pittman RN, Chakravarti D (2002) Ataxin-3 is a histone-binding protein with two independent transcriptional corepressor activities. J Biol Chem 277:45004–45012

Li H, Li SH, Cheng AL, Mangiarini L, Bates GP, Li XJ (1999) Ultrastructural localization and progressive formation of neuropil aggregates in Huntington's disease transgenic mice. Hum Mol Genet 8:1227–1236

Li H, Li SH, Johnston H, Shelbourne PF, Li XJ (2000) Amino-terminal fragments of mutant huntingtin show selective accumulation in striatal neurons and synaptic toxicity. Nat Genet 25:385–389

Li H, Li SH, Yu ZX, Shelbourne P, Li XJ (2001) Huntingtin aggregate-associated axonal degeneration is an early pathological event in Huntington's disease mice. J Neurosci 21:8473–8481

Li JL, Hayden MR, Almqvist EW, Brinkman RR, Durr A, Dode C, Morrison PJ, Suchowersky O, Ross CA, Margolis RL, Rosenblatt A, Gomez-Tortosa E, Cabrero DM, Novelletto A, Frontali M, Nance M, Trent RJ, McCusker E, Jones R, Paulsen JS, Harrison M, Zanko A, Abramson RK, Russ AL, Knowlton B, Djousse L, Mysore JS, Tariot S, Gusella MF, Wheeler VC, Atwood LD, Cupples LA, Saint-Hilaire M, Cha JH, Hersch SM, Koroshetz WJ, Gusella JF, MacDonald ME, Myers RH (2003a) A genome scan for modifiers of age at onset in Huntington disease: the HD MAPS study. Am J Hum Genet 73:682–687

Li JY, Plomann M, Brundin P (2003b) Huntington's disease: a synaptopathy? Trends Mol Med 9:414–420

Li M, Miwa S, Kobayashi Y, Merry DE, Yamamoto M, Tanaka F, Doyu M, Hashizume Y, Fischbeck KH, Sobue G (1998) Nuclear inclusions of the androgen receptor protein in spinal and bulbar muscular atrophy. Ann Neurol 44:249–254

Li SH, Li XJ (2004a) Huntingtin-protein interactions and the pathogenesis of Huntington's disease. Trends Genet 20:146–154

Li SH, Li XJ (2004b) Huntingtin and its role in neuronal degeneration. Neuroscientist 10:467–475

Li SH, Cheng AL, Zhou H, Lam S, Rao M, Li H, Li XJ (2002) Interaction of Huntington disease protein with transcriptional activator Sp1. Mol Cell Biol 22:1277–1287

Lieberman AP, Harmison G, Strand AD, Olson JM, Fischbeck KH (2002) Altered transcriptional regulation in cells expressing the expanded polyglutamine androgen receptor. Hum Mol Genet 11:1967–1976

Lievens JC, Rival T, Iche M, Chneiweiss H, Birman S (2005) Expanded polyglutamine peptides disrupt EGF receptor signaling and glutamate transporter expression in Drosophila. Hum Mol Genet 14:713–724

Lin X, Antalffy B, Kang D, Orr HT, Zoghbi HY (2000) Polyglutamine expansion downregulates specific neuronal genes before pathologic changes in SCA1. Nat Neurosci 3:157–163

Ludolph AC, Seelig M, Ludolph AG, Sabri MI, Spencer PS (1992) ATP deficits and neuronal degeneration induced by 3-nitropropionic acid. Ann NY Acad Sci 648:300–302

Lunkes A, Lindenberg KS, Ben-Haiem L, Weber C, Devys D, Landwehrmeyer GB, Mandel JL, Trottier Y (2002) Proteases acting on mutant huntingtin generate cleaved products that differentially build up cytoplasmic and nuclear inclusions. Mol Cell 10:259–269

Luthi-Carter R, Strand A, Peters NL, Solano SM, Hollingsworth ZR, Menon AS, Frey AS, Spektor BS, Penney EB, Schilling G, Ross CA, Borchelt DR, Tapscott SJ, Young AB, Cha JH, Olson JM (2000) Decreased expression of striatal signaling genes in a mouse model of Huntington's disease. Hum Mol Genet 9:1259–1271

Luthi-Carter R, Strand AD, Hanson SA, Kooperberg C, Schilling G, La Spada AR, Merry DE, Young AB, Ross CA, Borchelt DR, Olson JM (2002) Polyglutamine and transcription: gene expression changes shared by DRPLA and Huntington's disease mouse models reveal context-independent effects. Hum Mol Genet 11:1927–1937

Maltecca F, Filla A, Castaldo I, Coppola G, Fragassi NA, Carella M, Bruni A, Cocozza S, Casari G, Servadio A, De Michele G (2003) Intergenerational instability and marked anticipation in SCA-17. Neurology 61:1441–1443

Mangiarini L, Sathasivam K, Seller M, Cozens B, Harper A, Hetherington C, Lawton M, Trottier Y, Lehrach H, Davies SW, Bates GP (1996) Exon 1 of the HD gene with an expanded CAG repeat is sufficient to cause a progressive neurological phenotype in transgenic mice. Cell 87:493–506

Mantamadiotis T, Lemberger T, Bleckmann SC, Kern H, Kretz O, Martin Villalba A, Tronche F, Kellendonk C, Gau D, Kapfhammer J, Otto C, Schmid W, Schutz G (2002) Disruption of CREB function in brain leads to neurodegeneration. Nat Genet 31:47–54

Martianov I, Viville S, Davidson I (2002) RNA polymerase II transcription in murine cells lacking the TATA binding protein. Science 298:1036–1039

McCampbell A, Taye AA, Whitty L, Penney E, Steffan JS, Fischbeck KH (2001) Histone deacetylase inhibitors reduce polyglutamine toxicity. Proc Natl Acad Sci USA 98:15179–15184

McCampbell A, Taylor JP, Taye AA, Robitschek J, Li M, Walcott J, Merry D, Chai Y, Paulson H, Sobue G, Fischbeck KH (2000) CREB-binding protein sequestration by expanded polyglutamine. Hum Mol Genet 9:2197–2202

Michalik A, Van Broeckhoven C (2003) Pathogenesis of polyglutamine disorders: aggregation revisited. Hum Mol Genet: R173–186

Minamiyama M, Katsuno M, Adachi H, Waza M, Sang C, Kobayashi Y, Tanaka F, Doyu M, Inukai A, Sobue G (2004) Sodium butyrate ameliorates phenotypic expression in a transgenic mouse model of spinal and bulbar muscular atrophy. Hum Mol Genet 13:1183–1192

Morfini G, Pigino G, Brady ST (2005) Polyglutamine expansion diseases: failing to deliver. Trends Mol Med 11:64–70

Mrak RE, Griffin WS (2005) Glia and their cytokines in progression of neurodegeneration. Neurobiol Aging 26:349–354

Muller F, Tora L (2004) The multicoloured world of promoter recognition complexes. EMBO J 23:2–8

Murai K, Naruse Y, Shaul Y, Agata Y, Mori N (2004) Direct interaction of NRSF with TBP: chromatin reorganization and core promoter repression for neuron-specific gene transcription. Nucleic Acids Res 32:3180–3189

Nucifora FC Jr, Ellerby LM, Wellington CL, Wood JD, Herring WJ, Sawa A, Hayden MR, Dawson VL, Dawson TM, Ross CA (2003) Nuclear localization of a non-caspase truncation product of atrophin-1, with an expanded polyglutamine repeat, increases cellular toxicity. J Biol Chem 278:13047–13055

Obrietan K, Hoyt KR (2004) CRE-mediated transcription is increased in Huntington's disease transgenic mice. J Neurosci 24:791–796

Ordway JM, Tallaksen-Greene S, Gutekunst CA, Bernstein EM, Cearley JA, Wiener HW, Dure LSt, Lindsey R, Hersch SM, Jope RS, Albin RL, Detloff PJ (1997) Ectopically expressed CAG repeats cause intranuclear inclusions and a progressive late onset neurological phenotype in the mouse. Cell 91:753–763

Panov AV, Gutekunst CA, Leavitt BR, Hayden MR, Burke JR, Strittmatter WJ, Greenamyre JT (2002) Early mitochondrial calcium defects in Huntington's disease are a direct effect of polyglutamines. Nat Neurosci 5:731–736

Parker JA, Arango M, Abderrahmane S, Lambert E, Tourette C, Catoire H, Neri C (2005) Resveratrol rescues mutant polyglutamine cytotoxicity in nematode and mammalian neurons. Nat Genet 37:349–350

Perutz MF, Johnson T, Suzuki M, Finch JT (1994) Glutamine repeats as polar zippers: their possible role in inherited neurodegenerative diseases. Proc Natl Acad Sci USA 91:5355–5358

Piccioni F, Pinton P, Simeoni S, Pozzi P, Fascio U, Vismara G, Martini L, Rizzuto R, Poletti A (2002) Androgen receptor with elongated polyglutamine tract forms aggregates that alter axonal trafficking and mitochondrial distribution in motor neuronal processes. FASEB J 16:1418–1420

Poirier MA, Jiang H, Ross CA (2005) A structure-based analysis of huntingtin mutant polyglutamine aggregation and toxicity: evidence for a compact beta-sheet structure. Hum Mol Genet 14:765–774

Pollitt SK, Pallos J, Shao J, Desai UA, Ma AA, Thompson LM, Marsh JL, Diamond MI (2003) A rapid cellular FRET assay of polyglutamine aggregation identifies a novel inhibitor. Neuron 40:685–694

Ralser M, Albrecht M, Nonhoff U, Lengauer T, Lehrach H, Krobitsch S (2005) An integrative approach to gain insights into the cellular function of human ataxin-2. J Mol Biol 346:203–214

Reid SJ, Rees MI, van Roon-Mom WM, Jones AL, MacDonald ME, Sutherland G, During MJ, Faull RL, Owen MJ, Dragunow M, Snell RG (2003) Molecular investigation of TBP allele length: a SCA17 cellular model and population study. Neurobiol Dis 13:37–45

Rubinsztein DC (2002) Lessons from animal models of Huntington's disease. Trends Genet 18:202–209

Rubinsztein DC, Leggo J, Chiano M, Dodge A, Norbury G, Rosser E, Craufurd D (1997) Genotypes at the GluR6 kainate receptor locus are associated with variation in the age of onset of Huntington disease. Proc Natl Acad Sci USA 94:3872–3876

Saluja D, Vassallo MF, Tanese N (1998) Distinct subdomains of human TAFII130 are required for interactions with glutamine-rich transcriptional activators. Mol Cell Biol 18:5734–5743

Sanchez I, Mahlke C, Yuan J (2003) Pivotal role of oligomerization in expanded polyglutamine neurodegenerative disorders. Nature 421:373–379

Sapp E, Penney J, Young A, Aronin N, Vonsattel JP, DiFiglia M (1999) Axonal transport of N-terminal huntingtin suggests early pathology of corticostriatal projections in Huntington disease. J Neuropathol Exp Neurol 58:165–173

Sato K, Kashihara K, Okada S, Ikeuchi T, Tsuji S, Shomori T, Morimoto K, Hayabara T (1995) Does homozygosity advance the onset of dentatorubral-pallidoluysian atrophy? Neurology 45:1934–1936

Satyal SH, Schmidt E, Kitagawa K, Sondheimer N, Lindquist S, Kramer JM, Morimoto RI (2000) Polyglutamine aggregates alter protein folding homeostasis in Caenorhabditis elegans. Proc Natl Acad Sci USA 97:5750–5755

Schaffar G, Breuer P, Boteva R, Behrends C, Tzvetkov N, Strippel N, Sakahira H, Siegers K, Hayer-Hartl M, Hartl FU (2004) Cellular toxicity of polyglutamine expansion proteins: mechanism of transcription factor deactivation. Mol Cell 15:95–105

Scherzinger E, Lurz R, Turmaine M, Mangiarini L, Hollenbach B, Hasenbank R, Bates GP, Davies SW, Lehrach H, Wanker EE (1997) Huntingtin-encoded polyglutamine expansions form amyloid-like protein aggregates in vitro and in vivo. Cell 90:549–558

Scherzinger E, Sittler A, Schweiger K, Heiser V, Lurz R, Hasenbank R, Bates GP, Lehrach H, Wanker EE (1999) Self-assembly of polyglutamine-containing huntingtin fragments into amyloid-like fibrils: implications for Huntington's disease pathology. Proc Natl Acad Sci USA 96:4604–4609

Schilling G, Becher MW, Sharp AH, Jinnah HA, Duan K, Kotzuk JA, Slunt HH, Ratovitski T, Cooper JK, Jenkins NA, Copeland NG, Price DL, Ross CA, Borchelt DR (1999a) Intranuclear inclusions and neuritic aggregates in transgenic mice expressing a mutant N-terminal fragment of huntingtin. Hum Mol Genet 8:397–407

Schilling G, Wood JD, Duan K, Slunt HH, Gonzales V, Yamada M, Cooper JK, Margolis RL, Jenkins NA, Copeland NG, Takahashi H, Tsuji S, Price DL, Borchelt DR, Ross CA (1999b) Nuclear accumulation of truncated atrophin-1 fragments in a transgenic mouse model of DRPLA. Neuron 24:275–286

Schilling G, Jinnah HA, Gonzales V, Coonfield ML, Kim Y, Wood JD, Price DL, Li XJ, Jenkins N, Copeland N, Moran T, Ross CA, Borchelt DR (2001) Distinct behavioral and neuropathological abnormalities in transgenic mouse models of HD and DRPLA. Neurobiol Disord 8:405–418

Schmitz ML, Stelzer G, Altmann H, Meisterernst M, Baeuerle PA (1995) Interaction of the COOH-terminal transactivation domain of p65 NF-kappa B with TATA-binding protein, transcription factor IIB, and coactivators. J Biol Chem 270:7219–7226

Serra HG, Byam CE, Lande JD, Tousey SK, Zoghbi HY, Orr HT (2004) Gene profiling links SCA1 pathophysiology to glutamate signaling in Purkinje cells of transgenic mice. Hum Mol Genet 13:2535–2543

Shimohata T, Nakajima T, Yamada M, Uchida C, Onodera O, Naruse S, Kimura T, Koide R, Nozaki K, Sano Y, Ishiguro H, Sakoe K, Ooshima T, Sato A, Ikeuchi T, Oyake M, Sato T, Aoyagi Y, Hozumi I, Nagatsu T, Takiyama Y, Nishizawa M, Goto J, Kanazawa I, Davidson I, Tanese N, Takahashi H, Tsuji S (2000) Expanded polyglutamine stretches interact with TAFII130, interfering with CREB-dependent transcription. Nat Genet 26:29–36

Singhrao SK, Thomas P, Wood JD, MacMillan JC, Neal JW, Harper PS, Jones AL (1998) Huntingtin protein colocalizes with lesions of neurodegenerative diseases: An investigation in Huntington's, Alzheimer's, and Pick's diseases. Exp Neurol 150:213–222

Slow EJ, van Raamsdonk J, Rogers D, Coleman SH, Graham RK, Deng Y, Oh R, Bissada N, Hossain SM, Yang YZ, Li XJ, Simpson EM, Gutekunst CA, Leavitt BR, Hayden MR (2003) Selective striatal neuronal loss in a YAC128 mouse model of Huntington disease. Hum Mol Genet 12:1555–1567

Smith DL, Portier R, Woodman B, Hockly E, Mahal A, Klunk WE, Li XJ, Wanker E, Murray KD, Bates GP (2001) Inhibition of polyglutamine aggregation in R6/2 HD brain slices-complex dose-response profiles. Neurobiol Disord 8:1017–1026

Squitieri F, Gellera C, Cannella M, Mariotti C, Cislaghi G, Rubinsztein DC, Almqvist EW, Turner D, Bachoud-Levi AC, Simpson SA, Delatycki M, Maglione V, Hayden MR, Donato SD (2003) Homozygosity for CAG mutation in Huntington disease is associated with a more severe clinical course. Brain 126:946–955

Steffan JS, Kazantsev A, Spasic-Boskovic O, Greenwald M, Zhu YZ, Gohler H, Wanker EE, Bates GP, Housman DE, Thompson LM (2000) The Huntington's disease protein interacts with p53 and CREB-binding protein and represses transcription. Proc Natl Acad Sci USA 97:6763–6768

Steffan JS, Bodai L, Pallos J, Poelman M, McCampbell A, Apostol BL, Kazantsev A, Schmidt E, Zhu YZ, Greenwald M, Kurokawa R, Housman DE, Jackson GR, Marsh JL, Thompson LM (2001) Histone deacetylase inhibitors arrest polyglutamine-dependent neurodegeneration in Drosophila. Nature 413:739–743

Stenoien DL, Cummings CJ, Adams HP, Mancini MG, Patel K, DeMartino GN, Marcelli M, Weigel NL, Mancini MA (1999) Polyglutamine-expanded androgen receptors form aggregates that sequester heat shock proteins, proteasome components and SRC-1, and are suppressed by the HDJ-2 chaperone. Hum Mol Genet 8:731–741

Stenoien DL, Mielke M, Mancini MA (2002) Intranuclear ataxin1 inclusions contain both fast- and slow-exchanging components. Nat Cell Biol 4:806–810

Sugars KL, Rubinsztein DC (2003) Transcriptional abnormalities in Huntington disease. Trends Genet 19:233–238

Suhr ST, Senut MC, Whitelegge JP, Faull KF, Cuizon DB, Gage FH (2001) Identities of sequestered proteins in aggregates from cells with induced polyglutamine expression. J Cell Biol 153:283–294

Swope DL, Mueller CL, Chrivia JC (1996) CREB-binding protein activates transcription through multiple domains. J Biol Chem 271:28138–28145

Szebenyi G, Morfini GA, Babcock A, Gould M, Selkoe K, Stenoien DL, Young M, Faber PW, MacDonald ME, McPhaul MJ, Brady ST (2003) Neuropathogenic forms of huntingtin and androgen receptor inhibit fast axonal transport. Neuron 40:41–52

Tabrizi SJ, Workman J, Hart PE, Mangiarini L, Mahal A, Bates G, Cooper JM, Schapira AH (2000) Mitochondrial dysfunction and free radical damage in the Huntington R6/2 transgenic mouse. Ann Neurol 47:80–86

Takano H, Gusella JF (2002) The predominantly HEAT-like motif structure of huntingtin and its association and coincident nuclear entry with dorsal, an NF-kB/Rel/dorsal family transcription factor. BMC Neurosci 3:15

Takiyama Y, Sakoe K, Amaike M, Soutome M, Ogawa T, Nakano I, Nishizawa M (1999) Single sperm analysis of the CAG repeats in the gene for dentatorubral-pallidoluysian atrophy (DRPLA): the instability of the CAG repeats in the DRPLA gene is prominent among the CAG repeat diseases. Hum Mol Genet 8:453–457

Takiyama Y, Sakoe K, Soutome M, Namekawa M, Ogawa T, Nakano I, Igarashi S, Oyake M, Tanaka H, Tsuji S, Nishizawa M (1997) Single sperm analysis of the CAG repeats in the gene for Machado-Joseph disease (MJD1): evidence for non-Mendelian transmission of the MJD1 gene and for the effect of the intragenic CGG/GGG polymorphism on the intergenerational instability. Hum Mol Genet 6:1063–1068

Tanaka M, Machida Y, Niu S, Ikeda T, Jana NR, Doi H, Kurosawa M, Nekooki M, Nukina N (2004) Trehalose alleviates polyglutamine-mediated pathology in a mouse model of Huntington disease. Nat Med 10:148–154

Toyoshima Y, Yamada M, Onodera O, Shimohata M, Inenaga C, Fujita N, Morita M, Tsuji S, Takahashi H (2004) SCA17 homozygote showing Huntington's disease-like phenotype. Ann Neurol 55:281–286

Truant R, Xiao H, Ingles CJ, Greenblatt J (1993) Direct interaction between the transcriptional activation domain of human p53 and the TATA box-binding protein. J Biol Chem 268:2284–2287

Trushina E, Dyer RB, Badger JD 2nd, Ure D, Eide L, Tran DD, Vrieze BT, Legendre-Guillemin V, McPherson PS, Mandavilli BS, Van Houten B, Zeitlin S, McNiven M, Aebersold R, Hayden M, Parisi JE, Seeberg E, Dragatsis I, Doyle K, Bender A, Chacko C, McMurray CT (2004) Mutant huntingtin impairs axonal trafficking in mammalian neurons in vivo and in vitro. Mol Cell Biol 24:8195–8209

Tsai CC, Kao HY, Mitzutani A, Banayo E, Rajan H, McKeown M, Evans RM (2004) Ataxin 1, a SCA1 neurodegenerative disorder protein, is functionally linked to the silencing mediator of retinoid and thyroid hormone receptors. Proc Natl Acad Sci USA 101:4047–4052

Usdin MT, Shelbourne PF, Myers RM, Madison DV (1999) Impaired synaptic plasticity in mice carrying the Huntington's disease mutation. Hum Mol Genet 8:839–846

Warrick JM, Chan HY, Gray-Board GL, Chai Y, Paulson HL, Bonini NM (1999) Suppression of polyglutamine-mediated neurodegeneration in Drosophila by the molecular chaperone HSP70. Nat Genet 23:425–428

Wellington CL, Ellerby LM, Gutekunst CA, Rogers D, Warby S, Graham RK, Loubser O, van Raamsdonk J, Singaraja R, Yang YZ, Gafni J, Bredesen D, Hersch SM, Leavitt BR, Roy S, Nicholson DW, Hayden MR (2002) Caspase cleavage of mutant huntingtin precedes neurodegeneration in Huntington's disease. J Neurosci 22:7862–7872

Wellington CL, Ellerby LM, Hackam AS, Margolis RL, Trifiro MA, Singaraja R, McCutcheon K, Salvesen GS, Propp SS, Bromm M, Rowland KJ, Zhang T, Rasper D, Roy S, Thornberry N, Pinsky L, Kakizuka A, Ross CA, Nicholson DW, Bredesen DE, Hayden MR (1998) Caspase cleavage of gene products associated with triplet expansion disorders generates truncated fragments containing the polyglutamine tract. J Biol Chem 273:9158–9167

Wexler NS, Young AB, Tanzi RE, Travers H, Starosta-Rubinstein S, Penney JB, Snodgrass SR, Shoulson I, Gomez F, Ramos Arroyo MA et al (1987) Homozygotes for Huntington's disease. Nature 326:194–197

Wood JD, Nucifora FC Jr, Duan K, Zhang C, Wang J, Kim Y, Schilling G, Sacchi N, Liu JM, Ross CA (2000) Atrophin-1, the dentato-rubral and pallido-luysian atrophy gene product, interacts with ETO/MTG8 in the nuclear matrix and represses transcription. J Cell Biol 150:939–948

Wyttenbach A, Carmichael J, Swartz J, Furlong RA, Narain Y, Rankin J, Rubinsztein DC (2000) Effects of heat shock, heat shock protein 40 (HDJ-2), and proteasome inhibition on protein aggregation in cellular models of Huntington's disease. Proc Natl Acad Sci USA 97:2898–2903

Wyttenbach A, Swartz J, Kita H, Thykjaer T, Carmichael J, Bradley J, Brown R, Maxwell M, Schapira A, Orntoft TF, Kato K, Rubinsztein DC (2001) Polyglutamine expansions cause decreased CRE-mediated transcription and early gene expression changes prior to cell death in an inducible cell model of Huntington's disease. Hum Mol Genet 10:1829–1845

Yamada M, Hayashi S, Tsuji S, Takahashi H (2001) Involvement of the cerebral cortex and autonomic ganglia in Machado-Joseph disease. Acta Neuropathol (Berl) 101:140–144

Yamamoto A, Lucas JJ, Hen R (2000) Reversal of neuropathology and motor dysfunction in a conditional model of Huntington's disease. Cell 101:57–66

Yue S, Serra HG, Zoghbi HY, Orr HT (2001) The spinocerebellar ataxia type 1 protein, ataxin-1, has RNA-binding activity that is inversely affected by the length of its polyglutamine tract. Hum Mol Genet 10:25–30

Yvert G, Lindenberg KS, Devys D, Helmlinger D, Landwehrmeyer GB, Mandel JL (2001) SCA7 mouse models show selective stabilization of mutant ataxin-7 and similar cellular responses in different neuronal cell types. Hum Mol Genet 10:1679–1692

Yvert G, Lindenberg KS, Picaud S, Landwehrmeyer GB, Sahel JA, Mandel JL (2000) Expanded polyglutamines induce neurodegeneration and trans-neuronal alterations in cerebellum and retina of SCA7 transgenic mice. Hum Mol Genet 9:2491–2506

Zhang L, Fischbeck KH, Arnheim N (1995) CAG repeat length variation in sperm from a patient with Kennedy's disease. Hum Mol Genet 4:303–305

Zhang X, Smith DL, Meriin AB, Engemann S, Russel DE, Roark M, Washington SL, Maxwell MM, Marsh JL, Thompson LM, Wanker EE, Young AB, Housman DE, Bates GP, Sherman MY, Kazantsev AG (2005) A potent small molecule inhibits polyglutamine aggregation in Huntington's disease neurons and suppresses neurodegeneration in vivo. Proc Natl Acad Sci USA 102:892–897

Zhou H, Cao F, Wang Z, Yu ZX, Nguyen HP, Evans J, Li SH, Li XJ (2003) Huntingtin forms toxic NH2-terminal fragment complexes that are promoted by the age-dependent decrease in proteasome activity. J Cell Biol 163:109–118

Zoghbi HY, Orr HT (2000) Glutamine repeats and neurodegeneration. Annu Rev Neurosci 23:217–247

Zuccato C, Tartari M, Crotti A, Goffredo D, Valenza M, Conti L, Cataudella T, Leavitt BR, Hayden MR, Timmusk T, Rigamonti D, Cattaneo E (2003) Huntingtin interacts with REST/NRSF to modulate the transcription of NRSE-controlled neuronal genes. Nat Genet 35:76–83

Zuhlke C, Gehlken U, Hellenbroich Y, Schwinger E, Burk K (2003a) Phenotypical variability of expanded alleles in the TATA-binding protein gene. Reduced penetrance in SCA17? J Neurol 250:161–163

Zuhlke CH, Spranger M, Spranger S, Voigt R, Lanz M, Gehlken U, Hinrichs F, Schwinger E (2003b) SCA17 caused by homozygous repeat expansion in TBP due to partial isodisomy 6. Eur J Hum Genet 11:629–632

Nucleic Acids and Molecular Biology, Vol. 19
M. Fry/K. Usdin (Eds.)
Human Nucleotide Expansion Disorders

The Enigma of Spinocerebellar Ataxia Type 6

Marina Frontali

Istituto di Neurobiologia e Medicina Molecolare del CNR,
Consiglio Nazionale delle Ricerche, Via Fosso del Cavaliere, 00133 Rome, Italy
marina.frontali@artov.inmm.cnr.it

1 Introduction

The enigma of spinocerebellar ataxia (SCA) type 6 stems from the uncertainties about the disease pathogenesis. SCA6 shares the type of mutation, i.e. an expansion of a polyglutamine repeat, with SCA types 1, 2, 3, 7, and 17, as well as with other non-SCA disorders such as Huntington disease or spinal bulbar muscular atrophy. This suggests a SCA6 pathogenesis based on a toxic gain of function, similar to that of the aforementioned diseases. On the other hand, unlike other SCAs, the gene codes for a well-known protein, forming the main subunit of a calcium channel (Cav2.1) expressed in the brain and particularly in the cerebellum. Point mutations in this gene are responsible for two different channelopathies, episodic ataxia type 2 (EA2) and familial hemiplegic migraine type 1 (FHM1) (Ophoff et al. 1996), the former exhibiting striking similarities with SCA6. EA2 mutations cause a partial or complete loss of the channel function (Guida et al. 2001; Wappl et al. 2002). This raised the alternative hypothesis that polyglutamine expansions might have a role in channel activity, thus assigning SCA6 to the ion channel disorders rather than to the polyglutamine disorders.

2 Genetics

SCA6 is an autosomal dominant disorder due to small expansions of a d(CAG) trinucleotide repeat located in the COOH terminus of the α1A subunit of voltage-gated calcium channels type P/Q (Cav2.1) which is abundantly expressed in cerebellum.

The normal allele size of the polymorphic d(CAG) repeat ranges from four to 18 units (Zhuchenko et al. 1997; Shizuka et al. 1998a, b), while that of expanded alleles is from 20 to 30 repeats (Jodice et al. 1997; Matsuyama et al. 1997; Shizuka et al. 1998a, b; Katayama et al. 2000). Data on the intermediate allele with 19 repeats are contradictory. Both a homozygote and a heterozygote for a 19-repeat allele were found to be affected (Katayama et al. 2000; Mariotti et al. 2001), the latter, however, with unusual clinical features, i.e. autonomic and pyramidal characteristics and lacunar lesions in the pons and basal ganglia. However, several other heterozygotes for a 19-repeat allele were found to be unaffected even at old age (Ishikawa et al. 1997; Mariotti et al. 2001; Takahashi et al. 2004). Intermediate alleles with incomplete penetrance were described in polyglutamine disorders (Rubinzstein et al. 1996). One possible alternative explanation is that the 19-repeat allele exerts a pathogenic effect only when in double dose and the patient reported by Katayama et al. (2000) is not affected by SCA6.

Compared with other d(CAG) expansion disorders, which typically have more than 30 repeats, the size of SCA6 expanded alleles is significantly smaller, overlapping their wild-type allele distribution (Margolis 2003). As expected on the basis of the relatively low number of repeats, allele size is usually stable over successive generations, and no mosaicism is apparent in cells from different parts of the brain (Ishikawa et al. 1999b) or in sperm (Shizuka et al. 1998b). However, some degree of meiotic instability should be assumed since in three families an intergenerational jump of the expanded allele size has been reported (Jodice et al. 1997; Matsuyama et al. 1997; Mariotti et al. 2001).

Homozygous patients for an expansion larger than 19 repeats were reported to slightly differ from heterozygous ones, showing an earlier onset and a more rapid course (Geschwind et al. 1997; Matsumura et al. 1997; Takiyama et al. 1998; Kato et al. 2000; Fukutake et al. 2002); however, Ishikawa et al. (1997) could not detect any difference. An extensive study of the age at onset among SCA6 patients found that in three out of four homozygotes it fell within a 95% confidence interval with that of heterozygous patients (Takahashi et al. 2004).

3 The Cav2.1 α1A Isoforms

The α1A protein, encoded by the *CACNA1A* gene on chromosome 19p13, is the pore-forming subunit of voltage-gated Ca^{2+} channels type P/Q with a specific expression in neurons, in general, and in Purkinje and granule cells, in particular. The channel is part of the neuronal calcium signalling involved in neuron excitability, cell apoptosis and survival, release of neurotransmitters, synaptic plasticity and gene transcription (Pietrobon and Striessnig 2003). It responds to membrane depolarization with an influx of Ca^{2+} ions from the extracellular to the intracellular compartment. Auxiliary subunits, β and $\alpha_2\delta$, interact with α1A to modulate the channel activity. The protein is predicted to have four highly conserved homologous domains, each formed by six transmembrane segments connected through intracellular and extracellular linkers, and two intracytoplasmic, *N*-terminal and *C*-terminal, tails. The latter, containing the polyglutamine repeat in some of the isoforms, is actively involved in the channel activity. By undergoing conformational changes, under different states of the channel, it cooperates with the gating activity, particularly during the inactivation phase (Hering et al. 2000; Kobrinski et al. 2003). In addition, it contains the binding sites for calmodulin and other Ca^{2+}-sensing proteins, involved in the calcium-dependent regulation of the channel (Weiss and Burgoyne 2002), as well as a site for interaction with the β auxiliary subunit (Soong et al. 2002).

Alternative splicing generates several isoforms of the protein (Mori et al. 1991; Zhuchenko et al. 1997; Bourinet et al. 1999; Soong et al. 2002). In particular, a five-nucleotide stretch, d(GGCAG), between exons 46 and 47, is critical for the expression of the d(CAG) repeat stretch: when the five nucleotides are spliced out, a stop codon is encountered upstream to the $d(CAG)_n$ stretch and a short isoform is formed that is devoid of a polyglutamine tract. However, when the five nucleotides are left in place, the d(CAG) repeat is translated into a polyglutamine sequence and the 3′ tail is elongated by 244 exon nucleotides (Zhuchenko et al. 1997). Both long and short protein isoforms have been found in the cerebellar cortex (Ishikawa et al. 1999a). When expressed in cultured non-neuronal cells, the mouse long and short isoforms do not appear to significantly differ at the functional level (Tsunemi et al. 2002). So far, however, the functional characteristics of the different isoforms, the location, timing and regulation of their expression in human neurons are largely unknown.

4 Epidemiology

SCA6 is the second most frequent autosomal dominant cerebellar ataxia (ADCA) in Germany (Riess et al. 1997) and the Netherlands (van de Warrenburg et al. 2002), accounting for about 20% of all ADCAs. By contrast, it is almost absent in Portugal (Silveira et al. 2002) and Spain (Pujana et al. 1999), and has a low relative frequency (about 2%) in Italy and Finland (Brusco et al. 2004; Juvonen et al. 2005). Among non-Caucasian populations it is particularly frequent in Japan and is absent in India (Margolis 2003). Prevalence estimates of SCA6 are few: in northeast England, where SCA6 accounts for about 20% of all ADCAs, the minimum prevalence was found to be 1.59 in 100 000 subjects older than 16 years (about 1 : 60000) and 3.18 in 100 000 (about 1 : 30000) in adults over 45 years of age (Craig et al. 2004).

Sporadic SCA6 expansions have been detected in Caucasian and Japanese populations that have a relatively high frequency of SCA6 (Ikeuchi et al. 1997; Matsumura et al. 1997; Riess et al. 1997; Zhuchenko et al. 1997; Shizuka et al. 1998b). A new mutation, however, has been documented in only one patient (Shizuka et al. 1998a). Should all the reported sporadic cases be new mutations, the mutability of normal alleles would be very high. Takano et al. (1998) have proposed that the relative frequency of several SCAs, including SCA6, is correlated with the frequency of large normal alleles in the Japanese and Caucasian population, implying that a higher frequency is likely to be due to new mutations arising from the reservoir of larger normal alleles. It should be considered, however, that if a mutation/selection equilibrium is assumed, a high frequency of SCA6 de novo mutations would be in disagreement with the, presumably, small or absent selection against a disease such as this one, with a late onset and a long life span.

A possible explanation for the different frequencies of SCA6 in different countries arises from data showing a founder effect in Germany (Dichgans et al. 1999), the Netherlands (van de Warrenburg et al. 2002; Verbeek et al. 2004), northeastern England (Craig et al. 2004) and Japan (Mori et al. 2001; Terasawa et al. 2004). This would imply that the frequency of SCA6 does not depend on the rate of new mutations, but rather on a founder effect. According to this hypothesis the high number of SCA6 sporadic cases is explained either by incomplete penetrance or, more likely, by the presence of a neglected mild expression of the disorder in relatives of probands.

5 Clinical Features

SCA6 was initially reported as a multisystem deficit including cerebellum, brainstem and peripheral systems, similarly to SCA1, SCA2 and SCA3 (Zhuchenko et al. 1997; Geschwind et al. 1997; Stevanin et al. 1997). Later studies, however, described it as pure cerebellar ataxia (Ikeuchi et al. 1997; Ishikawa et al. 1997; Matsumura et al. 1997; Stevanin et al. 1997; Nagai et al. 1998; Watanabe et al. 1998; Garcia-Planells et al. 1999). Still other cases were reported to be preceded by episodes of the same type as in EA2 (Calandriello et al. 1996; Geschwindt et al. 1997; Jodice et al. 1997; Sinke et al. 2001; Koh et al. 2001). The issue of the similarity between the SCA6 and EA2 phenotypes has implications for their underlying pathologic mechanism. In fact, when different mutations cause the same phenotype, it is far more likely that they both lead to a loss-of-protein function rather than to the acquisition of new toxic activity.

Table 1 summarizes the main clinical features observed in 315 SCA6 patients as described in 16 independent studies (Geschwind et al. 1997; Calandriello et al. 1997; Jodice et al. 1997a; Ikeuchi et al. 1997; Ishikawa et al. 1997; Matsumura et al. 1997; Stevanin et al. 1997; Gomez et al. 1997; Jen et al. 1998; Nagai et al. 1998; Satoh et al. 1998; Shizuka et al. 1998b; Takiyama et al. 1998; Watanabe et al. 1998; Garcia-Planells et al. 1999; Kaseda et al. 1999), and compares them with those reported for 138 EA2 patients with ascertained *CACNA1A* point mutations. (Subramony et al. 1996; Yue et al. 1997, 1998; De-

Table 1 Frequency of clinical features in spinocerebellar ataxia type 6 (*SCA6*) and episodic ataxia type 2 (*EA2*) patients as reported in 16 and 12 studies respectively (see text for references)

Signs and symptoms	SCA6 $N = 315$		EA2 $N = 138$ [a]	
	Frequency	Percentage	Frequency	Percentage
Cerebellar signs				
Ataxia, slowly progressive	307/314	98	83/138	60
Nystagmus (often downbeat)	256/309	83	96/129	74
Dysarthria	274/305	90	12/138	9
Cerebellar atrophy	94/97	97	19/37	51
Vestibulo-cerebellar signs (episodic)				
Vertigo/ataxia episodes	53/86	62	112/138	81
Sensitivity to acetazolamide treatment	10/16	62	47/52	90
Extracerebellar signs				
Sensory loss	48/268	18	0/138	0
Pyramidal signs	46/302	16	0/138	0
Brainstem signs	26/150	17	1/138	< 1
Peripheral neuropathy	5/168	3	0/138	0
Cognitive deficit	9/224	4	11/138	8
Brainstem atrophy	6/97	6	0/138	0
Mean (± SE) age at onset of progressive ataxia [b]	46.3±1.2		46.75±4.7	
Mean (± SE) age at onset of episodes [b]			14.5±1.6	
Mean age at examination ± SE [c]	58.5±2.0		39.8±2.8	

SE standard error
[a] Only studies with a detailed phenotype were considered.
[b] Mean age at onset was calculated on the basis of studies that include reports for each patient. Data were available for 82 SCA6 and 39 EA2 patients.
[c] The mean ± SE refers to 251 SCA6 patients and to 32 EA2 patients for whom data were reported in the literature.

nier et al. 1999, 2001; Jen et al. 1999, 2004; Guida et al. 2001; Jouvenceau et al. 2001; Mantuano et al. 2004; Spacey et al. 2005; Imbrici et al. 2004; Kaunisto et al. 2003; Matsuyama et al. 2003; Wan et al. 2004).

Almost all SCA6 patients show a slowly progressive cerebellar involvement with trunk and limb ataxia, hypotonia, dysmetria, dysarthria, gaze-evoked nystagmus with or without a downbeat component, dysmetric saccades, saccadic pursuit, and hyperactive vestibulo-ocular reflex (Gomez et al. 1997). A minority of patients show additional extracerebellar signs, such as decrease in vibration sense, hyperreflexia with or without Babinski sign, dysphagia and/or ophtalmoplegia, brainstem atrophy, as occurs in patients with other

SCAs (Geschwind et al. 1997; Gomez et al. 1997; Ikeuchi et al. 1997; Kaseda et al. 1999; Matsumura et al. 1997; Shizuka et al. 1998b; Takiyama et al. 1998; Watanabe et al. 1998). In some studies the authors explicitly state that these latter features were found more frequently in older subjects (Ikeuchi et al. 1997; Stevanin et al. 1997) or in patients with other coexisting disorders, such as diabetes mellitus (Takiyama et al. 1998). In others, however, no mention was made of the age at which these symptoms were observed. Cognitive performance was found to be reduced in a small percentage of patients (Geschwind et al. 1997; Zhuchenko et al. 1997); however, Globas et al. (2003) did not find any significant cognitive impairment in 12 SCA6 patients as compared with normal controls.

A relatively high percentage of SCA6 patients may initially experience episodes, very similar to those described in EA2 patients, characterized by vertigo, gait and trunk ataxia, dysarthria, accompanied by visual disturbance (such as dyplopia or blurred vision), and tinnitus, lasting from minutes to days and triggered by head movements and physical or emotional stress (Calandriello et al. 1996; Geschwindt et al. 1997; Jodice et al. 1997; Koh et al. 2001; Sinke et al. 2001). These episodes, typically preceding the onset of progressive ataxia, responded to treatment with acetazolamide, a carbonic anhydrase inhibitor widely used to treat EA2 patients, in the few SCA6 patients for whom this therapy has been attempted. This early phase can have a variable duration, usually preceding the onset of progressive ataxia by a few years; however, in some cases the disease never progresses to a full-blown clinical presentation and maintains an episodic character with mild non-progressive interictal cerebellar signs (Calandriello et al. 1996; Jodice et al. 1997; Takiyama et al. 1998; Koh et al. 2001). In other patients the episodes may continue in the progressive phase, exacerbating the cerebellar signs (Yabe et al. 1998).

Age at onset is on average between 40 and 50 years and the life span is normal. As in other polyglutamine disorders, an inverse correlation between age at onset and the number of d(CAG) repeats has been consistently reported (Geschwind et al. 1997; Ikeuchi et al. 1997; Matsuyama et al. 1997; Matsumura 1997; Zhuchenko et al. 1997). Less than 50% of variance is accounted for by the expanded allele size, suggesting a major influence of other factors (van de Warrenburg et al. 2005). Larger normal alleles were reported to contribute to the age at onset (Takahashi et al. 2004; van de Warrenburg et al. 2005). An interaction between wild-type and expanded alleles in determining the age at onset was also suggested for Huntington disease (Djoussé et al. 2003). The anticipation of the age at onset over successive generations that characterizes expanded trinucleotide disorders was also reported in SCA6 families (Ikeuci et al. 1997; Matsumura et al. 1997; Matsuyama et al. 1997; Watanabe et al. 1998; Sinke et al. 2001). In this case, however, anticipation is rather surprising, particularly in studies in which no intergenerational variation of the expanded allele size was found. The phenomenon, hence, should be ascribed to an ascertainment bias, since one is more likely to observe a parent–child pair with the offspring affected earlier than the parent, rather than vice versa (Penrose 1948). Alternative explanations are exposure to different environmental factors affecting the age at onset in different generations, or the difficulty of assessing the age at onset in older patients.

Neuroimaging in SCA6 patients reveals a cerebellar vermis atrophy predominating in the anterior portion, which might later extend to cerebellar hemispheres, usually with preservation of brainstem (Calandriello et al. 1996; Gomez et al. 1997; Jodice et al. 1997; Nagai et al. 1998; Satoh et al. 1998; Shizuka et al. 1998a, b; Takiyama et al. 1998). Occasionally, on MRI a size reduction of the pons as well as of the red nucleus and the middle cerebellar peduncle has been reported (Murata et al. 1998; Arpa et al. 1999; Nakagawa et al. 1999). A widespread reduction of glucose metabolism on PET scan, with particularly low values in the brainstem and cerebellar hemispheres, was observed in a group of SCA6 patients as compared with normal controls (Soong et al. 2001b). Unfortunately, however, in this study the control subjects were younger than the patients, requiring further confirmation of these data.

6 Similar and Discrepant Clinical Features of SCA6 and EA2

As studies of patients with ascertained *CACNA1A* point mutations reveal, the clinical presentation of SCA6 differs in several criteria from that of EA2. A higher percentage of EA2 patients report vertigo/ataxia episodes which are sometimes accompanied by muscular weakness (Jen et al. 2001), epilepsy (Jouvenceau et al. 2001; Imbrici et al. 2004) or dystonia (Guida et al. 2001). The latter features have not been described so far in SCA6 patients. Acetazolamide treatment was found to be efficacious more frequently in treating episodes in EA2 individuals than in SCA6 patients (Yue et al. 1998; Jen et al. 1999; Denier et al. 1999, 2001; Guida et al. 2001; Matsuyama et al. 2003; Kaunisto et al. 2004; Mantuano et al. 2004; Jen et al. 2005; Spacey et al. 2005; Wan et al. 2005). The episodes, however, are not a hallmark of EA2 since a slowly progressive cerebellar ataxia is present in the majority of patients, although at a lower frequency than in SCA6 (Subramony et al. 1996; Yue et al. 1997, 1998; Denier et al. 1999; Jen et al. 1999; Guida et al. 2001; Jouvenceau et al. 2001; Imbrici et al. 2004; Kaunisto et al. 2004, 2005; Mantuano et al. 2004; Wan et al. 2005). Pyramidal, brainstem and peripheral deficits have not been reported in EA2, except for one patient (Denier et al. 1999), while learning difficulties and cognitive signs are present in a small percentage of patients, particularly those carrying a protein-truncating mutation (Mantuano et al. 2004).

The aforementioned differences between EA2 and SCA6 may reflect different pathogenic processes. It should be noted, however, that the age at examination in the two groups (Table 1) is markedly higher in the SCA6 group. This could influence many of the observed differences. An older age at examination may reveal the aspects of a more advanced phase of the disease and hence explain the higher percentage of patients with a progressive ataxia. In addition, in elderly patients other age-related neurological disorders could account for the higher frequency of extracerebellar signs. This hypothesis is consistent with the absence of extracerebellar signs in three studies of SCA6 patients with an age at examination below 70 years (Calandriello et al. 1997; Ishikawa et al. 1997; Satoh et al. 1998), and with the presence of brainstem signs in one EA2 patient over 70 (Denier et al. 1999).

Another discrepant finding in EA2, as compared with SCA6, is the age at onset which is usually reported in the first or second decade of life, i.e. an age markedly lower than that of SCA6 patients. In the latter case, however, the age at onset is defined as the age at onset of progressive ataxia, whereas in EA2 it refers to the age at onset of episodes. In the few EA2 patients for whom age at onset of progressive ataxia is available, the data are strikingly similar to those for SCA6 patients (Subramony et al. 1996; Kaunisto et al. 2004). The difference in the age at onset between the two disorders might, therefore, be less marked than appears at first glance.

The similarity between SCA6 and EA2 is further supported by the finding in the same family, carrying a $d(CAG)_n$ repeat expansion and no CACNA1A point mutation, of members with the typical episodic features of EA2 and others with SCA6 progressive ataxia (Jodice et al. 1997a; Koh et al. 2001).

7 Neuropathology

The brains of SCA6 patients at autopsy show a marked atrophy of the cerebellar vermis and, to a lesser extent, of the hemispheres. Histologically, the cerebellar cortex is characterized by a remarkable loss of Purkinje cells. Granule cells are also affected, although less severely (Subramony et al. 1996; Gomez et al. 1997; Sasaki et al. 1998; Ishikawa et al. 1999b; Tashiro et al. 1999). Loss of neurons in the dentate and inferior olivary nuclei was reported in one study (Subramony et al. 1996) and atrophy of brainstem has been occasionally described (Zhuchenko et al. 1997).

Non-ubiquitinated cytoplasmic protein aggregates were detected by anti-α1A antibodies (Ishikawa et al. 1999a, 2001) both in transfected cells and in cerebella of SCA6 patients. In addition, the cytoplasm and the nucleus of Purkinje cells showed small aggregates immunoreactive with 1C2, an antibody that detects polyglutamine sequences larger than 40 repeats; however, aggregates immunoreactive with the α1A subunit were not reactive with 1C2. The interpretation of this finding is not straightforward. It should be noted that protein aggregates in other polyglutamine disorders are most likely due to the tendency of polyglutamine stretches to form β-sheets by linking to one another through hydrogen bonds between their main chain and the side chain amides (Perutz 1994). This process is strongly dependent on the number of repeat units: no aggregate formation is observed with polyglutamine stretches below 27 repeats (Scherzinger et al. 1999). The structural transition permitting aggregation begins at 32–37 glutamine residues, corresponding to the lower limit of the expansion range seen in patients affected with non-SCA6 polyglutamine expansion disorders. In light of these data, the significance of aggregates in brains of SCA6 patients remains obscure. Are the anti-polyglutamine reactive aggregates due to the SCA6 expansion inducing the aggregation of other proteins with longer polyglutamine tracts, and if so, which are these proteins? Are the aggregates non-specific byproducts of a neurodegenerative process of Purkinje cells caused by a dysfunction of Ca^{2+} channels? On the other hand, the anti-α1A reactive aggregates are hardly comparable with those found in patients with other SCAs, since they are not ubiquitinated and are located only in the cytosol. It should be noted that the α1A subunit has many isoforms, only some of which contain the polyglutamine stretch. Are the patterns of expression of protein isoforms, their likelihood of being incorporated into the membrane or their turnover different from the normal as a result of expansion of the polyglutamine stretch? If this is the case, then these aggregates could be due to the degradation of unused or obsolete isoforms. The increased expression of the α1A subunit with a polyglutamine expansion has indeed been reported (Ishikawa et al. 1999, 2001; Piedras-Renteria et al. 2001). This implies that SCA6 is completely different from other polyglutamine disorders for which a comparable expression level was reported for wild-type and mutated genes (Persichetti et al. 1995; Servadio et al. 1995).

8
Toxic Gain of Function Versus Abnormal Channel Activity

Expanded polyglutamine stretches confer gain of function to nuclear or cytoplasmic proteins that are encoded by genes responsible for polyglutamine disorders. Although the pathogenic role of polyglutamine aggregates is highly controversial (Michalik and van Broekhoven 2003), evidence shows that the nuclear localization of the mutant proteins, or of part of them, plays a pathogenic role (Klement et al. 1998; Sandou et al. 1998; Walsh et al. 2005). The α1A subunit is a cell membrane protein that is unlikely to translocate into the nucleus. It has recently been proposed, however, that its carboxy-terminal polyglutamine-containing tail may undergo a proteolytic cleavage (Kubodera et al. 2003) and recent evidence suggests that this tail might indeed translocate into the nucleus (Agostoni et al. 2005; Kordasiewicz et al. 2006). These data would be consistent with the hypothesis that the tail with small expansions of the polyglutamine repeat exerts a toxic gain of function similar to that of other polyglutamine disorders. In addition, such a hypothesis would require an accurate study of the different α1A isoforms, the regulation of their expression and their functional roles.

Typically for polyglutamine disorders, the expression of the mutated protein in cultured cells leads to cell death (Suzuki and Koike 2005), supporting the hypothesis of a toxic gain of function, whatever the underlying process. Expression of the full-length mutated α1A subunit in cultured cells does not lead to cell death (Matsuyama et al. 1999), but rather it weakens the ability of the Ca^{2+} current through the P/Q channels to prevent cell death caused by other factors, such as serum starvation (Matsuyama et al. 2004). These data appear to indicate that the expansion, by reducing the calcium influx, decreases its protective effect against cell death. Kordasiewicz et al. (2006) reported, instead, an increased cell death when the *C*-terminal tail is expressed and translocated into the nucleus.

A loss of channel function as a basis for SCA6, like that believed to result from EA2-causing mutations (Guida et al. 2001), is also supported by electrophysiological data obtained through patch-clamp recording of cultured non-neuronal cells, usually renal embryonic HEK293 cells, transfected with α1A subunit complementary DNA with different numbers of d(CAG) units. In two studies, the mutated protein induced a hyperpolarizing shift in the voltage dependence of channel inactivation (Matsuyama et al. 1999; Toru et al. 2000). Such a change is predicted to exert a considerable decrease in channel availability at resting potentials, approximately halving the Ca^{2+} influx, which may in turn indirectly lead to cell death. This effect appears to be dependent on the type of isoform and on the number of repeat units (Toru et al. 2000). It is notable that in these experiments the current density in transfected cells was not reduced, implying that the mutated protein was normally transported to the membrane and is not sequestered into aggregates (Matsuyama et al. 1999).

However, other electrophysiological data, (Restituito et al. 2000), obtained under a different experimental setting, suggested an increased calcium influx as a consequence of a hyperpolarizing shift in the voltage-dependent channel activation and a slowed rate of inactivation. This effect was not obtained when the mutated protein had fewer than 30 glutamine units and when an auxiliary β subunit different from β4 was coexpressed. Still different results were obtained by Piedras-Renterìa et al. (2001), who observed a sharp increase in current density in cells expressing the mutated protein. The effect, however, was not dependent on the number of polyglutamine residues. No significant alteration of channel function was observed either in the activation or in the inactivation kinetics. Although confirmation of the functional analysis through single-channel recording is necessary, these data suggest that SCA6 is induced by an abnormal expression or turnover of the mutated protein rather than by a channel malfunction as previously suggested.

Although difficult to reconcile, most electrophysiological data favour the hypothesis of a SCA6 pathogenesis based on a channel malfunctioning rather than on a toxic gain of function of the polyglutamine stretch per se, as envisaged for other d(CAG) expansion disorders.

9
Conclusions

Data favouring either one of the two hypotheses concerning the pathogenic mechanism of SCA6 are few, often controversial and seldom replicated. On one hand, gain of function is supported by analogies with expanded polyglutamine disorders that include the following: (1) polyglutamine aggregate formation in brains of patients with SCA6, although these aggregates differ from those found in patients with other SCAs; (2) cleavage of the α1A COOH tail and its translocation into the nucleus; and (3) clinical features that, although widely overlapping those resulting from channel loss of function in EA2, may include some extracerebellar symptoms reminiscent of symptoms of other SCAs. Such evidence should be weighted against contradicting evidence such as the size of the expansion mutation being too small to produce stable aggregates or the absence of cytotoxicity in cultured cells that express the full-length mutated protein.

On the other hand, evidence in favour of loss of function as a result of channel malfunction derives mainly from the similarities between clinical phenotypes of SCA6 and EA2 and the electrophysiological analysis of Ca^{2+} currents in non-neuronal cultured cells expressing the mutated channel. However, the latter data consist of often-contradictory results demonstrating either loss or gain of Ca^{2+} influx.

The enigma of SCA6, therefore, is far from being solved. Its solution must await the gathering of less controversial and more consistent data. These will hopefully come from the study of neuronal cellular models, which proved to be essential in the electrophysiological analyses of the α1A subunit (Tottene et al. 2002), and from SCA6 transgenic models.

References

Agostoni E, Jodice C, Patore A (2005) Nuclear expression of the c-terminal region of alpha1 calcium channel subunit. In: 2nd meeting on molecular mechanisms of neurodegeneration, Milan, Italy, 7–10 May, abstracts, p 89. (http://users.unimi.it/triplet)

Arpa J, Cuesta A, Cruz-Martinez A, Santiago S, Sarrià J, Palau F (1999) Clinical features and genetic analysis of a Spanish family with spinocerebellar ataxia 6. Acta Neurol Scand 99:43–47

Baloh RW, Qing Yue MD, Furman JM, Nelson SF (1997) Familial episodic ataxia. Ann Neurol 41:8–16

Bourinet E, Soong TW, Sutton K, Slaymaker S, Mathews E, Monteil A, Zamponi GW, Nargeot J, Snutch TP (1999) Splicing of alpha 1A subunit gene generates phenotypic variants of P- and Q-type calcium channels. Nat Neurosci 2:407–415

Brusco A, Gellera C, Cagnoli C, Saluto A, Castucci A, Michielotto C, Fetoni V, Mariotti C, Migone N, Di Donato S, Taroni F (2004) Molecular genetics of hereditary spinocerebellar ataxia: mutation analysis of spinocerebellar ataxia genes and CAG/CTG repeat expansion detection in 225 Italian families. Arch Neurol 61:727–733

Calandriello L, Veneziano L, Francia A, Sabbadini G, Colonnese C, Mantuano E, Jodice C, Trettel F, Viviani P, Manfredi M, Frontali M (1996) Acetazolamide-responsive episodic ataxia in an Italian family refines gene mapping on chromosome 19p13. Brain 120:805–812

Craig K, Keers SM, Archibald K, Curtis A, Chinnery PF (2004) Molecular epidemiology of spinocerebellar ataxia type 6. Ann Neurol 55:752–755

Denier C, Ducros A, Vahedi K, Joutel A, Thierry P, Rotz A, Castelnovo G, Deonna T, Gerard P, Devoize JL, Gayou A, Perrouty B, Soisson T, Autret A, Warter JM, Vighetto A, Van Bogaert P, Alamowitch S, Roullet E, Tournier-Lasserve E (1999) High prevalence of CACNA1A truncations and broader clinical spectrum in episodic ataxia type 2. Neurology 52:1816–1821

Denier C, Ducros A, Durr A, Eymard B, Chassande B, Tournier-Lasserve E (2001) Missense CACNA1A mutation causing episodic ataxia type 2. Arch Neurol 58:292–295

Dichgans M, Schols L, Herzog J, Stevanin G, Weirich-Schwaiger H, Rouleau G, Burk K, Klockgether T, Zuhlke C, Laccone F, Riess O, Gasser T (1999) Spinocerebellar ataxia type 6 evidence for a strong founder effect among german families. Neurology 52:849–851

Djousse L, Knowlton B, Hayden M, Almqvist EW, Brinkman R, Ross C, Margolis R, Rosenblatt A, Durr A, Dode C, Morrison PJ, Novelletto A, Frontali M, Trent RJ, McCusker E, Gomez-Tortosa E, Mayo D, Jones R, Zanko A, Nance M, Abramson R, Suchowersky O, Paulsen J, Harrison M, Yang Q, Cupples LA, Gusella JF, MacDonald ME, Myers RH (2003) Interaction of normal and expanded CAG repeat sizes influences age at onset of Huntington disease. Am J Med Genet A 119:279–282

Fukutake T, Kamitsukasa T, Arai K, Hattori T, Nakajima T (2002) A patient homozygous for the SCA6 gene with retinitis pigmentosa. Clin Genet 61:375–379

Garcia-Planells J, Cuesta A, Vilchez JJ, Martinez F, Prieto F, Palau F (1999) Genetics of the SCA6 gene in a large family segregating an autosomal dominant "pure" cerebellar ataxia. J Med Genet 36:148–151

Geschwind DH, Perlman S, Figueroa BS, Karrim BS, Baloh RW, Pulst SM (1997) Spinocerebellar ataxia type 6. Frequency of the mutation and genotype-phenotype correlations. Neurology 49:1247–1251

Globas C, Bosch S, Zuhlke Ch, Daum I, Dichgans J, Burk K (2003) The cerebellum and cognition. Intellectual function in spinocerebellar ataxia type 6 (SCA6). J Neurol 250:1482–1487

Gomez CM, Thompson M, Gammack JT, Perlman SL, Dobyns WD, Truwit CL, Zee DS, Clark HB, Anderson JH (1997) Spinocerebellar ataxia type 6: gaze-evoked and vertical nystagmus, Purkinje cell degeneration, and variable age of onset. Ann Neurol 42:933–950

Guida S, Trettel F, Pagnutti S, Mantuano E, Tottene A, Veneziano L, Fellin T, Spadaro M, Stauderman KA, Williams ME, Volsen S, Ophoff RA, Frants RR, Jodice C, Frontali M, Pietrobon D (2001) Complete loss of P/Q calcium channel activity caused by a CACNA1A missense mutation carried by episodic ataxia type 2 patients. Am J Hum Genet 68:759–764

Hering S, Berjukov S, Sokolov S, Marksteiner R, Weiss RG, Kraus R, Timin EN (2000) Molecular determinants of inactivation in voltage-gated Ca^{2+} channels. J Physiol 528:237–249

Ikeuchi T, Takano H, Koide R, Horikawa Y, Honma Y, Onisji Y, Igarashi S, Tanaka H, Nakao N, Sahashi K, Tsukagoshi H, Inoue K, Takahashi H, Tsuji S (1997) Spinocerebellar ataxia type 6: CAG repeat expansion in alpha1A voltage-dependent calcium channel gene and clinical variations in Japanese population. Ann Neurol 42:879–884

Imbrici P, Jaffe SL, Eunson LH, Davies NP, Herd C, Robertson R, Kullmann DM, Hanna MG (2004) Dysfunction of the brain calcium channel CaV2.1 in absence epilepsy and episodic ataxia. Brain 127:2682–2692

Ishikawa K, Tanaka H, Saito M, Ohkoshi N, Fujita T, Yoshizawa K, Ikeuchi T, Watanabe M, Hayashi A, Takiyama Y, Nishizawa M, Nakano I, Matsubayashi K, Miwa M, Shoji S, Kanazawa I, Tsuji S, Mizusawa H (1997) Japanese families with autosomal dominant pure cerebellar ataxia map to chromosome 19p13.1–p13.2 and are strongly associated with mild CAG expansions in the spinocerebellar ataxia type 6 gene in chromosome 19p13.1. Am J Hum Genet 61:336–346

Ishikawa K, Fujigasaki H, Saegusa H, Ohwada K, Fujita T, Iwamoto H, Komatsuzaki Y, Toru S, Toriyama H, Watanabe M, Ohkoshi N, Shoji S, Kanazawa I, Tanabe T, Mizusawa H (1999a) Abundant expression and cytoplasmic aggregations of [alpha]1A voltage-dependent calcium channel protein associated with neurodegeneration in spinocerebellar ataxia type 6. Hum Mol Genet 8:1185–1193

Ishikawa K, Watanabe M, Yoshizawa K, Fujita T, Iwamoto H, Yoshizawa T, Harada K, Nakamagoe K, Komatsuzaki Y, Satoh A, Doi M, Ogata T, Kanazawa I, Shoji S, Mizusawa H (1999b) Clinical, neuropathological, and molecular study in two families with spinocerebellar ataxia type 6 (SCA6). J Neurol Neurosurg Psychiatry 67:86–89

Ishikawa K, Ohwada K, Ishida K, Fujigasaki H, Shun Li M, Tsunemi T, Ohkoshi N, Toru S, Mizutani T, Hayashi M, Arai N, Hasegawa K, Kawanami T, Kato T, Makifuchi T, Shoji S, Tanabe T, Mizusawa H (2001) Cytoplasmic and nuclear polyglutamine aggregates in SCA6 Purkinje cells. Neurology 56:1753–1756

Jen J, Kim GW, Baloh RW (2004) Clinical spectrum of episodic ataxia type 2. Neurology 62:17–22

Jen JC, Yue Q, Karrim J, Nelson SF, Baloh RW (1998) Spinocerebellar ataxia type 6 with positional vertigo and acetazolamide responsive episodic ataxia. J Neurol Neurosurg Psychiatry 65:565–568

Jen JC, Yue Q, Nelson SF, Yu H, Litt M, Nutt J, Baloh RW (1999) A novel nonsense mutation in CACNA1A causes episodic ataxia and hemiplegia. Neurology 53:34–37

Jen JC, Wan J, Graves M, Yu H, Mock AF, Coulin CJ, Kim G, Yue Q, Papazian DM, Baloh RW (2001) Loss of function EA2 mutations are associated with impaired neuromuscular transmission. Neurology 57:1843–1848

Jodice C, Mantuano E, Veneziano L, Trettel F, Sabbadini G, Calandriello L, Francia A, Spadaro M, Pierelli F, Salvi F, Ophoff RA, Frants RR, Frontali M (1997) Episodic ataxia type 2 (EA2) and spinocerebellar ataxia type 6 (SCA6) due to CAG repeat expansion in the CACNA1A gene on chromosome 19p. Hum Mol Genet 6:1973–1978

Jouvenceau A, Eunson LH, Spauschus A, Ramesh V, Zuberi SM, Kullmann DM, Hanna MG (2001) Human epilepsy associated with dysfunction of the brain P/Q-type calcium channel. Lancet 358:801–807

Juvonen V, Hietala M, Kairisto V, Savontaus ML (2005) The occurrence of dominant spinocerebellar ataxias among 251 Finnish ataxia patients and the role of predisposing large normal alleles in a genetically isolated population. Acta Neurol Scand 111:154–162

Kaseda Y, Kawakami H, Matsuyama Z, Kumagai R, Toji M, Komure O, Nishimura M, Izumi Y, Udaka F, Kameyama M, Nishio T, Sunohara N, Kuroda Y, Nakamura S (1999) Spinocerebellar ataxia type 6 in relation to CAG repeat length. Acta Neurol Scand 99:209–212

Katayama T, Ogura Y, Aizawa H, Kuroda H, Suzuki Y, Kuroda K, Kikuchi K (2000) Nineteen CAG repeats of the SCA6 gene in a Japanese patient presenting with ataxia. J Neurol 415:711–712

Kato T, Tanaka F, Yamamoto M, Yosida E, Indo T, Watanabe H, Yoshiwara T, Doyu M, Sobue G (2000) Sisters homozygous for the spinocerebellar ataxia type 6 (SCA6)/CACNA1A gene associated with different clinical phenotypes. Clin Genet 58:69–73

Kaunisto MA, Harno H, Kallela M, Somer H, Sallinen R, Hamalainen E, Miettinen PJ, Vesa J, Orpana A, Palotie A, Farkkila M, Wessman M (2004) Novel splice site CACNA1A mutation causing episodic ataxia type 2. Neurogenetics 5:69–73

Kobrinski E, Schwartz E, Abernathy DR, Soldatov NM (2003) Voltage-gated mobility of the Ca^{2+} channel cytoplasmic tails and its regulatory role. J Biol Chem 278:5021–5028

Koh SH, Kim HT, Kim SH, Lee GY, Kim J, Kim MH (2001) Spinocerebellar Ataxia type 6 and Episodic Ataxia type 2 in a Korean Family. J Korean Med Sci 16:809–813

Kordasiewicz HB, Thompson RM, Clark HB, Gomez CM (2006) C-termini of P/Q-type Ca^{2+} channel (alpha)1A subunits translocate to nuclei and promote polyglutamine-mediated toxicity. Hum Mol Genet 15:1587–1599

Kubodera T, Yokota T, Ohwada K, Ishikawa K, Miura H, Matsuoka T, Mizusawa H (2003) Proteolytic cleavage and cellular toxicity of the human alpha1A calcium channel in spinocerebellar ataxia type 6. Neurosci Lett 341:74–78

Mantuano E, Veneziano L, Spadaro M, Giunti P, Guida S, Leggio MG, Verriello L, Wood N, Jodice C, Frontali M (2004) Clusters of non-truncating mutations of P/Q type $Ca2^{+}$ channel subunit Ca(v)2.1 causing episodic ataxia 2. J Med Genet 41:e82

Margolis RL (2003) Dominant spinocerebellar ataxias: a molecular approach to classification, diagnosis, pathogenesis and the future. Expert Rev Mol Diagn 3:715–732

Mariotti C, Gellera C, Grisoli M, Mineri R, Castucci A, Di Donato S (2001) Pathogenic effect of an intermediate-size SCA-6 allele (CAG)(19) in a homozygous patient. Neurology 57:1502–1504

Matsumura R, Futamura N, Fujimoto Y, Yanagimoto S, Horikawa H, Suzumura A, Takayanagi T (1997) Spinocerebellar ataxia type 6. Molecular and clinical features of 35 Japanese patients including one homozygous for the CAG repeat expansion. Neurology 49:1238–1243

Matsuyama Z, Kawakami H, Maruyama H, Izumi Y, Komure O, Udaka F, Kameyama M, Nishio T, Kuroda Y, Nishimura M, Nakamura S (1997) Molecular features of the CAG repeats of spinocerebellar ataxia 6 (SCA6). Hum Mol Genet 6:1283–1287

Matsuyama Z, Wakamori M, Mori Y, Kawakami H, Nakamura S, Imoto K (1999) Direct alteration of the P/Q-type Ca^{2+} channel property by polyglutamine expansion in spinocerebellar ataxia 6. J Neurosci 19:RC14

Matsuyama Z, Murase M, Shimizu H, Aoki Y, Hayashi M, Hozumi I, Inuzuka T (2003) A novel insertion mutation of acetazolamide-responsive episodic ataxia in a Japanese family. J Neurol Sci 210:91–93

Matsuyama Z, Yanagisawa NK, Aoki Y, Black JL 3rd, Lennon VA, Mori Y, Imoto K, Inuzuka T (2004) Polyglutamine repeats of spinocerebellar ataxia 6 impair the cell-death-preventing effect of CaV2.1 Ca^{2+} channel-loss-of-function cellular model of SCA6. Neurobiol Disord 17:198–204

Michalik A, Van Broeckhoven C (2003) Pathogenesis of polyglutamine disorders: aggregation revisited. Hum Mol Genet 12 Spec No 2:R173–186

Mori Y, Friedrich T, Kim MS, Mikami A, Nakai J, Ruth P, Bosse E, Hofmann F, Flockerzi V, Furuichi T, Mikishiba K, Imoto K, Tanabe T, Numa S (1991) Primary structure of functional expression from complementary DNA of a brain calcium channel. Nature 350:398–402

Mori M, Adachi Y, Kusumi M, Nakashima K (2001) Spinocerebellar ataxia type 6: founder effect in Western Japan. J Neurol Sci 185:43–47

Murata Y, Kawakami H, Yamaguchi S, Nishimura M, Kohriyama T, Ishizaki F, Matsuyama Z, Mimori Y, Nakamura S (1998) Characteristic magnetic resonance imaging findings in spinocerebellar ataxia 6. Arch Neurol 55:1348–1352

Nagai Y, Azuma T, Unauchi M, Fujita M, Umi M, Hirano M, Matsubara T, Ueno S (1998) Clinical and molecular genetic study in seven Japanese families with spinocerebellar ataxia type 6. J Neurol Sci 157:52–59

Nakagawa N, Katayama T, Makita Y, Kuroda K, Aizawa H, Kikuchi K (1999) A case of spinocerebellar ataxia type 6 mimicking olivopontocerebellar atrophy. Neuroradiology 41:501–503

Ophoff RA, Terwindt GM, Vergouwe MN, van Eijk R, Oefner PJ, Hoffman SMG, Lamerdin JE, Mohrenweiser HW, Bulman DE, Ferrari M, Haan J, Lindhout D, van Ommen GJB, Hofker MH, Ferrari MD, Frants RR (1996) Familial hemiplegic migraine and episodic ataxia type-2 are caused by mutations in the Ca^{2+} channel gene CACNL1A4. Cell 87:543–552

Penrose LS (1948) The problem of anticipation in pedigrees of Dystonia Myotonica. Ann Eugenics 14:125–132

Persichetti F, Ambrose CM, Ge P, McNeil SM, Srinidhi J, Anderson MA, Jenkins B, Barnes GT, Duyao MP, Kanaley L, Wexler NS, Myers RH, Bird ED, Vonastetel JP, MacDonald ME, Gusella JF (1995) Normal and expanded Huntington's disease gene alleles produce distinguishable proteins due to translation across the CAG repeat. Mol Med 1:374–383

Perutz MF (1994) Polar zippers: their role in human disease. Protein Sci 3:1629–1637

Piedras-Renterìa ES, Watase K, Harata N, Zhuchenko O, Zoghbi H, Lee CC, Tsien RW (2001) Increased expression of α_{1A} Ca^{2+} channel currents arising from expanded trinucleotide repeats in Spinocerebellar ataxia type 6. J Neurosci 21:9185–9193

Pietrobon D, Striessnig J (2003) Neurobiology of migraine. Nat Rev Neurosci 4:386–398

Pujana MA, Corral J, Gratacos M, Combarros O, Berciano J, Genis D, Banchs I, Estivill X, Volpini V (1999) Spinocerebellar Ataxia in Spanish patients: genetic analysis of familial and sporadic cases. The Ataxia Study Group. Hum Genet 104:516–522

Restituito S, Thomson RM, Eliet J, Raike RS, Riedl M, Charnet P, Gomez C (2000) The polyglutamine expansion in spinocerebellar ataxia type 6 causes a β subunit-specific enhanced activation of P/Q-type calcium channels in Xenopus oocytes. J Neurosci 20:6394–6430

Riess O, Schols L, Bottger H, Nolte D, Vieira-Saecker AM, Schimming C, Kreuz F, Macek M Jr, Krebsovà A, Macek M Sen, Klockgether T, Zuhlke C, Laccone FA (1997) SCA6 is caused by moderate CAG expansion in the alpha1A-voltage-dependent calcium channel gene. Hum Mol Genet 6:1289–1293

Rubinsztein DC, Leggo J, Coles R, Almqvist E, Biancalana V, Cassiman JJ, Chotai K, Connarty M, Crauford D, Curtis A, Curtis D, Davidson MJ, Differ AM, Dode C, Dodge A, Frontali M, Ranen NG, Stine OC, Sherr M, Abbott MH, Franz ML, Graham CA, Harper PS, Hedreen JC, Jackson A, Kaplan JC, Losekoot M, MacMillan JC, Morrison P, Trottier Y, Novelletto A, Simpson SA, Theilmann J, Whittaker JL, Folstein SE, Ross CA, Hayden MR (1996) Phenotypic characterization of individuals with 30–40 CAG repeats in the Huntington disease (HD) gene reveals HD cases with 36 repeats and apparently normal elderly individuals with 36–39 repeats. Am J Hum Genet 59:16–22

Sasaki H, Kojima H, Yabe I, Hamada T, Sawa H, Hiraga H, Nagashima K (1998) Neuropathological and molecular studies of spinocerebellar ataxia type 6 (SCA6). Acta Neuropathol 95:199–204

Satoh JI, Tokumoto H, Yukitake M, Matsui M, Matsuyama Z, Kawakami H, Nakamura S, Kuroda Y (1998) Spinocerebellar ataxia type 6: MRI of three Japanese patients. Neuroradiology 40:222–227

Scherzinger E, Sittler A, Schweiger K, Heiser V, Lurz R, Hasenbank R, Bates GP, Lehrach H, Wanker EE (1999) Self assembly of polyglutamine containing huntingtin fragments into amiloid-like fibrils: implication for Huntington's disease pathology. Proc Natl Acad Sci USA 96:4604–4608

Servadio A, Koshy B, Armstrong D, Antalffy B, Orr HT, Zoghbi HY (1995) Expression analysis of the ataxin-1 protein in tissues from normal and spinocerebellar ataxia type 1 individuals. Nat Genet 10:94–98

Shizuka M, Watanabe H, Ikeda Y, Mizushima K, Okamoto K, Shoji M (1998a) Molecular analysis of a de novo mutation for spinocerebellar ataxia type 6 and $(CAG)_n$ repeat units in normal elder controls. J Neurol Sci 161:85–87

Shizuka M, Watanabe H, Ikeda Y, Mizushima K, Kanai M, Tsuda T, Abe K, Okamoto K, Shoji M (1998b) Spinocerebellar ataxia type 6: CAG trinucleotide expansion, clinical characteristics and sperm analysis. Eur J Neurol 5:381–387

Silveira I, Miranda C, Guimaraes L, Moreira CC, Alonso I, Mendonca P, Ferro A, Pinto-Basto J, Coelho J, Ferreirinha F, Poirier J, Parreira E, Vale J, Januario C, Barbot C, Tuna A, Barros J, Koide R, Tsuji S, Holmes SE, Margolis RL, Jardim L, Pandolfo M, Coutinho P, Sequeiros J (2002) Trinucleotide repeats in 202 Families with Ataxia. A small expanded $(CAG)_n$ allele at the SCA17 locus. Arch Neurol 59:623–629

Sinke RJ, Ippel EF, Diepstraten CM, Beemer FA, Wokke JHJ, Van Hilten BJ, Knoers NVAM, Ploos van Amstel HC, Kremer HPH (2001) Clinical and molecular correlations in Spinocerebellar Ataxia 6. Arch Neurol 58:1839–1844

Soong BW, Lu YC, Choo KB, Lee HY (2001a) Frequency analysis of Autosomal Dominant Cerebellar Ataxias in Taiwanese patients and clinical molecular characterization of Spinocerebellar Ataxia type 6. Arch Neurol 58:1105–1109

Soong BW, Liu RS, Wu LC, Lu YC, Lee HY (2001b) Metabolic characterization of Spinocerebellar Ataxia type 6. Arch Neurol 58:300–304

Soong TW, DeMaria CD, Alvania RS, Zweifel LS, Liang MC, Mittman S, Agnew WS, Yue DT (2002). Systematic identification of splice variants in human P/Q-type channel alpha1(2.1) subunits: implications for current density and Ca^{2+} dependent inactivation. J Neurosci 22:10142–10152

Spacey SD, Materek LA, Szczygielski BI, Bird TD (2005) Two novel CACNA1A gene mutations associated with episodic ataxia type 2 and interictal dystonia. Arch Neurol 62:314–316

Stevanin G, Durr A, David G, Didierjean O, Cancel G, Rivaud S, Tourbah A, Warter JM, Agid Y, Brice A (1997) Clinical and molecular features of spinocerebellar ataxia type 6. Neurology 49:1243–1246

Subramony SH, Fratkin JD, Manyam BV, Currier RD (1996) Dominantly inherited cerebello-olivary atrophy is not due to a mutation at the spinocerebellar ataxia 1, Machado-Joseph disease, or dentato-rubro-pallido-luysian atrophy locus. Mov Disord 11:174–180

Subramony SH, Schott K, Raike RS, Callahan J, Langford LR, Christova PS, Anderson JH, Gomez CM (2003) Novel CACNA1A mutation causes febrile episodic ataxia with interictal cerebellar deficits. Ann Neurol 54:725–731

Suzuki M, Koike T (2005) Early apoptotic and late necrotic components associated with altered Ca(2+) homeostasis in a peptide-delivery model of polyglutamine-induced neuronal death. J Neurosci Res 80:549–561

Takahashi H, Ishikawa K, Tsutsumi T, Fujigasaki H, Kawata A, Okiyama R, Fujita T, Yoshizawa K, Yamaguchi S, Tomiyasu H, Yoshii F, Mitani K, Shimizu N, Yamazaki M,

Miyamoto T, Orimo T, Shoji S, Kitamura K, Mizusawa H (2004) A clinical and genetic study in a large cohort of patients with spinocerebellar ataxia type 6. J Hum Genet 49:256–264

Takano H, Cancel G, Ikeuchi T, Lorenzetti D, Mawad R, Stevanin G, Didierijean O, Durr A, Oyake M, Shimohata T, Sasaki R, Koide R, Igarashi S, Hayashi S, Takiyama Y, Nishizawa M, Tanaka H, Zoghbi H, Brice A, Tsuji S (1998) Close associations between prevalences of dominant inherited ataxias with CAG-repeat expansions and frequencies of large normal CAG alleles in Japanese and Caucasian population. Am J Hum Genet 63:1060–1066

Takiyama Y, Sakoe K, Nemekawa M, Soutome M, Esumi E, Ogawa T, Ishikawa K, Mizusawa H, Nakano I, Nishizawa M (1998) A Japanese family with spinocerebellar ataxia type 6 which includes three individuals homozygous for an expanded CAG repeat in the SCA6/CACNL1A4 gene. J Neurol Sci 158:141–147

Tashiro H, Suzuki SO, Hitotsumatsu T, Iwaki T (1999) An autopsy case of spinocerebellar ataxia type 6 with mental symptoms of schizophrenia and dementia. Clin Neuropathol 18:198–204

Terasawa H, Oda M, Morino H, Miyachi T, Izumi Y, Maruyama H, Matsumoto M, Kawakami H (2004) A novel haplotype of spinocerebellar ataxia type 6 contributes to the highest prevalence in western Japan. Neurosci Lett 358:107–110

Toru S, Murakoshi T, Ishikawa K, Saegusa H, Fujigasaki H, Uchihara T, Nagayama S, Osanai M, Mizusawa H, Tanabe T (2000) Spinocerebellar ataxia type 6 mutation alters P-type calcium channel function. J Biol Chem 275:10893–10898

Tottene A, Fellin T, Pagnutti S, Luvisetto S, Striessnig J, Fletcher C, Pietrobon D (2002) Familial hemiplegic migraine mutations increase Ca^{2+} influx through single human CaV2.1 channels and decrease maximal CaV2.1 current density in neurons. Proc Natl Acad Sci USA 99:13284–13289

Tsunemi T, Saegusa H, Ishikawa K, Nagayama S, Murakoshi T, Mizusawa H, Tanabe T (2002) Novel Cav2.1 splice variants isolated from Purkinje cells do not generate P-type Ca^{2+} current. J Biol Chem 277:7214–7221

Yabe I, Sasaki H, Yamashita I, Takei A, Tashiro K (2001) Clinical trial of acetazolamide in SCA6, with assessment using ataxia rating scale and body stabilometry. Acta Neurol Scand 104:44–47

Yue Q, Jen CJ, Nelson SF, Baloh RW (1997) Progressive ataxia due to a missense mutation in a calcium-channel gene. Am J Hum Genet 61:1078–1087

Yue Q, Jen JC, Thwe MM, Nelson SF, Baloh RW (1998) De novo mutation in CACNA1A caused acetazolamide-responsive episodic ataxia. Am J Med Genet 77:298–301

Van de Warrenburg BPC, Sinke RJ, Verschuuren-Bemelmans CC, Scheffer H, Brunt ER, Ippel PF, Maat-Kievit JA, Dooijes D, Notermans NC,. Lindhout D, Knoers NVAM, Kremer HPH (2002) Spinocerebellar ataxias in the Netherlands. Neurology 58:702–708

Verbeek DS, Piersma SJ, Hennekam EF, Ippel EF, Pearson PL, Sinke RJ (2004) Haplotype study in Dutch SCA3 and SCA6 families: evidence for common founder mutations. Eur J Hum Genet 12:441–446

Walsh R, Storey E, Stefani D, Kelly L, Turnbull V (2005) The roles of proteolysis and nuclear localisation in the toxicity of the polyglutamine diseases. Neurotox Res 7:43–57

Wan J, Carr JR, Baloh RW, Jen JC (2005) Nonconsensus intronic mutations cause episodic ataxia. Ann Neurol 57:131–135

Wappl E, Koschak A, Poteser M, Sinnegger MJ, Walter D, Eberhart A, Grosachner K, Glossman H, Kraus RL, Grabner M, Striessnig J (2002) Functional consequences of P/Q-type Ca^{2+} channel Cav2.1 missense mutations associated with Episodic Ataxia type 2 and progressive ataxia. J Biol Chem 277:6960–6966

Watanabe H, Tanaka F, Matsumoto M, Doyu M, Ando T, Mitsuma T, Sobue G (1998) Frequency analysis of autosomal dominant ataxias in Japanese patients and clinical characterization of spinocerebellar ataxia type 6. Clin Genet 53:13–19

Weiss JL, Burgoyne RD (2002) Sense and sensibility in the regulation of voltage-gated Ca^{2+} channels. Trends Neurosci 25:489–491

Zhuchenko O, Bailey J, Bonnen P, Ashizawa T, Stockton DW, Amos C, Dobyns WB, Subramony SH, Zoghbi HY, Chi LC (1997) Autosomal dominant cerebellar ataxia (SCA6) associated with small polyglutamine expansion in the α_{1A}-voltage-dependent calcium channel. Nat Genet 15:62–69

Part IV
Disorders Associated with Repeats in an Undetermined Location

Part IV
Disorders Associated with Repeats
[illegible]

Nucleic Acids and Molecular Biology, Vol. 19
M. Fry/K. Usdin (Eds.)
Human Nucleotide Expansion Disorders

Spinocerebellar Ataxia Type 12 and Huntington's Disease-Like 2: Clues to Pathogenesis

Russell L. Margolis[1,2,5] (✉) · Susan E. Holmes[1] · Elizabeth O'Hearn[2,3] · Dobrila D. Rudnicki[1] · John Hwang[1] · Natividad Cortez-Apreza[1] · Olga Pletnikova[4] · Juan C. Troncoso[2,4]

[1]Laboratory of Genetic Neurobiology, Division of Neurobiology, Department of Psychiatry, CMSL 8-121, 600 N. Wolfe Street, Johns Hopkins University School of Medicine, Baltimore, MD 21287, USA
rmargoli@jhmi.edu

[2]Department of Neurology, Johns Hopkins University School of Medicine, CMSL 8-121, 600 N. Wolfe Street, Baltimore, MD 21287, USA

[3]Department of Neuroscience, Johns Hopkins University School of Medicine, CMSL 8-121, 600 N. Wolfe Street, Baltimore, MD 21287, USA

[4]Division of Neuropathology, Department of Pathology, Johns Hopkins University School of Medicine, CMSL 8-121, 600 N. Wolfe Street, Baltimore, MD 21287, USA

[5]Program in Cellular and Molecular Medicine, Johns Hopkins University School of Medicine, CMSL 8-121, 600 N. Wolfe Street, Baltimore, MD 21287, USA

1 The Identification of Spinocerebellar Ataxia Type 12 and Huntington's Disease-Like 2

As part of the long-standing program in Huntington's disease (HD) at Johns Hopkins University, our group developed an interest in investigating the genetic etiology and pathogenesis of disorders similar to HD. After the discovery of the HD mutation in 1993 (Huntington's Disease Collaborative Research Group 1993), it became clear that a portion of the patients followed in our clinic with presumed HD did not have the HD mutation. With careful review of all available records, reexamination, and additional genetic testing, four cases emerged that had a familial progressive neurodegenerative disorder resembling HD (Rosenblatt et al. 1998). Using a modified version of the repeat expansion detection (RED) assay (Schalling et al. 1993), we found a d(CAG)/d(CTG) repeat expansion in genomic DNA from one of these four cases that could not be explained by any known repeat expansion, and that segregated with disease in affected family members. The RED assay was then used to isolate a clone of a section of genomic DNA containing the repeat (Holmes et al. 2001b), following a method similar to that used by Ranum and colleagues (Koob et al. 1998). This technique allows isolation of novel expansion mutations directly from DNA of a single patient without the necessity of a positional cloning approach. From this clone, the locus of the repeat expansion was readily apparent, and a simple PCR assay was developed for examining repeat length in other individuals. We used the PCR assay to determine that the repeat expansion segregated perfectly with the illness. We also used the PCR assay to determine that one of the other individuals from our clinic with an HD-like familial disorder was from another branch of the proband's family. The disorder was named HD-like 2 (HDL2).

Spinocerebellar ataxia type 12 (SCA12) was identified through a collaborative arrangement between Johns Hopkins University and Athena Diagnostics, a commercial laboratory providing genetic tests for neurodegenerative and other diseases. After obtaining informed consent, DNA from patients with ataxias for which no mutation was detected by Athena Diagnostics was sent to our laboratory along with brief clinical descriptions. The RED assay was applied to these cases, and a case was identified with an unexplained d(CAG)/d(CTG) repeat expansion that, as in (HDL2), segregated perfectly with affected status in family members. This expansion was cloned, and the flanking sequence again revealed the locus of the expansion and enabled us to develop a PCR assay of repeat length (Holmes et al. 1999). The pattern of segregation was somewhat less clear than in HDL2, a function of the phenotype, which is relatively mild, variable, and often resembles essential tremor. Though we were reasonably confident that the expansion mutation was the cause of the disease, the proof that the mutation was causal only emerged with the finding of multiple other families, as described below.

2 HDL2 and SCA12 Clinical Phenotype

At least at the level of individual patients, HDL2 is clinically indistinguishable from HD; most individuals with HDL2 were initially diagnosed as having HD, often by clinicians experienced in the differential diagnosis of movement disorders. HDL2 cases generally appear to have one of two presentations, though these presentations probably reflect opposite ends of a continuum rather than discrete disease subtypes (Table 1). The first HDL2 presentation is similar to the juvenile onset or Westphal variant of HD, and was observed in all affected members of the index family, and in some members of another American HDL2 family (Walker et al. 2002, 2003). Onset, typically of weight loss (despite increased appetite) and poor coordination, is usually in the fourth decade. Subsequently, a variety of movement abnormalities develop, most prominently rigidity, dysarthria, hyperreflexia, bradyki-

Table 1 Signs and symptoms in Huntington's disease-like 2 (*HDL2*) and Spinocerebellar ataxia type 12 (*SCA12*)

	HDL2	SCA12
Cerebellar signs	0	+–++
Action tremor	0	+++
Parkinsonism	+–+++	0 –++
Chorea	+–+++	0
Dystonia	0 –++	0 –+
Hyperreflexia	++	+
Dysarthria	++	+
Abnormal eye movements	+	+
Ataxia	+–++	+–++
Sensory neuropathy	0	0 –++
Weight loss	+–++	0
Dementia	+++	+
Psychiatric syndromes	+++	+

nesia, and tremor. Most patients develop dystonia and/or chorea, though chorea is typically quite mild. Cerebellar signs are absent, and eye movements are minimally affected. All patients gradually become demented and develop psychiatric disturbances such as depression, irritability, and apathy. The disease is relentlessly progressive, resulting in a profound dementia and rigidity in 10–15 years, with death from nonspecific complications following thereafter.

The second presentation resembles typical HD. Onset may be somewhat later than in the index family, perhaps by as much as 10 years, though with interfamily variability. Chorea, as in most cases of HD, is quite pronounced, while dystonia, bradykinesia, tremor, hyperreflexia, and dysarthria are less prominent. Eye movement abnormalities, particularly dysmetric saccades, seem more common. Psychiatric and cognitive disturbances may be milder or at least slower to develop, and overall the disease may progress more slowly.

The similarity between HD and HDL2 certainly justifies the name HDL2. In fact, given that HDL1 is a familial form of a prion disorder, stemming from an insertion into the *PRPN* gene and with a pathology unlike HD (Moore et al. 2001), a more appropriate name for HDL2 might have been HD2. If the HDL2 mutation had been discovered prior to the HD mutation, the designations HD and HDL2 would be reversed, and it is likely that more emphasis would now be placed on the genotypic diversity of the HD phenotype.

The HDL2 phenotype is striking. Individuals from known HDL2 families with even mild signs and symptoms can be diagnosed with a high degree of confidence. When all affected individuals are considered, SCA12 has an equally characteristic course, but in any one individual diagnosis may be more difficult. The mean onset age in the index American family was 34 years, with a range of 8–55 years (Holmes et al. 1999; O'Hearn et al. 2001). In the two published series of SCA12 patients from India, the mean onset age was 40 years (Bahl et al. 2005) and 38 years (Sinha et al. 2004). Most typically, disease begins with an action tremor of the arms or head followed by development of mild ataxia, limb dysmetria, and/or generalized hyperreflexia. Some individuals develop action tremor (postural and/or intention tremor) of the head, limbs, lips, or tongue. Some members of the index American pedigree develop bradykinesia or dementia, and others have psychiatric syndromes of uncertain relationship to the SCA12 mutation. Individuals from India with SCA12 tend to develop less parkinsonism but more frequent pyramidal signs, and as many as 50% develop a sensory neuropathy. Signs of cerebellar dysfunction tend to be less prominent and less disabling in individuals with SCA12 than in individuals with other SCAs. While the designation SCA12 was applied to the disease on the basis of particularly dramatic presentations of a few members of the index family, cerebellar involvement is much less prominent than in the other SCAs. Overall, SCA12 is among the least disabling of all the SCAs, and may not have a major impact on longevity in many affected individuals. A few affected individuals have remained employed throughout adulthood.

3 HDL2 and SCA12 Pathological Phenotype

HDL2 neuroimaging consistently reveals prominent caudate atrophy, moderate atrophy of the cerebral cortex, and little abnormality in other brain structures. HDL2 and HD MRI images cannot be distinguished from each other (Fig. 1). Examination of the first autopsied HDL2 case (Fig. 2) was notable for mild atrophy of cortical gray matter, ventricular dilatation, and severe atrophy of the head of the caudate and the putamen. Severe neuronal degeneration and reactive astrocytosis, with vacuolation of the neuropil, was observed throughout the caudate, with more severe involvement of dorsal than ventral regions and a selective loss of medium-sized neurons. Degeneration was somewhat less marked in the putamen, with the same dorsal-to-ventral gradient. Neuronal loss and astrocytosis was more moderate in the globus pallidus, and moderate neuronal degeneration accompanied by pigment incontinence was observed in the substantia nigra, but no Lewy bodies. No β amyloid deposits or neurofibrillary tangles were detected. Immunohistochemically, the key finding was the presence of intranuclear aggregates that stained with 1C2 (somewhat specific for expanded polyglutamine tracts) and anti-ubiquitin antibodies, but not anti-huntingtin antibodies. Inclusions were more common in the cortex than in the striatum. TorsinA (Walker et al. 2002) and TATA-box binding protein (TBP) have also been detected in these inclusions.

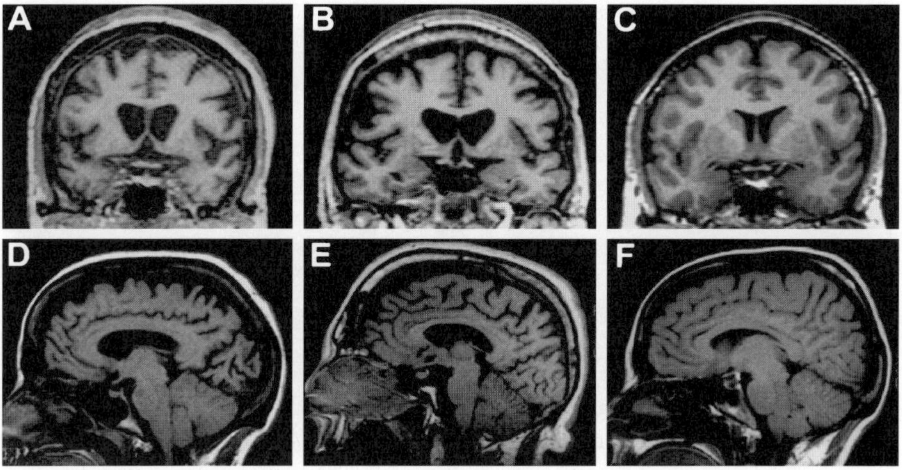

Fig. 1 Huntington's disease-like 2 (*HDL2*) is indistinguishable from Huntington's disease (*HD*) by MRI scan. **a, d** HDL2 case, MRI at age 36, 10-year disease duration. **b, e** Typical HD case, age 48 years, 12-year disease duration. **c, f** Normal control, age 43 years. Note the atrophy of the striatum and cerebral cortex in the HDL2 and HD cases, with relative sparing of the cerebellum and brain stem. (Reprinted with permission, *Annals of Neurology*, copyright 2001; Margolis et al. 2001)

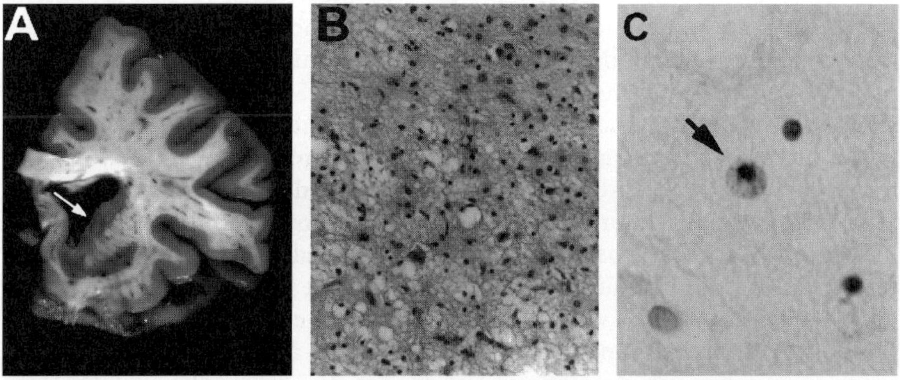

Fig. 2 HDL2 pathology. **a** Gross pathology, with prominent striatal atrophy (*arrow*) and moderate cortical atrophy. **b** Microscopic pathology of the caudate, with neuronal degeneration, astrocytic gliosis, and vacuolization. **c** Intranuclear inclusions stained by the 1C2 antibody (*arrow*). (Reprinted with permission, *Annals of Neurology*, copyright 2001; Margolis et al. 2001)

Recent evidence suggests that nuclear RNA inclusions may also be present in HDL2 brain (see below). A total of four HDL2 brains have now been examined in some detail, with consistent findings.

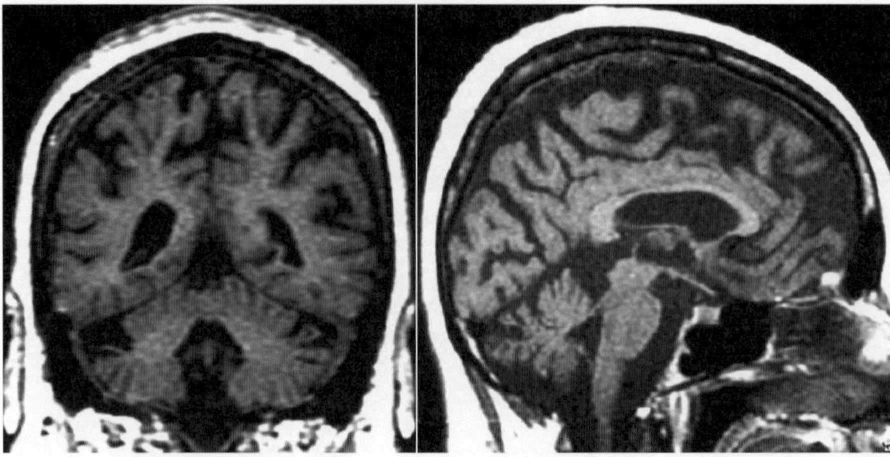

Fig. 3 Spinocerebellar ataxia type 12 neuroimaging. MRI of a 59-year-old woman demonstrating moderate cerebral cortical and cerebellar atrophy affecting the vermis and lateral cerebellar hemispheres. (Reprinted with permission, *Neurology*, copyright 2001, Lippincott Williams & Wilkins; O'Hearn et al. 2001)

The variability of SCA12 neuroimages is consistent with the variability in the SCA12 phenotype. MRI of most affected individuals reveals atrophy of the cerebral cortex and less marked atrophy of the cerebellum, with somewhat greater atrophy in the vermis than in the hemispheres; atrophy in other brain regions is less consistent (Fig. 3). A single SCA12 brain has come to autopsy. This brain was characterized by bilateral diffuse moderate atrophy of the cerebral cortex, most prominent in the parietal lobe; mild atrophy of the cerebellum, pons, and corpus calosum; and mild ventricular enlargement. Microscopic examination revealed moderate loss of Purkinje cells and mild neuronal loss in the substantia nigra, dentate nucleus, and inferior olivary nucleus. Neuronal intranuclear inclusions staining for ubiquitin and resembling Marinesco bodies were found in substantia nigra neurons and, more rarely, in Purkinje cells and motor cortical neurons. No Lewy bodies, neuronal tangles, or inclusions staining for tau or with 1C2 were detected (O'Hearn et al. 2004).

4 Epidemiology

HDL2 and SCA12 are both rare diseases found in restricted ethnic groups. In carefully examined individuals with an HD-like syndrome or pathology in Europe and North America, as few as 1% do not have the HD mutation (Andrew et al. 1994; Persichetti et al. 1994; Xuereb et al. 1996; Stevanin et al. 2003).

In North America, a systematic examination of 538 individuals with HD-like phenotypes who did not have the HD mutation found only six cases of HDL2 (Margolis et al. 2004a). Every case of known ethnicity was either of definite or probable African origin, with preliminary indications that all cases have the same haplotype (Krause et al. 2002). No cases have been detected in Japan (Margolis et al. 2004a; Shimohata et al. 2004) or in Europe (Stevanin et al. 2002; Bauer et al. 2002). Most remarkably, among individuals with HD-like syndromes in South Africa, preliminary evidence suggests that the frequency of HDL2 is about 25%, nearly as common as HD (Krause et al. 2002). The frequency of HDL2 in Asia outside of Japan remains unknown.

SCA12, unlike HDL2, has been detected in two separate populations. The index family, American and said to be of German origin, is the only SCA12 pedigree ever detected in North America (Holmes et al. 1999; Cholfin et al. 2001; O'Hearn et al. 2001). No SCA12 cases of European (Holmes et al. 1999; Fujigasaki et al. 2001; Worth and Wood 2001; Brusco et al. 2004; Hellenbroich et al. 2004), Japanese (Maruyama et al. 2002; Matsumura et al. 2003; Sasaki et al. 2003), or Chinese (Tsai et al. 2004; Zhao et al. 2002) origins have been ascertained. The frequency of SCA12 in African populations has never been examined. However, SCA12 has been detected in approximately 25 families from India by two independent groups. SCA12 appears to be the second most common form of SCA in India (though ascertainment has mostly been in northern India), accounting for about 8% of all cases of ataxia (Srivastava et al. 2001; Sinha et al. 2004; Bahl et al. 2005). All cases come from a single endogamous ethnic group originating in Haryana state in northern India and share a common haplotype, which differs from the haplotype of the American family (Bahl et al. 2005). It therefore appears that the SCA12 mutation has arisen independently at least twice, compared with the single origin of HDL2. Origins of each disease only once or twice in human history are consistent with the low prevalence of each disease compared with HD or the other SCAs.

5 Phenotype–Genotype Relationship

Establishing the minimum length of expansion sufficient to cause disease in SCA12 and HDL2 is certainly critical for providing accurate genetic diagnosis, but this information, in comparison with that for other repeat expansion diseases, may also provide insight into pathogenic processes. In HDL2, normal allele length varies from six to 28 triplets, with a mode of 14 triplets. Repeat lengths associated with disease range from 40 to 57 triplets. It is possible that expansions in the 40–45 triplet range may be incompletely penetrant, as one individual from an HDL2 pedigree did not have clear evidence of HDL2 at the age of 65 despite a repeat expansion of 44 triplets, though subtle manifestations of HDL2 may have been obscured by a stroke in this person. It is also not possible to exclude phenotypic manifestations of shorter alleles. A 48-year-old woman with alleles of 12 and 33 triplets developed an acute atypical cerebellar disorder after a hospitalization for uncontrolled diabetes mellitus type II (Margolis et al. 2004a). Five years later, cerebellar signs had only minimally progressed. Her son, with alleles of 35 and 14 triplets, developed Cogan's syndrome at age 25, an autoimmune disease resulting in complete hearing loss. Disease onset was notable for tinnitus, occasional lapses of concentration, and difficulty with balance. A detailed neurological examination at age 30 revealed only a lack of smoothness in horizontal and vertical gaze and saccades, possible dysdiadochokinesis, and moderate unsteadiness with tandem walk. While the relationship between repeat expansion and disease is uncertain in these two individuals, they do provide evidence for unstable vertical transmission of repeat lengths as short as 33 triplets (Margolis et al. 2004a). In summary, the range of normal and expanded repeat lengths in HDL2 is remarkably consistent with that seen in HD and most of the other polyglutamine diseases.

Normal alleles at the SCA12 locus range in length from four to 32 triplets, with a mode of ten triplets in all samples studied to date (Holmes et al. 1999; Fujigasaki et al. 2000; Cholfin et al. 2001; Srivastava et al. 2001; Worth and Wood 2001; Zhao et al. 2002; Brusco et al. 2004; Sulek et al. 2004; Tsai et al. 2004). Alleles associated with SCA12 range in length from 51 to 78 triplets (Holmes et al. 1999; Bahl et al. 2005). Two northern Germans with ataxia had repeats of 40 and 41 triplets, but family and phenotypic information was insufficient to determine if these are coincidental findings (Hellenbroich et al. 2004). An Iranian woman with unipolar depression and her monozygotic twin sons with schizophrenia each had an expansion of 53 triplets at the SCA12 locus (Holmes et al. 2003). The implication was that expanded alleles exist in Iran, but a clear relationship between the repeat expansions and the psychiatric disorders could not be established. An individual with typical Creutzfeld–Jacob disease (CJD) had an SCA12 allele of 49 triplets; the relationship between the CJD and the expansion is unknown (Hellenbroich et al. 2004). An asymptomatic individual from India who is homozygous for SCA12 expansions (repeat lengths 52 and 59 triplets) suggests that homozygosity does not lead to devastating disease or to very early disease onset (Bahl et al. 2005). Overall, the repeat lengths typically associated with SCA12 are somewhat longer than in the polyglutamine diseases (with the possible exception of SCA3). Also, repeat lengths over 70 triplets, seen in a number of SCA12 patients, generally result in juvenile onset of polyglutamine diseases. The combination of relatively mild phenotype and relatively long repeat expansions compared with the polyglutamine diseases suggests that SCA12 is unlikely to be a polyglutamine disorder.

Repeat instability during vertical transmission and the correlation between longer repeat length and younger age of onset are characteristic features of repeat expansion diseases. In HDL2 there is now clear evidence that longer

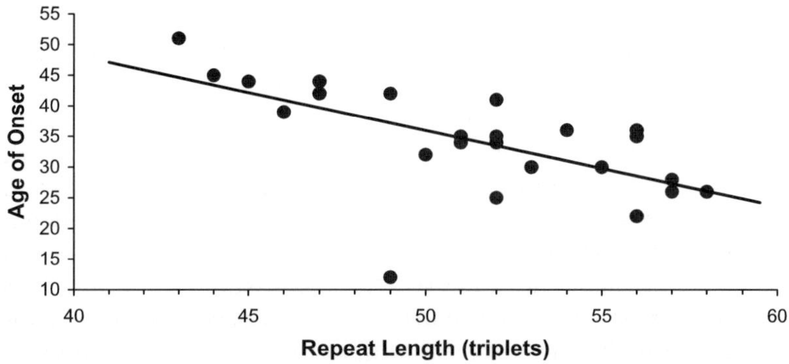

Fig. 4 Younger age of HDL2 onset is associated with longer repeat lengths. $N = 24$, $R = 0.62$, $r^2 = 0.39$, $p = 0.0011$. The relationship is quantitatively similar to that observed in HD. (Reprinted with permission, *Annals of Neurology*, copyright 2004; Margolis et al. 2004)

repeat length correlates with an earlier onset age, with a relationship similar to that seen in HD (Fig. 4) (Margolis et al. 2004a). Preliminary evidence suggests the same phenomenon occurs in SCA12 (Margolis et al. 2004b). The few examples in which the length of an HDL2 expanded repeat is known in both parent and child suggest a tendency towards a modest increase in repeat length during vertical transmission, consistent with the suggestion of anticipation in the index pedigree (Margolis and Ross 2001; Margolis et al. 2004a). While slight variations in repeat length among siblings with SCA12 suggest some change in repeat length during vertical transmission, there is not a clear direction to the change. This is somewhat surprising, given the relatively long and variable repeat lengths observed in SCA12. As would be expected given the minimal change in repeat length during vertical transmission, there is little evidence for anticipation in SCA12.

6
The HDL2 and the SCA12 Repeats Do Not Encode Polyglutamine

Our immediate assumption after detection of expansions by the RED assay was that HDL2 and SCA12 would be polyglutamine disorders. Most obviously, in all previous neurodegenerative diseases associated with d(CAG)/d(CTG) repeat expansions except SCA8 (and myotonic dystrophy, in which neurodegeneration is less prominent), the repeat is in-frame to encode polyglutamine. Unlike SCA8, the length of the repeat expansions in HDL2 is in exactly the same range as in the known polyglutamine diseases. The repeat length for SCA12, as noted before, was less consistent with the that for the polyglutamine disorders, but was not far from the range. Further, in HDL2, intranuclear inclusions in brain from affected individuals, as noted before, stain with an antibody with some specificity for polyglutamine expansions. However, for each disease, this initial assumption appears to be wrong.

The HDL2 locus maps to 16q24.3. In the d(CAG) orientation, an open reading frame exists in which the repeat is in-frame to encode polyglutamine, with 115 residues from the first d(ATG) to the repeat, and 27 amino acids from the *C*-terminal end of the repeat before the first stop codon. However, there is no known gene, predicted gene, or expressed sequence tag (EST) that contains this open reading frame and the predicted protein sequence is not homologous to that of any known proteins. Our experimental efforts to identify a transcript containing this open reading frame expressed in brain using reverse transcription (RT) PCR failed. In addition, multitissue Northern blots using antisense oligonucleotide probes to sequence flanking either side of the d(CAG) repeat failed to detect a transcript in brain tissue, as did complementary DNA (cDNA) library screening (Holmes et al. 2001b). Preliminary Western blots of human brain, probed with polyclonal antibodies generated against the putative sequence flanking the d(CAG) repeat, indicate that the open reading frame is either untranscribed or untranslated.

In the d(CTG) orientation, the HDL2 repeat is located 760 nucleotides downstream of the 3′ end of exon 1 of the gene junctophilin-3 (*JPH3*), and more than 36 kb upsteam of exon 2 (Holmes et al. 2001b) (Fig. 5). A polyadenylation signal is located 281 nucleotides 3′ to the end of the repeat, GENSCAN predicts a transcript in which exon 1 of *JP-3* is spliced to an exon containing the d(CTG) repeat, and multiple ESTs exist in which exon 1 is

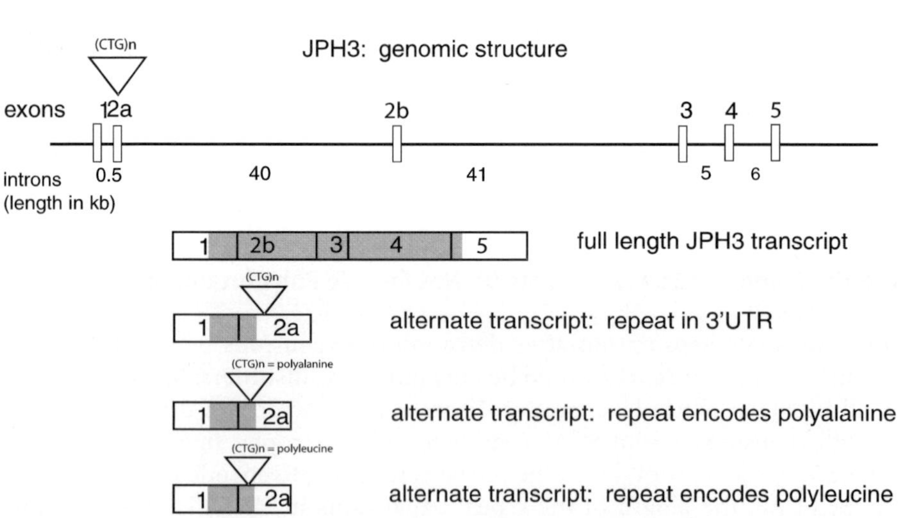

Fig. 5 *JPH3* genomic structure and transcripts. Several other exons 3′ to exon 2b may also exist which are only infrequently included in transcripts. Additional transcripts missing one or more middle exons also exist. Not to scale. See text for details

spliced to an alternate terminal exon containing the repeat. We therefore concluded that the d(CTG) repeat is contained within an alternate exon of *JPH3*, which we refer to as 2A. The use of multiple different canonical splice acceptor sites in the various exon 1–exon 2A ESTs indicated that the repeat could be alternatively in-frame to encode polyleucine or polyalanine, or could fall in the 3′ untranslated region (UTR). Using RT-PCR, we experimentally confirmed the existence of these transcripts in cerebral cortex, and confirmed the presence of multiple splice acceptor sites leading to different reading frames.

In SCA12, unlike HDL2, there is little ambiguity about which strand is relevant for disease pathogenesis, as there is no gene or open reading frame predicted on the strand in which the repeat is in the d(CTG) orientation. In the d(CAG) orientation, the repeat is 133 nucleotides upstream of the 5′ end of a cDNA of the initially published version of the gene *PPP2R2B* (Mayer et al. 1991) on 5q31–33. However, the precise relationship between the repeat and the *PPP2R2B* transcript remains ambiguous. A multiple start-site element downstream-1 (MED-1) sequence d(GCTCCC) (Ince and Scotto 1995) exists 65 nucleotides 3′ to the repeat. This motif is typically found downstream of transcription initiation sites in genes that contain multiple initiation sites. Consistent with the presence of the MED-1 sequence, it appears that this transcript has variable start sites. A combination of experimental and bioinformatic evidence places the repeat, alternatively, as close as nine nucleotides upstream of the transcription initiation site or within the 5′ UTR region of this exon. The structure of this exon is further complicated by an internal alternative, variable intron, spliced out beginning 28 or 141 nucleotides 3′ of the d(CAG) repeat.

The relationship between *PPP2R2B* and the SCA12 repeat locus is emerging as much more complicated than we had assumed when we first described the SCA12 mutation. At least six exons have been detected upstream of the 5′ exon of the originally described *PPP2R2B* transcript. These upstream exons are spliced in a variety of combinations to exons 9–16, which appear to be invariant. The result is multiple different *N*-terminal regions of the encoded protein (Fig. 6).

Initially, no experimental or bioinformatic evidence suggested that the SCA12 d(CAG) repeat might encode polyglutamine, either within an unidentified open reading frame of *PPP2R2B*, or within another unidentified adjacent or overlapping gene (Holmes et al. 1999, 2001a). However, recently available ESTs indicate that a small fraction of PPP2R2B transcripts may contain the d(CAG) repeat; several ESTs from human and mouse brain, as well as other tissues, begin within the repeat or 5′ to the repeat. With excision of the internal intron, the repeat is contained within an open reading frame contiguous with the normal *PPP2R2B* open reading frame. These transcripts contain various insertions and deletions, such that the actual reading frame for the repeat is uncertain, and the repeat could encode polyglutamine, polyserine,

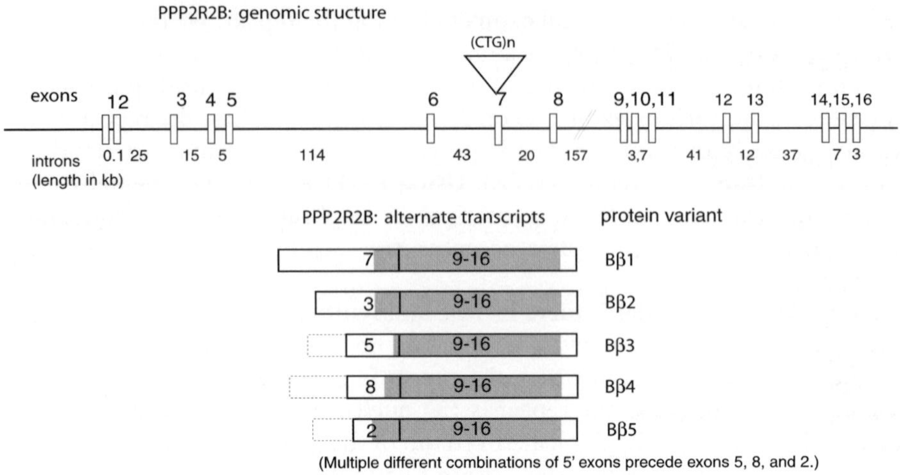

Fig. 6 *PPP2R2B* genomic structure and splicing. Alternative splice variants of *PPP2R2B* are shown in simplified form. *Darkened areas* of transcripts represent open reading frames. Additional upstream exons are present in transcripts in which the open reading frame begins with exon 2, 5, or 8. Exons 1 and 7 have multiple alternative splice accepter sites. Not to scale. See text for details

or polyalanine. However, no transcripts were detected on Northern blots of messenger RNA from brain and the LA-N-1 neuroblastoma cell line (which expresses high levels of *PPP2R2B*) probed with antisense oligonucleotides targeted to sequence immediately 5′ and 3′ to the d(CAG) repeat. Similarly, no protein was detected on Western blots derived from lymphoblastoid cells of affected SCA12 patients probed with the 1C2 antibody. The absence of 1C2-positive inclusions in SCA12 brain is further evidence against a pathogenic role of polyglutamine.

In sum, unlike most of the other neurodegenerative diseases caused by a d(CAG)/d(CTG) repeat expansion, we hypothesize that neither the SCA12 nor the HDL2 repeat is likely to encode polyglutamine, at least at a significant level. A similar conclusion has been reached for SCA8 (Koob et al. 1999). One implication of this conclusion is that there is dissociation between the nature of the mutation at the molecular level and the mode of disease pathogenesis; the presence of a d(CAG)/d(CTG) repeat expansion does not necessarily imply that disease is caused by a polyglutamine expansion. In addition, the selective forces favoring the existence of expandable d(CAG)/d(CTG) repeats within the human genome must not operate only through pressure on the length of polyglutamine tracts. It seems plausible that polymorphism in d(CAG)/d(CTG) repeat length may function as a general mechanism for introducing variability into the genome, affecting genomic DNA structure, transcript expression, RNA splicing, or protein sequence, depending on the context of the repeat.

7
HDL2 and SCA12: Clues to Pathogenesis

7.1
HDL2 and Gain of Function

If the repeat expansions in SCA12 and HDL2 do not encode polyglutamine, then what alternative pathogenic mechanisms may account for the development of these diseases? One possibility is a toxic gain of function of a mutant protein product, parallel to that of the polyglutamine diseases. In SCA12, transcripts containing the repeat in an open reading frame are very uncommon, diminishing, though not completely eliminating, the potential for a pathogenic role of an expanded amino acid repeat. In HDL2, the alternative transcript of *JPH3* containing exon 2A could potentially give rise to a protein containing expanded polyalanine or polyleucine. Both polyalanine and polyleucine expansions are toxic in cell models (Rankin et al. 2000; Dorsman et al. 2002), and the disease oculopharyngeal muscular dystrophy, an adult onset neuromuscular degenerative disorder, is caused by an alanine expansion (Brais et al. 1998). If HDL2 was the result of either polyalanine or polyleucine toxicity, we would predict that the intranuclear inclusions in HDL2 would stain with antibodies raised against epitopes flanking either the polyalanine or polyleucine tract, and that Western blots of HDL2 brain probed with these antibodies would show a band corresponding to the alternative form of *JPH3* containing an expanded polyalanine or polyleucine tract. Our preliminary data suggest that these predictions are false.

If not toxic gain of function at the protein level, is toxicity possible at the RNA level? Here the precedent is myotonic dystrophy, types 1 and 2 (DM1, DM2) (Ranum and Day 2004). DM1 is caused by a d(CTG) repeat expansion in the 3′ UTR of the gene myotonic dystrophy protein kinase 1 (*DMPK1*) (Brook et al. 1992; Mahadevan et al. 1992; Fu et al. 1992). Brain and muscle tissue from patients with DM1, cell models in which DM1 is overexpressed, and mouse models in which the d(CTG) repeat is overexpressed all indicate that expression of the transcript containing the r(CUG) repeat results in small nuclear RNA aggregates containing the expanded repeat (Taneja et al. 1995; Davis et al. 1997; Jiang et al. 2004; Mankodi et al. 2000, 2003). The aggregates, or the unaggregated transcripts with the r(CUG) expansion, may exert a dominant effect on RNA splicing through the dysregulation of RNA binding proteins, most prominently CUG-BP1 and muscleblind (Timchenko et al. 1996; Miller et al. 2000). DM2, with a very similar phenotype to DM1, is caused by an intronic d(CCTG) repeat expansion in *ZNF9* (Liquori et al. 2001). The findings that RNA nuclear aggregation and dysregulation of RNA binding proteins are similar in DM1 and DM2 provides strong support for a causal role of RNA transcripts in these diseases, and is a prime example of how analysis of a rare form of a disease can shed light on a more common form (Liquori et al. 2001; Mankodi et al. 2001; Fardaei et al. 2002). There is no evidence that expansions of repeats in the d(CAG) orientation result in RNA toxicity or inclusions, so it seems less likely, though untested, that RNA toxicity is central to SCA12 pathogenesis. On the other hand, the HDL2 repeat is found in alternate *JPH3* transcripts. Preliminary data suggest that the HDL2 expansion can lead to RNA inclusions, with some properties similar to those seen in DM1 and DM2. This finding provides tantalizing though unconfirmed evidence that a toxic transcript may play a role in HDL2 pathogenesis.

7.2 The Function of *JPH3* and *PPP2R2B*

Could the HDL2 or SCA12 repeats have an impact on gene expression? This possibility is intriguing because of the known functions of *JPH3* and *PPP2R2B*. *JPH3* encodes one member of a four protein family (Takeshima et al. 2000; Nishi et al. 2000, 2003). Whereas *JPH1* is expressed in skeletal muscle and *JPH2* is expressed in cardiac and skeletal muscle, *JPH3*, along with *JPH4* (Nishi et al. 2003), is expressed in brain and to a much lesser extent in testes. While detailed studies of *JPH3* expression have not been performed, JP3 (the mouse orthologue of *JPH3*) is widely expressed in the brain, with high density in hippocampus, cerebral cortex, striatum, and cerebellar cortex (Nishi et al. 2003).

The N-terminus of each junctophilin contains a series of repeating tracts, each 14 amino acids in length, termed membrane orientation and recognition nexus (MORN) motifs that serve to anchor the N-terminus of the protein to the plasma membrane. The junctophilins also have an endoplasmic reticulum (ER)/sarcoplasmic reticulum (SR) transmembrane domain, which serves to anchor the C-terminus to the ER or the SR (Takeshima et al. 2000). This structure suggested that the junctophilins may tether plasma membrane to the ER/SR, placing plasma membrane voltage sensors adjacent to ER/SR ion channels, particularly ryanodine and IP3 receptor-gated calcium channels (Fig. 7). Consistent with this hypothesis, the junctional complexes in *JP1* knockout mice (Ito et al. 2001), compared with controls, were fewer in number and structurally abnormal, muscle response to electrical stimulation was impaired, and muscle response to calcium was abnormal. The implication is that loss of *JPH3* expression in brain could similarly impair neuronal function, and perhaps survival, through destabilization of calcium flux.

Like *JPH3*, the function of *PPP2R2B* has been partially established. It is one of multiple alternative regulatory units of the ubiquitous enzyme protein phosphatase 2A (PP2A, also termed PP2). PP2A has been implicated in a plethora of cellular processes, including oncogenesis, growth and differentiation, DNA replication, morphogenesis and cytokinesis, regulation of kinase cascades, ion channel function, neurotransmitter release, microtubule assembly, and apoptosis (Janssens et al. 2005; Price and Mumby 1999; Santoro and

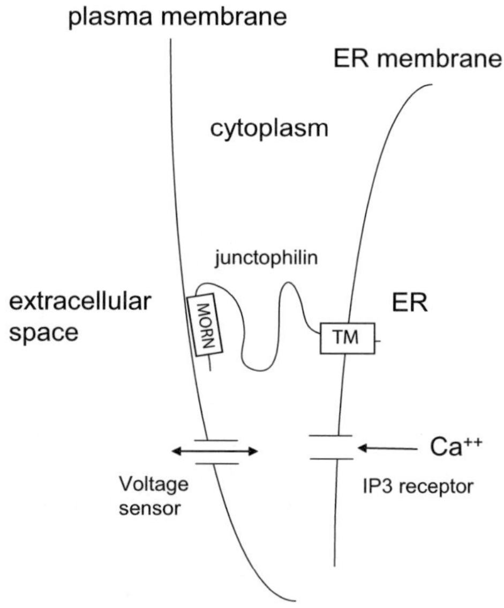

Fig. 7 Junctophilin function. Putative role of junctophilin in bridging the gap between the plasma membrane and the endoplasmic reticulum (*ER*), bringing plasma membrane voltage sensors into the proximity of IP3 receptors that modulate calcium release from the ER. Junctophilin membrane orientation and recognition nexus motifs (*MORN*), which serve to anchor junctophilin to the inner plasma membrane, and the ER transmembrane domain (*TM*) are indicated. (Adapted from Takeshima et al. 2000)

Grummt 2001; Virshup 2000). PP2A is trimeric in structure, and includes a catalytic subunit (PP2Ac or subunit C; two known isoforms), a structural subunit (PR65 or subunit A; two known isoforms), and a regulatory subunit (more than 12 known isoforms divided into three families) (Fig. 8). The structural and catalytic subunits form a complex to which, in a highly regulated process, one of the regulatory subunits is recruited. As many as 75 different trimeric combinations may exist. The structural and catalytic subunits are constitutively expressed in all mammalian cells, whereas expression of the regulatory subunits is spatially and temporally restricted. It is the regulatory subunits that confer much of the substrate specificity and intracellular targeting to the entire enzyme complex (McCright et al. 1996; Millward et al. 1999; Tehrani et al. 1996; Virshup, 2000).

PPP2R2B encodes a protein usually termed PR55β or Bβ (Mayer et al. 1991) that is widely and specifically expressed in neurons throughout the brain (Strack et al. 1998). The *PPP2R2* family of subunits, which includes *PPP2R2B*, regulate PP2A dephosphorylation of, among other substrates, vimentin (Turowski et al. 1999), histone-1 (Ferrigno et al. 1993), and tau (Sontag et al. 1996). Mutations of different *PPP2R2* subunits produce deleterious effects in yeast and *Drosophila* (Shiomi et al. 1994). Downregulation of PP2A contain-

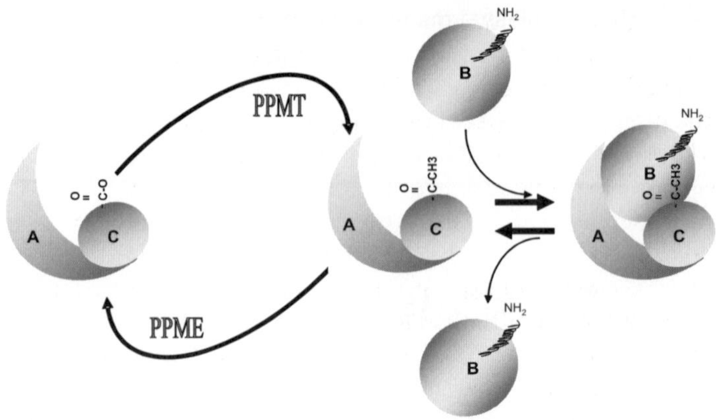

Fig. 8 PP2A holoenzyme structure. After a complex forms between structural (*A*) and catalytic (*C*) subunits, methylestification of the catalytic unit by a 38-kDa AdoMet-dependent PP2A methyltransferase (*PPMT*) favors the incorporation of a regulatory unit (*B*). Demethylation by a 46-kDa PP2A methylesterase (*PPME*) destabilizes the trimeric structure (Tolstykh et al. 2000). The regulatory unit forms a β-propeller structure. The protruding N-terminus of the regulatory unit targets the PP2A holoenzyme to intracellular domains and substrates. N-terminus variability among the multiple regulatory subunits provides spatial and temporal specificity for PP2A activity

ing either *PPP2R2A* or *PPP2R2B* in cultured mammalian brain slices results in Alzheimer's-like accumulation of hyperphosphorylated tau (Sontag et al. 1999; Gong et al. 2000).

Adding to the complexity of understanding the physiological role of *PPP2R2B*, it, like other regulatory units, has multiple alternative splice forms. In the mouse, at least three splice forms have been examined (Schmidt et al. 2002), each with a variable first exon spliced to invariant downstream exons (our preliminary examination of human splice variants demonstrates an even greater complexity). This is of particular interest because each alternative first exon is under the control of a different promoter and encodes a different *N*-terminal signal. Structurally, the *N*-terminal portion of each regulatory subunit is exposed, with the functional consequence that this N-terminus provides the targeting information for PP2A.

7.3
Expression Level and Pathogenesis

Could loss of expression of the mutated *JPH3* allele lead to the HDL2 phenotype? So far, direct evidence of a decrease in expression of a full-length *JPH3* transcript in HDL2 brain has proven elusive. On the other hand, Takeshima and colleagues have generated JP-3 knockout mice. At 6–12 weeks of age, these mice show motor incoordination but no gross brain abnormalities and no electrophysiological abnormalities of cerebellar function (Ito et al. 2001). Our preliminary data suggest that even mice with one deleted copy of *JP3* may have motor abnormalities and a shorter life span than littermate controls. Overall, however, the phenotype appears relatively mild compared with the severity of HDL2, perhaps reflecting a redundancy in function between *JP3* and *JP4*. It seems reasonable to propose that loss of *JPH3* expression may contribute to the HDL2 phenotype, but that it is not sufficient to cause the disease.

Could SCA12 stem from an effect of the repeat expansion on *PPP2R2B* expression? As noted before, the SCA12 expansion mutation is found immediately flanking, and at times within the 5′ UTR or even the open reading frame, of one of the alternatively spliced first exons of the *PPP2R2B* transcript. Strack and colleagues compared the transcript and protein product (termed Bβ1) generated with the rat orthologue that includes this exon with the transcript and protein product (termed Bβ2) generated by including an alternative first exon located 150 kb upstream. Transcripts containing the alternative first exons are divergently regulated during development, with Bβ1 levels declining from late fetal to adult stages, while Bβ2 levels increased. Bβ2, but not Bβ1, localized to mitochondria, and increased cell death following serum deprivation in a PC6-3 cell line (Dagda et al. 2003).

This line of investigation demonstrates that the ratio of Bβ species containing alternative *N*-terminal regions might considerably affect cell function and survival. The SCA12 repeat expansion could alter this ratio in a number of ways. For instance, expansion could influence splicing, as exemplified by the effect on exon skipping in bovine *c-myb* with and without an insertion of a dodecamer repeat in exon 9 (Shinagawa et al. 1997). Such a mechanism could be significant in HDL2 as well, as a shift from the full-length to a truncated *JPH3* transcript would result in less junctophilin-3 protein capable of tethering plasma membrane to the ER. Alternatively, given its location, the SCA12 repeat could affect transcriptional activity, as exemplified by the dodecamer repeat expansion in the promoter of *CTSB* (encoding cystatin B) that reduces gene expression in association with myotonic epilepsy type 1 (Lalioti et al. 1997). Preliminary evidence from a reporter gene assay suggests, contrary to expectations, that SCA12 repeat expansions increase transcription of the human equivalent of the Bβ1 isoform described by Strack. Should additional experiments confirm this result, it is possible that SCA12 may arise, at least in part, from excess expression of *PPP2R2B* variant Bβ1 that results in a shift in PP2A target specificity.

8 Conclusion

HDL2 and SCA12, the most recent of the d(CAG)/d(CTG) expansion diseases to be discovered, share intriguing similarities. Each is rare, and primarily limited to one ethnic group, as one might expect for diseases identified after the more common Mendelian causes of the HD and SCA phenotypes had been determined. HDL2 and SCA12 both demonstrate the genetic factors shared by most other repeat diseases: instability in vertical transmission, association of longer repeat length with earlier onset age, intermediate repeat lengths of uncertain significance, and incomplete penetrance of shorter expansions. Both, like all the other d(CAG)/d(CTG) disorders, are primarily neurodegenerative (DM1 and DM2 less markedly so than the others). It is not difficult to conceptualize a common mechanism through which any protein with a polyglutamine expansion might selectively affect the brain; however, it is less clear how the broader class of d(CAG)/d(CTG) expansion mutations, several of which do not encode polyglutamine, might also be relatively specific for neurodegeneration (again less so for DM1 and DM2). One possibility is ascertainment bias, in that this type of mutation has been sought in brain diseases but not others. Another explanation is that neurons are selectively vulnerable to several different classes of insults, including long polyglutamine tracts, long tracts of proteins encoded by other reading frames, and transcripts with long r(CUG) or other repeating units. If this is the case, then the pathogenesis for SCA12, HDL2, and SCA8 should be sought in abnormal transcripts or proteins that have gained a toxic function, rather than (or in addition to) the effect of the expansion on expression of the gene in which it is located. DM1 serves as an illustrative example of this concept, as pathogenesis was illuminated by the recognition that overexpressing transcripts with r(CUG) expansions could recapitulate many of the features of the disease in various model systems.

On the other hand, differences between SCA12 and HDL2 suggest other conclusions. While the complete SCA12 syndrome does not resemble other known disorders, the similarity of HDL2 to HD is striking. This presents a golden opportunity to search for common mechanisms of disease pathogenesis, especially as played out in the medium spiny neurons of the striatum. Perhaps HDL2 can serve a clarifying role for HD pathogenesis in the same way that DM2 illuminated the pathogenesis of DM1. In addition, the possibility that HDL2 may arise from toxic transcripts with expanded r(CUG) repeats raises important issues for the pathogenesis of DM: Why is the phenotype of DM1 and DM2 so different from that of HDL2? Does this imply that the mutation in each gene has other effects, or is it a matter of relative expression levels in different brain regions?

The variability of the SCA12 phenotype, and its relatively benign course, differ from those of HDL2 and most of the other d(CAG)/d(CTG) expansion diseases. This is consistent with an effect of the mutation on gene expression levels. If so, this will be one of the few diseases caused by a mutation directly affecting a phosphatase, and could therefore provide a portal into novel mechanisms of neurodegeneration.

Finally, HDL2 and SCA12 raise the question of the value of searching for and characterizing rare disorders. Given the minimal public health impact of such rare diseases, would scarce resources be better applied to the direct study of more common diseases? Here we put forth a clear answer: rare diseases provide unique opportunities to understand more common disorders.

Acknowledgements The authors thank Christopher A. Ross, Adam Rosenblatt, Stefan Strack, Achal K. Srivastava, Satish Jain, K.K. Sinha, Mark Moliver, and Mitali Mukerji for valuable insight and guidance; Abhijit Agarwal and Alka Ahuja for research assistance; Marie Sonderman for technical assistance; and the individuals with SCA12 and HDL2, and their families for their patience and cooperation. This work was supported by NIH NS38054, NS16375, and MH01275, and the Hereditary Disease Foundation.

References

Andrew SE, Goldberg YP, Kremer B, Squitieri F, Theilmann J, Zeisler J, Telenius H, Adam S, Almquist E, Anvret M, Lucotte G, Stoessl AJ, Campanella G, Hayden MR (1994) Huntington disease without CAG expansion: Phenocopies or errors in assignment? Am J Hum Genet 54:852–863

Bahl S, Virdi K, Mittal U, Sachdeva M, Kalla A, Holmes SE, O'Hearn E, Margolis RL, Jain S, Srivastava AK, Mukerji M (2005) Evidence of a common founder for SCA12 in the Indian population. Ann Hum Genet 69:528–534

Bauer I, Gencik M, Laccone F, Peters H, Weber BH, Feder EH, Weirich H, Morris-Rosendahl DJ, Rolfs A, Gencikova A, Bauer P, Wenning GK, Epplen JT, Holmes SE, Margolis RL, Ross CA, Riess O (2002) Trinucleotide repeat expansions in the junctophilin-3 gene are not found in Caucasian patients with a Huntington's disease-like phenotype. Ann Neurol 51:662

Brais B, Bourchard JP, Xie YG, Rochefort DL, Chretien N, Tome FM, LaFreniere RG, Rommens JM, Uyama E, Nohira O, Blumen S, Korcyn AD, Heutink P, Mathieu J, Duranceau A, Codere F, Fardeau M, Rouleau GA (1998) Short GCG expansions in the PABP2 gene cause oculopharyngeal muscular dystrophy. Nat Genet 18:164–167

Brook JD, McCurrach ME, Harley HG, Buckler AJ, Church D, Aburatani H, Hunter K, Davies J, Shelbourne P, Buxton J, Jones C, Juvonen V, Johnson K, Harper PS, Shaw DJ, Housman DE (1992) Molecular basis of myotonic dystrophy: expansion of a trinucleotide (CTG) repeat at the 3′ end of a transcript encoding a protein kinase family member. Cell 68:799–808

Brusco A, Gellera C, Cagnoli C, Saluto A, Castucci A, Michielotto C, Fetoni V, Mariotti C, Migone N, Di Donato S, Taroni F (2004) Molecular genetics of hereditary spinocerebellar ataxia: mutation analysis of spinocerebellar ataxia genes and CAG/CTG repeat expansion detection in 225 Italian families. Arch Neurol 61:727–733

Cholfin JA, Sobrido MJ, Perlman S, Pulst SM, Geschwind DH (2001) The SCA12 mutation as a rare cause of spinocerebellar ataxia. Arch Neurol 58:1833–1835

Dagda RK, Zaucha JA, Wadzinski BE, Strack S (2003) A developmentally regulated, neuron-specific splice variant of the variable subunit Bbeta targets protein phosphatase 2A to mitochondria and modulates apoptosis. J Biol Chem 278:24976–24985

Davis BM, McCurrach ME, Taneja KL, Singer RH, Housman DE (1997) Expansion of a CUG trinucleotide repeat in the 3′ untranslated region of myotonic dystrophy protein kinase transcripts results in nuclear retention of transcripts. Proc Natl Acad Sci USA 94:7388–7393

Dorsman JC, Pepers B, Langenberg D, Kerkdijk H, Ijszenga M, Den Dunnen JT, Roos RA, van Ommen GJ (2002) Strong aggregation and increased toxicity of polyleucine over polyglutamine stretches in mammalian cells. Hum Mol Genet 11:1487–1496

Fardaei M, Rogers MT, Thorpe HM, Larkin K, Hamshere MG, Harper PS, Brook JD (2002) Three proteins, MBNL, MBLL and MBXL, co-localize in vivo with nuclear foci of expanded-repeat transcripts in DM1 and DM2 cells. Hum Mol Genet 11:805–814

Ferrigno P, Langan TA, Cohen P (1993) Protein phosphatase 2A1 is the major enzyme in vertebrate cell extracts that dephosphorylates several physiological substrates for cyclin-dependent protein kinases. Mol Biol Cell 4:669–677

Fu Y-H, Pizzuti A, Fenwick RG, King J, Rajnarayan S, Dunne PW, Dubel J, Nasser GA, Ashizawa T, De Jong P, Wieringa B, Korneluk R, Perryman MB, Epstein HF, Caskey CT (1992) An unstable triplet repeat in a gene related to myotonic muscular dystrophy. Science 255:1256–1258

Fujigasaki H, Uchihara T, Koyano S, Iwabuchi K, Yagishita S, Makifuchi T, Nakamura A, Ishida K, Toru S, Hirai S, Ishikawa K, Tanabe T, Mizusawa H (2000) Ataxin-3 is translocated into the nucleus for the formation of intranuclear inclusions in normal and Machado-Joseph disease brains. Exp Neurol 165:248–256

Fujigasaki H, Verma IC, Camuzat A, Margolis RL, Zander C, Lebre AS, Jamot L, Saxena R, Anand I, Holmes SE, Ross CA, Durr A, Brice A (2001) SCA12 is a rare locus for autosomal dominant cerebellar ataxia: a study of an Indian family. Ann Neurol 49:117–121

Gong CX, Lidsky T, Wegiel J, Zuck L, Grundke-Iqbal I, Iqbal K (2000) Phosphorylation of microtubule-associated protein tau is regulated by protein phosphatase 2A in mammalian brain. Implications for neurofibrillary degeneration in Alzheimer's disease. J Biol Chem 275:5535–5544

Hellenbroich Y, Kaulich M, Opitz S, Schwinger E, Zuhlke C (2004) No association of the SCA1 (CAG)31 allele with Huntington's disease, myotonic dystrophy type 1 and spinocerebellar ataxia type 3. Psychiatr Genet 14:61–63

Holmes SE, O'Hearn EE, McInnis MG, Gorelick-Feldman DA, Kleiderlein JJ, Callahan C, Kwak NG, Ingersoll-Ashworth RG, Sherr M, Sumner AJ, Sharp AH, Ananth U, Seltzer WK, Boss MA, Vieria-Saecker AM, Epplen JT, Riess O, Ross CA, Margolis RL (1999) Expansion of a novel CAG trinucleotide repeat in the 5′ region of PPP2R2B is associated with SCA12. Nat Genet 23:391–392

Holmes SE, Hearn EO, Ross CA, Margolis RL (2001a) SCA12: an unusual mutation leads to an unusual spinocerebellar ataxia. Brain Res Bull 56:397–403

Holmes SE, O'Hearn E, Rosenblatt A, Callahan C, Hwang HS, Ingersoll-Ashworth RG, Fleisher A, Stevanin G, Brice A, Potter NT, Ross CA, Margolis RL (2001b) A repeat expansion in the gene encoding junctophilin-3 is associated with Huntington disease-like 2. Nat Genet 29:377–378

Holmes SE, O'Hearn E, Brahmachari SK, Choudhry S, Srivastava AK, Jain S, Ross CA, Margolis RL (2003) SCA12. In: Pulst S-M (ed) Genetics of movement disorder. Academic, San Diego, pp 121–130

Huntington's Disease Collaborative Research Group (1993) A novel gene containing a trinucleotide repeat that is expanded and unstable on Huntington's disease chromosomes. Cell 72:971–983

Ince TA, Scotto KW (1995) A conserved downstream element defines a new class of RNA polymerase II promoters. J Biol Chem 270:30249–30252

Ito K, Komazaki S, Sasamoto K, Yoshida M, Nishi M, Kitamura K, Takeshima H (2001) Deficiency of triad junction and contraction in mutant skeletal muscle lacking junctophilin type 1. J Cell Biol 154:1059–1067

Janssens V, Goris J, van Hoof C (2005) PP2A: the expected tumor suppressor. Curr Opin Genet Dev 15:34–41

Jiang H, Mankodi A, Swanson MS, Moxley RT, Thornton CA (2004) Myotonic dystrophy type 1 is associated with nuclear foci of mutant RNA, sequestration of muscleblind proteins and deregulated alternative splicing in neurons. Hum Mol Genet 13:3079–3088

Koob MD, Benzow KA, Bird TD, Day JW, Moseley ML, Ranum LP (1998) Rapid cloning of expanded trinucleotide repeat sequences from genomic DNA. Nat Genet 18:72–75

Koob MD, Moseley ML, Schut LJ, Benzow KA, Bird TD, Day JW, Ranum LPW (1999) An untranslated CTG expansion causes a novel form of spinocerebellar ataxia (SCA8). Nat Genet 21:379–384

Krause A, Temlett J, van der Meyden J, Ross CA, Callahan C, Margolis R (2002) CAG/CTG repeat expansions at the HDL2 locus are a common cause of Huntington disease in Black South Africans. Am J Hum Genet, p 528

Lalioti MD, Scott HS, Buresi C, Rossier C, Bottani A, Morris MA, Malafosse A, Antonarakis SE (1997) Dodecamer repeat expansion in cystatin B gene in progressive myoclonus epilepsy. Nature 386:847–851

Liquori CL, Ricker K, Moseley ML, Jacobsen JF, Kress W, Naylor SL, Day JW, Ranum LP (2001) Myotonic dystrophy type 2 caused by a CCTG expansion in intron 1 of ZNF9. Science 293:864–867

Mahadevan M, Tsilfidis C, Sabourin L, Shutler G, Amemiya C, Jansen G, Nelville C, Narang M, Barcelo J, O'Hoy K, Leblond S, Earle-MacDonald J, DeJong PJ, Wieringa B, Korneluk RG (1992) Myotonic dystrophy mutation: an unstable CTG repeat in the 3′ untranslated region of the gene. Science 255:1253–1255

Mankodi A, Logigian E, Callahan L, McClain C, White R, Henderson D, Krym M, Thornton CA (2000) Myotonic dystrophy in transgenic mice expressing an expanded CUG repeat. Science 289:1769–1773

Mankodi A, Teng-umnuay P, Krym M, Henderson D, Swanson M, Thornton CA (2003) Ribonuclear inclusions in skeletal muscle in myotonic dystrophy types 1 and 2. Ann Neurol 54:760–768

Mankodi A, Urbinati CR, Yuan QP, Moxley RT, Sansone V, Krym M, Henderson D, Schalling M, Swanson MS, Thornton CA (2001) Muscleblind localizes to nuclear foci of aberrant RNA in myotonic dystrophy types 1 and 2. Hum Mol Genet 10:2165–2170

Margolis RL, Ross CA (2001) Expansion explosion: new clues to the pathogenesis of repeat expansion neurodegenerative diseases. Trends Mol Med 7:479–482

Margolis RL, O'Hearn E, Rosenblatt A, Willour V, Holmes SE, Franz ML, Callahan C, Hwang HS, Troncoso JC, Ross CA (2001) A disorder similar to Huntington's disease is associated with a novel CAG repeat expansion. Ann Neurol 50:373–380

Margolis RL, Holmes SE, Rosenblatt A, Gourley L, O'Hearn E, Ross CA, Seltzer WK, Walker RH, Ashizawa T, Rasmussen A, Hayden M, Almqvist EW, Harris J, Fahn S, MacDonald ME, Mysore J, Shimohata T, Tsuji S, Potter N, Nakaso K, Adachi Y, Nakashima K, Bird T, Krause A, Greenstein P (2004a) Huntington's disease-like 2 (HDL2) in North America and Japan. Ann Neurol 56:670–674

Margolis RL, O'Hearn E, Holmes SE, Srivastava AK, Mukherji M, Sinha KK (2004b) Spinocerebellar ataxia type 12. Gene Clinics www.geneclinics.org

Maruyama H, Izumi Y, Morino H, Oda M, Toji H, Nakamura S, Kawakami H (2002) Difference in disease-free survival curve and regional distribution according to subtype of spinocerebellar ataxia: a study of 1286 Japanese patients. Am J Med Genet 114:578–583

Matsumura R, Futamura N, Ando N, Ueno S (2003) Frequency of spinocerebellar ataxia mutations in the Kinki district of Japan. Acta Neurol Scand 107:38–41

Mayer RE, Hendrix P, Cron P, Mattheis R, Stone SR, Goris J, Merlevede W, Hofsteenge J, Hemmings BA (1991) Structure of the 55-kDa regulatory subunit of protein phosphatase 2A: evidence for a neuronal-specific isoform. Biochem J 30:3589–3597

McCright B, Rivers AM, Audlin S, Virshup DM (1996) The B56 family of protein phosphatase 2A (PP2A) regulatory subunits encodes differentiation-induced phosphoproteins that target PP2A to both nucleus and cytoplasm. J Biol Chem 271:22081–22089

Miller JW, Urbinati CR, Teng-umnuay P, Stenberg MG, Byrne BJ, Thornton CA, Swanson MS (2000) Recruitment of human muscleblind proteins to (CUG)(n) expansions associated with myotonic dystrophy. EMBO J 19:4439–4448

Millward TA, Zolnierowicz S, Hemmings BA (1999) Regulation of protein kinase cascades by protein phosphatase 2A. Trends Biochem Sci 24:186–191

Moore RC, Xiang F, Monaghan J, Han D, Zhang Z, Edstrom L, Anvret M, Prusiner SB (2001) Huntington disease phenocopy is a familial prion disease. Am J Hum Genet 69:1385–1388

Nishi M, Mizushima A, Nakagawara K, Takeshima H (2000) Characterization of human junctophilin subtype genes. Biochem Biophys Res Commun 273:920–927

Nishi M, Sakagami H, Komazaki S, Kondo H, Takeshima H (2003) Coexpression of junctophilin type 3 and type 4 in brain. Brain Res Mol Brain Res 118:102–110

O'Hearn E, Holmes SE, Calvert PC, Ross CA, Margolis RL (2001) SCA-12: Tremor with cerebellar and cortical atrophy is associated with a CAG repeat expansion. Neurology 56:299–303

O'Hearn E, Pletnikova O, Holmes SE, Trojanowski JQ, Margolis RL (2004) SCA12 Neuropathology: Cerebral cortical and cerebellar atrophy, Purkinje cell loss, and neuronal intranuclear inclusions. Mov Disord 19:1124

Persichetti F, Srinidhi J, Kanaley L, Ge P, Myers RH, D'Arrigo K, Barnes GT, MacDonald ME, Vonsattel JP, Gusella JF (1994) Huntington's disease CAG trinucleotide repeats in pathologically confirmed post-mortem brains. Neurobiol Disord 1:159–166

Price NE, Mumby MC (1999) Brain protein serine/threonine phosphatases. Curr Opin Neurobiol 9:336–342

Rankin J, Wyttenbach A, Rubinsztein DC (2000) Intracellular green fluorescent protein-polyalanine aggregates are associated with cell death. Biochem J 348(Pt 1):15–19

Ranum LP, Day JW (2004) Pathogenic RNA repeats: an expanding role in genetic disease. Trends Genet 20:506–512

Rosenblatt A, Ranen NG, Rubinsztein DC, Stine OC, Margolis RL, Wagster MV, Becher MW, Rosser AE, Leggo J, Hodges JR, French-Constant CK, Sherr M, Franz ML, Abbott MH, Ross CA (1998) Patients with features similar to Huntington's disease, without CAG expansion in huntingtin. Neurology 51:215–220

Santoro R, Grummt I (2001) Molecular mechanisms mediating methylation-dependent silencing of ribosomal gene transcription. Mol Cell 8:719–725

Sasaki H, Yabe I, Tashiro K (2003) The hereditary spinocerebellar ataxias in Japan. Cytogenet Genome Res 100:198–205

Schalling M, Hudson TJ, Buetow KW, Housman DE (1993) Direct detection of novel expanded trinucleotide repeats in the human genome. Nat Genet 4:135–139

Schmidt K, Kins S, Schild A, Nitsch RM, Hemmings BA, Gotz J (2002) Diversity, developmental regulation and distribution of murine PR55/B subunits of protein phosphatase 2A. Eur J Neurosci 16:2039–2048

Shimohata T, Onodera O, Honma Y, Hirota K, Nunomura Y, Kimura T, Kawachi I, Sanpei K, Nishizawa M, Tsuji S (2004) Gene diagnosis of patients with chorea. Rinsho Shinkeigaku 44:149–153

Shinagawa T, Ishiguro N, Horiuchi M, Matsui T, Okada K, Shinagawa M (1997) Deletion of c-myb exon 9 induced by insertion of repeats. Oncogene 14:2775–2783

Shiomi K, Takeichi M, Nishida Y, Nishi Y, Uemura T (1994) Alternative cell fate choice induced by low-level expression of a regulator of protein phosphatase 2A in the Drosophila peripheral nervous system. Development 120:1591–1599

Sinha KK, Worth PF, Jha DK, Sinha S, Stinton VJ, Davis MB, Wood NW, Sweeney MG, Bhatia KP (2004) Autosomal dominant cerebellar ataxia: SCA2 is the most frequent mutation in eastern India. J Neurol Neurosurg Psychiatry 75:448–452

Sontag E, Nunbhakdi-Craig V, Lee G, Bloom GS, Mumby MC (1996) Regulation of the phosphorylation state and microtubule-binding activity of Tau by protein phosphatase 2A. Neuron 17:1201–1207

Sontag E, Nunbhakdi-Craig V, Lee G, Brandt R, Kamibayashi C, Kuret J, White CL3, Mumby MC, Bloom GS (1999) Molecular Interactions among Protein Phosphatase 2A, Tau, and Microtubules. Implications for the regulation of tau phosphorylation and the development of tauopathies. J Biol Chem 274:25490–25498

Srivastava AK, Choudhry S, Gopinath MS, Roy S, Tripathi M, Brahmachari SK, Jain S (2001) Molecular and clinical correlation in five Indian families with spinocerebellar ataxia 12. Ann Neurol 50:796–800

Stevanin G, Camuzat A, Holmes SE, Julien C, Sahloul R, Dode C, Hahn-Barma V, Ross CA, Margolis RL, Durr A, Brice A (2002) CAG/CTG repeat expansions at the Huntington's disease-like 2 locus are rare in Huntington's disease patients. Neurology 58:965–967

Stevanin G, Fujigasaki H, Lebre AS, Camuzat A, Jeannequin C, Dode C, Takahashi J, San C, Bellance R, Brice A, Durr A (2003) Huntington's disease-like phenotype due to trinucleotide repeat expansions in the TBP and JPH3 genes. Brain 126:1599–1603

Strack S, Zaucha JA, Ebner FF, Colbran RJ, Wadzinski BE (1998) Brain protein phosphatase 2A: developmental regulation and distinct cellular and subcellular localization by B subunits. J Comp Neurol 392:515–527

Sulek A, Hoffman-Zacharska D, Bednarska-Makaruk M, Szirkowiec W, Zaremba J (2004) Polymorphism of trinucleotide repeats in non-translated regions of SCA8 and SCA12 genes: allele distribution in a Polish control group. J Appl Genet 45:101–105

Takeshima H, Komazaki S, Nishi M, Iino M, Kangawa K (2000) Junctophilins: a novel family of junctional membrane complex proteins. Mol Cell 6:11–22

Taneja KL, McCurrach M, Schalling M, Housman D, Singer RH (1995) Foci of trinucleotide repeat transcripts in nuclei of myotonic dystrophy cells and tissues. J Cell Biol 128:995–1002

Tehrani MA, Mumby MC, Kamibayashi C (1996) Identification of a novel protein phosphatase 2A regulatory subunit highly expressed in muscle. J Biol Chem 271:5164–5170

Timchenko LT, Timchenko NA, Caskey CT, Roberts R (1996) Novel proteins with binding specificity for DNA CTG repeats and RNA CUG repeats: implications for myotonic dystrophy. Hum Mol Genet 5:115–121

Tolstykh T, Lee J, Vafai S, Stock JB (2000) Carboxyl methylation regulates phosphoprotein phosphatase 2A by controlling the association of regulatory B subunits. EMBO J 19:5682–5691

Tsai HF, Liu CS, Leu TM, Wen FC, Lin SJ, Liu CC, Yang DK, Li C, Hsieh M (2004) Analysis of trinucleotide repeats in different SCA loci in spinocerebellar ataxia patients and in normal population of Taiwan. Acta Neurol Scand 109:355–360

Turowski P, Myles T, Hemmings BA, Fernandez A, Lamb NJ (1999) Vimentin dephosphorylation by protein phosphatase 2A is modulated by the targeting subunit B55. Mol Biol Cell 10:1997–2015

Virshup DM (2000) Protein phosphatase 2A: a panoply of enzymes. Curr Opin Cell Biol 12:180–185

Walker RH, Morgello S, Davidoff-Feldman B, Melnick A, Walsh MJ, Shashidharan P, Brin MF (2002) Autosomal dominant chorea-acanthocytosis with polyglutamine-containing neuronal inclusions. Neurology 58:1031–1037

Walker RH, Jankovic J, O'Hearn E, Margolis RL (2003) Phenotypic features of Huntington's disease-like 2. Mov Disord 18:1527–1530

Worth PF, Wood NW (2001) Spinocerebellar ataxia type 12 is rare in the United Kingdom. Neurology 56:419–420

Xuereb JH, MacMillan JC, Snell R, Davies P, Harper PS (1996) Neuropathological diagnosis and CAG repeat expansion in Huntington's disease. J Neurol Neurosurg Psychiatry 60:78–81

Zhao Y, Tan EK, Law HY, Yoon CS, Wong MC, Ng I (2002) Prevalence and ethnic differences of autosomal-dominant cerebellar ataxia in Singapore. Clin Genet 62:478–481

Part V
Postscript

Nucleic Acids and Molecular Biology, Vol. 19
M. Fry/K. Usdin (Eds.)
Human Nucleotide Expansion Disorders

Current Issues and Therapeutic Prospects

Karen Usdin[1] (✉) · Michael Fry[2]

[1]Gene Structure and Disease Section, Laboratory of Molecular & Cellular Biology
National Institute of Diabetes, Digestive and Kidney Diseases,
National Institutes of Health, Bethesda, MD 20892, USA
ku@helix.nih.gov

[2]Department of Biochemistry, Rappaport Faculty of Medicine Technion,
Israel Institute of Technology, Haifa, Israel
mickey@tx.technion.ac.il

1 Current Issues

1.1 Mechanisms of Nucleotide Repeat Expansion

As discussed by Sinden and Pytlos in this volume, the disease-causing repeats form a variety of unusual DNA structures, including hairpins, triplexes, and tetraplexes. Formation of such structures by the repeats has informed much of our current thinking about the expansion mechanism. Work in vitro and in bacteria, yeast and mice has implicated a variety of enzymes in this process, including those involved in DNA replication, recombination and repair. Part of the lack of clarity of the underlying expansion mechanism probably arises from the overlapping role that many of these proteins have in the maintenance of genome integrity. Contradictory evidence from different organisms also complicates the picture. An additional source of difficulty is that to date no model system fully recapitulates expansion in humans. One school of thought is that expansion is a uniquely human phenomenon; however, the recent description of an expansion disease in dogs, a form of progressive myoclonus epilepsy (Gredal et al. 2003), suggests that this is not the case. In addition, since all known folate-sensitive fragile sites in human chromosomes consist of long CGG · CCG-repeat tracts like those responsible for fragile X syndrome and FRAXE mental retardation, it has been suggested that this repeat might be a general feature of all such sites (Sutherland 2003). These sorts of sites, which are also seen in mice (Djalali et al. 1987), may thus be remnants of past repeat expansions. Their presence in mice raises the possibility that mice may in fact be suitable animals for examining expansion mechanisms. Clearly much more work is still required to resolve this question.

1.2 Consequences of Repeat Expansion

As can be gathered from chapters in this volume, research is beginning to clarify the relationship between nucleotide expansions and their consequences. Perhaps not surprisingly, these consequences depend on some combination of the properties of the repeat itself, its location in the affected gene, and the function of that gene. When the repeat is located in an open reading frame, the relationship between expansion and disease pathology is superficially quite straightforward; nucleotide expansion causes an increase in the length of a run of a particular amino acid, which in all instances to date has been glutamine, and the resulting polyglutamine (polyQ) stretch is toxic. However, the basis of the toxicity of the polyQ tract remains the subject of vigorous debate and, as discussed by Friedman et al. in this volume, mechanisms still under consideration include mitochondrial dysfunction, transcriptional dysregulation, axonal transport defects, and cytoskeletal abnormalities. The role of protein context in toxicity is becoming increasingly apparent, and in the case of SCA6 a lack of agreement exists as to how much of the pathology is due to polyQ toxicity and how much results from loss of normal protein function (Frontali, this volume).

The effects of expansion of repeats that are located outside the open reading frame seem to be remarkably varied. Some expansions are thought to constitute loss-of-function mutations, while others represent a gain of function. Loss-of-function expansions include those in which the primary promoter structure is disrupted, repeat-induced epigenetic changes lead to gene silencing, or blocks are formed to either transcription or translation. While definitive proof of the actual disease mechanism is often lacking, in many instances available evidence implicates unusual mechanisms. For example, the transcription block that has been suggested to cause Friedreich ataxia (FA or FRDA) may involve the formation of an unusual DNA structure, such as a triplex or "sticky DNA", that impedes RNA polymerase (Pandolfo, this volume). All the known gain-of-function expansions involve a novel disease paradigm: RNA-mediated pathology (Teng-umnuay and Swanson, and Tassone and Hagerman, this volume), which has just recently begun to be elucidated.

It has also been appreciated recently that expansion can have different consequences depending on the number of repeats. For example, as discussed by Tassone and Hagerman in this volume, carriers of *FMR1* alleles with 59–200 repeats are at risk for fragile X associated tremor and ataxia syndrome (FXTAS) and fragile X associated premature ovarian failure (FXPOF). However, carriers of alleles with more than 200 repeats have a completely different disorder, fragile X mental retardation syndrome (FXS). FXTAS and FXPOF are thought to represent gain-of-function disorders resulting from the effects of the expanded repeats in the transcript, while FXS results from loss-of-gene-function, specifically the absence of FMRP, the protein product of *FMR1* due to silencing of the gene. FXTAS has been suggested to be a functional laminopathy resulting from sequestering of lamin A/C (Arocena et al. 2005); however, it remains to be seen whether such an effect accounts for all FXTAS symptoms and whether the same mechanism is responsible for FXPOF. The mechanism of gene silencing in FXS is also unknown. Interestingly, the expanded *FMR1* alleles are transcribed in early embryogenesis and the r(CGG) repeats form RNA hairpins that are substrates for the ribonuclease Dicer (Handa et al. 2003). This raises the possibility (Handa et al. 2003; Jin et al. 2004) that silencing of *FMR1* may be mediated in some instances through the RNA interference pathway (Matzke and Birchler 2005).

RNA-mediated pathology is also thought to be responsible for most cases of myotonic dystrophy (DM1) (Teng-umnuay and Swanson, this volume). However, in congenital DM1, a very severe form of DM1 seen only when the repeat tract is extremely large, another novel disease mechanism may contribute to disease symptoms. In most cases of DM1, the repeats are heterochromatinized owing to the antisense effect of a bidirectional promoter of the downstream *SIX5* gene. Heterochromatin spreading is blocked by a CTCF-dependent insulator element. In congenital DM1, CTCF binding is lost, resulting in heterochromatin spreading into adjacent genes (Cho et al. 2005). The heterochromatization may affect the expression of adjacent genes, thereby accounting for some of the unusual features of this form of DM1. The cases of FXS, FXTAS, and FXPOF and of the different types of DM1 raise the possibility that other repeat expansions may also have more than one pathological mechanism.

While the broad outlines of many nucleotide repeat disorders are beginning to emerge, many of them are still bereft of molecular details. Large gaps in our knowledge are most apparent for particular diseases. The relative rarity of spinocerebellar ataxias (SCA) types 8 and 10, and the limited access to affected tissues has slowed progress toward elucidation of the disease mechanism (Dick et al., and Lin and Ashizawa, this volume). In the case of Huntington disease-like 2 (HDL2) and SCA12, both of which involve expansions of the repeat CAG · CTG, whether the repeats are translated is not even known (Margolis et al., this volume). While it is possible that the repeat falls into an open reading frame in a subset of alternative transcripts, it is unlikely that a polyQ tract is responsible for either disorder. It remains possible in the case of HDL2 that a polyalanine- or a polyleucine-containing protein is involved. Such amino acid tracts are toxic in cell models and a polyalanine tract has been shown to be responsible for oculopharyngeal muscular dystrophy (Brais et al. 1998). RNA toxicity also remains a viable explanation for both disorders, although in the case of SCA12 it would probably have to involve CAG-RNA rather than CUG-RNA. If this turns out to be the case, it would represent the first disease resulting from CAG-RNA-mediated pathology. However, the muscleblind proteins thought to be involved in the pathology responsible for DM1 and DM2 bind to RNA containing CHHG and CHG repeats, where H is A, U, or C (Kino et al. 2004). Thus, it is possible that the net effect of the SCA12 repeat is very similar to that of DM1 and DM2.

A number of diseases whose genetic bases are not yet known show anticipation, one of the hallmarks of the repeat expansion disorders. It may thus be that other diseases will be added to this group as their genetic basis becomes known. Available evidence suggests that some of these diseases may result from protein toxicity that is not associated with polyQ or RNA-mediated pathology that involves repeats other than r(CGG) or r(CUG).

2 Therapeutic Prospects

Despite the large gaps in our understanding of the molecular details of the nucleotide repeat disorders, what has been learned thus far is already paving the way for rational approaches to treating these diseases. Some of these approaches are aimed at correcting the underlying genetic defect and might be applicable for a wide range of repeat expansion disorders, whilst others target the downstream consequences of expansion and may sometimes be disease-specific.

2.1 Gene Correction/Replacement

Gene replacement strategies may be suitable for diseases resulting from loss-of-function mutations. The introduction of a normal frataxin gene using adeno-associated virus and lentivirus vectors had some positive effects in cells from individuals with FRDA (Fleming et al. 2005). However, many technical problems would need to be overcome before this approach could be effectively applied in patients. In disorders involving gain-of-function mutations, strategies that eliminate the affected protein/messenger RNA are potentially useful. Successful antisense knockdown of mutant huntingtin has been achieved in cultured cells (Hasholt et al. 2003), and an RNA interference strategy has been shown to slow disease progression in a mouse model of Huntington disease (HD) (Rodriguez-Lebron et al. 2005). However, for such therapy to be effective, expression of the normal allele would have to be maintained if the consequences of complete loss-of-gene expression are severe. Otherwise, this approach would be limited to those disorders where haploinsufficiency could be tolerated.

Cell transplantation offers an alternate therapeutic avenue for diseases involving neurodegeneration. Indeed, some success has been achieved in grafting of fetal striatal tissue in mice and humans with HD (reviewed in Dunnett and Rosser 2004).

Certain small molecules may allow gene correction. For example, cytosine arabinoside, ethidium bromide, 5-azacytidine, and aspirin facilitate contraction of repeat tracts in dividing DM1 cells (Gomes-Pereira and Monckton 2004). Ribozymes have also been used to remove expanded nucleotide repeats in DM cells (Phylactou 2004). However, it remains to be seen how these findings can be translated into actual therapies. First, problems with drug delivery and toxicity must be overcome. Second, in the case of disorders involving postmitotic cells like neurons the feasibility of this approach needs to be demonstrated in nondividing cells.

2.2 Targeting the Downstream Consequences of Expansion

2.2.1 Epigenetic Modification

Partial reversal of *FMR1* gene silencing in fragile X syndrome cells has been achieved using the DNA demethylating agent 5-azadeoxycytidine (Chiurazzi et al. 1998). Furthermore, a synergistic effect was obtained when this compound was used together with a histone deacetylase inhibitor (Chiurazzi et al. 1999). Histone deacetylase inhibitors have also been shown to reduce motor impairment in a mouse model of HD presumably because they reduce the histone hypoacetylation seen in the polyQ disorders (Hockly et al. 2003). Similar strategies might be useful in treating other disorders such as congenital DM where the most severe symptoms may also arise from aberrant DNA methylation and heterochromatinization (Cho et al. 2005). However, once again a number of obstacles remain in the way of translating such approaches into effective therapies. First, these drugs are relatively nonspecific and 5-azadeoxycytidine, in particular, is highly toxic. Second, DNA demethylation since it is thought to require DNA replication in postmitotic cells like neurons provides an additional technical challenge. In addition, in the case of fragile X syndrome increased *FMR1* transcription may predispose patients to the ovarian and cerebellar dysfunction that is thought to be due to toxicity of RNA with long CGG-repeat tracks. Furthermore, to be effective, treatments would have to overcome the block to translation posed by large numbers of repeats in the transcripts. The development of less toxic DNA methyltransferase inhibitors that do not require DNA replication for their activity as well as compounds that improve FMR1 translation is required.

2.2.2 Countering Protein Misfolding

Evidence suggests that both the polyQ disorders and FXTAS may result from some sort of altered protein folding. For example, overexpression of heat shock protein 70, a chaperone protein, ameliorated symptoms of neurodegeneration in a fly model of FXTAS (Jin et al. 2003), and in a mouse model of the polyQ disorder spinal and bulbar muscular atrophy (SBMA) (Adachi et al. 2003). Trehalose (Tanaka et al. 2004) and Congo red (Sanchez et al. 2003), which both reduce protein misfolding, decrease disease severity in mouse models of polyQ disorders. Thus, compounds that lower protein misfolding may also be useful in treating FXTAS.

2.2.3 Stimulating Neurogenesis

Paroxetine (Paxil) (Duan et al. 2004), fluoxetine (Prozac) (Grote et al. 2005), and FGF-2 (Jin et al. 2005) have shown some clinical promise in HD patients, perhaps because of their ability to compensate for the impaired neurogenesis and increased neuronal death in this disorder. A similar effect may be beneficial in countering the neuronal loss in FXTAS, while at the same time providing some relief in the case of Paxil and Prozac from the symptoms of depression commonly seen in affected individuals.

2.2.4 Reducing Oxidative Stress

Increased oxidative stress in response to expansion was reported in a variety of disorders, including FRDA and the polyQ disorders. Clinical trials with antioxidants such as idebenone, an analog of coenzyme Q (ubiquinone) have led to decreased cardiac hypertrophy in a mouse model of FRDA (Seznec et al. 2004) and humans (Rustin 2003). However, additional double-blind placebo controlled studies are needed to assess neurological outcomes. Antioxidants also attenuate the HD phenotype in mouse models of this disorder (Beal 2002). Since in both disorders mitochondrial dysregulation is observed, the development of agents targeted to the mitochondria might be desirable. One such compound, mitoquinone, a ubiquinone derivative, has been shown to be more effective in protecting cells from FRDA patients from oxidative stress than idebenone (Jauslin et al. 2003), and is currently being tested in phase I clinical trials.

2.2.5 Modulating Specific Downstream Targets

In some disorders very specific changes in gene expression have been noticed that may be relevant for disease symptoms. For example, altered group 1 metabotropic glutamate (mGluR) receptor signaling is seen in both *Fmr1* knockout mice (Huber et al. 2002) and in the polyQ disorders (Anborgh et al. 2005). Accordingly, drugs that target these receptors may have promise in alleviating disease symptoms. An mGluR5 antagonist, 2-methyl-6-(phenylethynyl)pyridine (MPEP), decreased audiogenic seizures and macroorchidism seen in *Fmr1* knockout mice (Yan et al. 2005) and corrected the neuroanatomical and behavioral defects resulting from an FMRP deficit in the fly (McBride et al. 2005). MPEP also increased survival of HD mice (Schiefer et al. 2004). However, while MPEP has high potency, it is relatively nonselective and is not very soluble in cerebrospinal fluid. Better outcomes may result from the use of more selective mGluR5 receptor antagonists such as 3-[(2-methyl-1,3-thiazol-4-yl)ethynyl]pyridine (MTEP) (Busse et al. 2004).

In conclusion, while it is clear that much more remains to be understood, enough has been learned in the relatively short time since the nucleotide repeat expansion disorders were first described to take the first steps toward developing mechanism-based approaches to their treatment. Indeed, in some instances, a very real possibility exists for the development of effective treatments in the not too distant future.

References

Adachi H, Katsuno M, Minamiyama M, Sang C, Pagoulatos G, Angelidis C, Kusakabe M, Yoshiki A, Kobayashi Y, Doyu M, Sobue G (2003) Heat shock protein 70 chaperone overexpression ameliorates phenotypes of the spinal and bulbar muscular atrophy transgenic mouse model by reducing nuclear-localized mutant androgen receptor protein. J Neurosci 23:2203–2211

Anborgh PH, Godin C, Pampillo M, Dhami GK, Dale LB, Cregan SP, Truant R, Ferguson SS (2005) Inhibition of metabotropic glutamate receptor signaling by the huntingtin-binding protein optineurin. J Biol Chem 280:34840–34848

Arocena DG, Iwahashi CK, Won N, Beilina A, Ludwig AL, Tassone F, Schwartz PH, Hagerman PJ (2005) Induction of inclusion formation and disruption of lamin A/C structure by premutation CGG-repeat RNA in human cultured neural cells. Hum Mol Genet 14:3661–3671

Beal MF (2002) Coenzyme Q10 as a possible treatment for neurodegenerative diseases. Free Radic Res 36:455–460

Brais B, Bouchard JP, Xie YG, Rochefort DL, Chretien N, Tome FM, Lafreniere RG, Rommens JM, Uyama E, Nohira O, Blumen S, Korczyn AD, Heutink P, Mathieu J, Duranceau A, Codere F, Fardeau M, Rouleau GA (1998) Short GCG expansions in the PABP2 gene cause oculopharyngeal muscular dystrophy. Nat Genet 18:164–167

Busse CS, Brodkin J, Tattersall D, Anderson JJ, Warren N, Tehrani L, Bristow LJ, Varney MA, Cosford ND (2004) The behavioral profile of the potent and selective mGlu5 receptor antagonist 3-[(2-methyl-1,3-thiazol-4-yl)ethynyl]pyridine (MTEP) in rodent models of anxiety. Neuropsychopharmacology 29:1971–1979

Chiurazzi P, Pomponi MG, Willemsen R, Oostra BA, Neri G (1998) In vitro reactivation of the FMR1 gene involved in fragile X syndrome. Hum Mol Genet 7:109–113

Chiurazzi P, Pomponi MG, Pietrobono R, Bakker CE, Neri G, Oostra BA (1999) Synergistic effect of histone hyperacetylation and DNA demethylation in the reactivation of the FMR1 gene. Hum Mol Genet 8:2317–2323

Cho DH, Thienes CP, Mahoney SE, Analau E, Filippova GN, Tapscott SJ (2005) Antisense transcription and heterochromatin at the DM1 CTG repeats are constrained by CTCF. Mol Cell 20:483–489

Djalali M, Adolph S, Steinbach P, Winking H, Hameister H (1987) A comparative mapping study of fragile sites in the human and murine genomes. Hum Genet 77:157–162

Duan W, Guo Z, Jiang H, Ladenheim B, Xu X, Cadet JL, Mattson MP (2004) Paroxetine retards disease onset and progression in huntingtin mutant mice. Ann Neurol 55:590–594

Dunnett SB, Rosser AE (2004) Cell therapy in Huntington's disease. NeuroRx 1:394–405

Fleming J, Spinoulas A, Zheng M, Cunningham SC, Ginn SL, McQuilty RC, Rowe PB, Alexander IE (2005) Partial correction of sensitivity to oxidant stress in Friedreich ataxia patient fibroblasts by frataxin-encoding adeno-associated virus and lentivirus vectors. Hum Gene Ther 16:947–956

Gomes-Pereira M, Monckton DG (2004) Chemically induced increases and decreases in the rate of expansion of a CAG*CTG triplet repeat. Nucleic Acids Res 32:2865–2872

Gredal H, Berendt M, Leifsson PS (2003) Progressive myoclonus epilepsy in a beagle. J Small Anim Pract 44:511–514

Grote HE, Bull ND, Howard ML, van Dellen A, Blakemore C, Bartlett PF, Hannan AJ (2005) Cognitive disorders and neurogenesis deficits in Huntington's disease mice are rescued by fluoxetine. Eur J Neurosci 22:2081–2088

Handa V, Saha T, Usdin K (2003) The fragile X syndrome repeats form RNA hairpins that do not activate the interferon-inducible protein kinase, PKR, but are cut by Dicer. Nucleic Acids Res 31:6243–6248

Hasholt L, Abell K, Norremolle A, Nellemann C, Fenger K, Sorensen SA (2003) Antisense downregulation of mutant huntingtin in a cell model. J Gene Med 5:528–538

Hockly E, Richon VM, Woodman B, Smith DL, Zhou X, Rosa E, Sathasivam K, Ghazi-Noori S, Mahal A, Lowden PA, Steffan JS, Marsh JL, Thompson LM, Lewis CM, Marks PA, Bates GP (2003) Suberoylanilide hydroxamic acid, a histone deacetylase inhibitor, ameliorates motor deficits in a mouse model of Huntington's disease. Proc Natl Acad Sci USA 100:2041–2046

Huber KM, Gallagher SM, Warren ST, Bear MF (2002) Altered synaptic plasticity in a mouse model of fragile X mental retardation. Proc Natl Acad Sci USA 99:7746–7750

Jauslin ML, Meier T, Smith RA, Murphy MP (2003) Mitochondria-targeted antioxidants protect Friedreich Ataxia fibroblasts from endogenous oxidative stress more effectively than untargeted antioxidants. FASEB J 17:1972–1974

Jin K, Lafevre-Bernt M, Sun Y, Chen S, Gafni J, Crippen D, Logvinova A, Ross CA, Greenberg DA, Ellerby LM (2005) FGF-2 promotes neurogenesis and neuroprotection and prolongs survival in a transgenic mouse model of Huntington's disease. Proc Natl Acad Sci USA 102:18189–18194

Jin P, Zarnescu DC, Zhang F, Pearson CE, Lucchesi JC, Moses K, Warren ST (2003) RNA-mediated neurodegeneration caused by the fragile X premutation rCGG repeats in Drosophila. Neuron 39:739–747

Jin P, Alisch RS, Warren ST (2004) RNA and microRNAs in fragile X mental retardation. Nat Cell Biol 6:1048–1053

Kino Y, Mori D, Oma Y, Takeshita Y, Sasagawa N, Ishiura S (2004) Muscleblind protein, MBNL1/EXP, binds specifically to CHHG repeats. Hum Mol Genet 13:495–507

Matzke MA, Birchler JA (2005) RNAi-mediated pathways in the nucleus. Nat Rev Genet 6:24–35

McBride SM, Choi CH, Wang Y, Liebelt D, Braunstein E, Ferreiro D, Sehgal A, Siwicki KK, Dockendorff TC, Nguyen HT, McDonald TV, Jongens TA (2005) Pharmacological rescue of synaptic plasticity, courtship behavior, and mushroom body defects in a Drosophila model of fragile X syndrome. Neuron 45:753–764

Phylactou LA (2004) Repair of myotonic dystrophy protein kinase (DMPK) transcripts by trans-splicing ribozymes. Methods Mol Biol 252:373–383

Rodriguez-Lebron E, Denovan-Wright EM, Nash K, Lewin AS, Mandel RJ (2005) Intrastriatal rAAV-mediated delivery of anti-huntingtin shRNAs induces partial reversal of disease progression in R6/1 Huntington's disease transgenic mice. Mol Ther 12:618–633

Rustin P (2003) The use of antioxidants in Friedreich's ataxia treatment. Expert Opin Investig Drugs 12:569–575

Sanchez I, Mahlke C, Yuan J (2003) Pivotal role of oligomerization in expanded polyglutamine neurodegenerative disorders. Nature 421:373–379

Schiefer J, Sprunken A, Puls C, Luesse HG, Milkereit A, Milkereit E, Johann V, Kosinski CM (2004) The metabotropic glutamate receptor 5 antagonist MPEP and the mGluR2 agonist LY379268 modify disease progression in a transgenic mouse model of Huntington's disease. Brain Res 1019:246–254

Seznec H, Simon D, Monassier L, Criqui-Filipe P, Gansmuller A, Rustin P, Koenig M, Puccio H (2004) Idebenone delays the onset of cardiac functional alteration without correction of Fe-S enzymes deficit in a mouse model for Friedreich ataxia. Hum Mol Genet 13:1017–1024

Sutherland GR (2003) Rare fragile sites. Cytogenet Genome Res 100:77–84

Tanaka M, Machida Y, Niu S, Ikeda T, Jana NR, Doi H, Kurosawa M, Nekooki M, Nukina N (2004) Trehalose alleviates polyglutamine-mediated pathology in a mouse model of Huntington disease. Nat Med 10:148–154

Yan QJ, Rammal M, Tranfaglia M, Bauchwitz RP (2005) Suppression of two major Fragile X Syndrome mouse model phenotypes by the mGluR5 antagonist MPEP. Neuropharmacology 49:1053–1066

Subject Index

Aberrant cell cycle entry, 191
Aberrant replication initiation, 190
AF4, 91
AF5q31, 91
Alternative DNA conformations, see Unusual DNA structures
Alternative RNA splicing, 152
Alu element, 106
Anticipation, 189, 261, 282
Armadillo repeat, 191
Astrocytosis, 256
Ca^{2+}-ATPase (SERCA), 153
Axonal transport, 214, 280
- fast (FAT), 215
- inhibition, 215

β-catenin, 191

CACNA1A Gene, 234
Cardiomyopathy, 152
Cataracts, 144, 145
Cathepsins, 134
CELF protein, 154
CEPH families, 126
Cerebellar ataxia, 121, 235, 236
Checkpoint control, 27, 30, 57
CLCN1, 153
Congenital Myotonic Dystrophy (CDM), 144, 157
CTCF, 281
Cystatin B (CSTB), 122
- knock out mouse, 134
- mutations, 123, 126
- polymorphisms, 126
- stefin B, 122
- promoter, 133
- reporter gene assays, 133
Cysteine proteinase inhibitor, 123

d(CAG) repeat
- expansion, 200
- length heterogeneity, 200
d(CCCCGCCCCGCG) repeat
- hypomethylation, 134
- linkage, 122
- meiotic instability, 129
- mitotic instability, 129
d(CGG) repeat
- animal models, 61
- *FMR1* gene, 57
- *FMR2* gene, 88
- *FAM11A* gene, 88
- folate-sensitive fragile sites, 21, 88, 279
- fragile X syndrome (FXS), 57
- FRAXE disease, 105
- FRAXF, 105
- fragile X associated tremor and ataxia syndrome (FXTAS), 57, 59, 72, 73
- full mutation, 57
- gray zone, 57
- normal length, 57, 64
- premature ovarian failure (POF), 57, 59
- premutation, 57, 65, 72
- r(CGG) mRNA, 57
d(CTG) repeat
- conservation, 169
- d(CTA) tract, 168, 175, 176
- interruptions, 175, 176, 180
- instability, 167, 175, 176, 178
- intergenerational changes, 176
- maternal bias, 176, 177, 179
- maternal transmission, 176, 177
- paternal transmission, 176, 177
- sperm, 178, 179
d(GAA) repeat, 103–114
d(CTTCT) repeat, 185
DM, see Myotonic Dystrophy

DMPK, 143, 147
DMWD, 147
DNA
- amplification, 31
- damage checkpoint, 30
- directed mutations, 13
- double-strand break repair, 27
- excision repair, 16
- gap repair, 25
- hairpins, 5, 277
- helicase, 14
- mismatch repair, 16
- nuclear matrix attachment, 190
- onion-skin amplification, 32
- parallel-stranded, 11
- primer-template misalignment, 13, 18
- quadruplex, 10
- recombination, 15, 26
- slipped intermediate, 25
- slipped-stranded, 8
- strand switching, 14, 20
- torsional stress, 190
- triplex, 9, 108–110, 277
- unwound, 10
DNA methyltransferase inhibitors, 283
DNA repair
- excision, 16, 25
- gap, 25
- mismatch, 16, 23
DNA replication, 12, 13, 18, 30, 31
- blockage, 12, 13
- slippage, 13
- restart, 17, 27, 28
DNA Unwinding Elements (DUE), 190
Dentatorubral-pallidoluysian Atrophy (DRPLA), 202–204, 212, 213, 217

EA2, see Episodic Ataxia type 2
E46L, 191
Episodic Ataxia type 2 (EA2), 233
- age at onset, 240
- clinical features, 239
EPM1, see Progressive Myoclonus Epilepsy type 1
Excision repair, 16, 25

FAM11A, 88
Flap endonuclease, 17, 24
Fluoxetine (Prozac), 284
FMR1 gene, 57, 63, 87
- acetylation, 66
- binding, 64, 66, 67, 69
- chromatin, 64, 66, 67
- exons, 63
- histones, 66, 67
- instability, 57, 61, 62
- intron, 63
- isoforms, 63
- methylation, 58, 62, 64, 66, 67
- mRNA, 65, 68, 72
- premutation, 68
- protein product (FMRP), 58, 59, 66, 68, 69, 71, 72
- regulation, 64, 66, 68
- silencing, 57, 58, 66, 67, 69
- transcription, 63–68
- translation, 68–72
FMR2 gene
- characteristics, 91
- mouse model, 93
FMR3 gene, 95
Folate-sensitive fragile sites, 279
- *FRAXA*, 87
- *FRAXE*, 87
- *FRAXF*, 87
Fragile X syndrome (FXS), 58
- autism, 57–59
- clinical involvement, 58
- FMRP, 58, 59, 65, 69, 72, 73
- mental retardation, 58, 59
- prevalence, 58, 59
- primary protein (FMRP) deficiency
- premutation, 59, 68
- premature ovarian failure (POF), 57, 59, 65
Frataxin, 104, 108–114
FRAXE disease
- learning deficit, 87
- mental retardation, 87
- repeat dynamics, 90
Friedreich ataxia (FA or FRDA)
- altered iron metabolism, 113
- animal models, 112
- clinical variability, 103, 104
- chromosomal location of affected gene, 104
- Coenzyme Q, 114
- d(GAA) repeat
  - instability, 105, 108
  - origin, 106

- diagnostic criteria, 103
- DNA triplexes, 108–110
- free radical damage, 111, 113
- frataxin
  - expression, 104, 111
  - iron chaperone, 112
  - structure and function, 111, 112
- idebenone, 114
- inheritance, 105
- iron accumulation, 111, 113
- iron chaperone, 112
- iron-sulfur clusters (ISCs), 112, 113
- mitochondrial dysfunction, 111, 113
- oxidative stress, 113
- pathogenic mechanisms, 108–110, 113
- point mutations, 110, 111
- "Sticky DNA", 108–110
- symptoms, 103

FXTAS, 57, 59, 60, 65, 72, 73
- animal models, 72
- inclusions, 60, 72, 73
- laminA/C, 73
- comparisons with myotonic dystrophy, 60
- premutation, 60
- RNA toxicity, 60, 73

FXS, see Fragile X syndrome

Glutamate receptor, 153
Group 1 metabotropic Glutamate (mGluR) receptor signaling, 284

Huntington Disease (HD), 202–205, 207–209, 212–218
- mitochondrial defects, 216
- neuropil aggregates, 206, 214
- synaptic function, 214, 215
- Westphal variant, 254

Huntington Disease-Like 1 (HDL1) and 2 (HDL2)
- aggregates, 256
- allele length, 259
- chorea, 255
- dementia, 255
- epidemiology, 258
- ethnicity, 259
- gain-of-function, 265
- haplotype, 259
- juvenile onset, 254
- mice, 268
- neuroimaging, 256
- pathology, 257
- penetrant, 259
- repeat instability, 260
- RNA, 265
- signs and symptoms, 255

Friedreich Ataxia (FA or FRDA), 103–114
Heterochromatin spreading, 233, 281
Heterotrimeric GTP-binding protein, 192
Histone modification
- acetylation, 66
- deacetylation, 68
- HDAC inhibitors, 218
- histone acetyltransferases (HATs), 213
- hypoacetylation, 213
- methylation, 67, 68

Histone deacetylase inhibitors, 218, 283

Insulin Receptor (IR), 153

Junctophilin-3 (*JPH3*), 262, 266

LAF4, 91
Lamin A/C, 73
Laminopathy, 281

Marinesco bodies, 258
Matrix attachment, 190
2-methyl-6-(phenylethynyl)pyridine (MPEP), 285
Mismatch repair, 16, 23
Molecular chaperones, 207, 217
- Hsp40, 207, 217, 218
- Hsp70, 207, 217, 218
- overexpression, 218

Muscleblind (MBNL), 150
MyoD, 158
Myotonia, 144
Myotonic Dystrophy (DM)
- arrhythmia, 145
- cardiomyopathy, 152
- cataracts, 144, 145
- congenital myotonic dystrophy (CDM), 144, 157
- DM1, 143
- DM2, 143
- gain-of-function model, 146
- pathogenesis, 144
- hypersomnia, 145
- hypogonadism, 145
- hypotonia, 144, 157

- pathogenesis, 144
- ribonuclear foci, 147

NMDA R1, 153
Non-B-DNA structures, see Unusual DNA structures
Nuclear Foci, 147, 193

Oculopharyngeal muscular dystrophy, 265
Oxidative stress, 284

Paroxetine (Paxil), 284
Polyalanine, 281
Polyglutamine (PolyQ)
- aggregates, 204, 206
  - colocalization with ubiquitin, 206, 207
  - neuronal distribution, 204, 205
  - neuronal intranuclear inclusion (NII), 204
  - nucleation-dependent polymerization, 205
  - polar zipper, 205
  - protein context, 205
  - protein misfolding, 205, 207
  - subcellular localization, 204, 206, 207
  - ubiquination, 206
  - toxicity, 207, 280
- diseases
  - age of onset, 202, 203
  - autosomal dominance, 202
  - clinical heterogeneity, 202
  - disease severity, 202, 203
  - excitotoxicity, 217
  - genetic modifiers, 203
  - glial dysfunction, 216, 217
  - haplotype analysis, 189
  - incomplete dominance, 202
  - meiotic drive, 203
  - meiotic instability, 203
  - neurotic degeneration, 215
  - paternal transmission, 203
  - reduced penetrance, 202
  - selective neurodegeneration, 204, 214
Polyleucine, 281
Progressive Myoclonus Epilepsy type 1, (EPM1)
- age of onset, 131
- cerebellar ataxia, 121
- chromosomal locus, 122
- contractions, 129
- dementia, 121
- founder effect, 122
- linkage, 122
- knockout mouse model, 134
- mental deterioration, 121
- nucleotide expansion, 128
- seizures, 134
- stimulus-sensitive myoclonus, 121
- tonic–clonic seizures, 121
- Unverricht–Lundborg type, 121
Protein misfolding, 284

Quasi-palindrome correction mutation, 23

Reiterative DNA synthesis, 31
Repeat expansion detection (RED), 253, 261
Repeat-induced epigenetic changes, 280
Ribozymes, 283
RNA, 69, 72, 146
- binding proteins, 148, 154, 192
- elevation, 72
- gain-of-function, 192
- hairpins, 281
- interference, 281
- mediated pathology, 280, 281
- splicing, 151, 265
- toxic gain-of-function, 57
- toxicity, 57, 60, 72, 73, 281
RNA-mediated disease, see RNA toxicity
RNA-mediated pathology, see RNA toxicity
RNA toxicity, 72, 73, 143, 266, 278–281
Ryanodine receptor RyR1, 153

Sarcoplasmic/Endoplasmic Reticulum Ca^{2+}-ATPase (SERCA), 153
Spinal and Bulbar Muscular Atrophy (SBMA), 203, 204, 218
- X-linked recessive, 202
SCA1, 206, 218
SCA2, 204, 206, 214
SCA3, 202, 203, 213
SCA6, 202, 204, 206, 214
- age at onset, 238, 240
- aggregate formation, 240, 241
- allele size, 233, 241
- clinical features, 236, 237, 239
- epidemiology, 235
- founder effect, 236
- gain-of-function vs abnormal channel activity, 241

- genetics, 233
- homozygous patients, 234
- neuroimaging, 239
- neuropathology, 240
- nuclear localization, 241
- repeat instability, 234
- treatment, 238
SCA7, 204, 218
SCA8
- asymptomatic, 171–173, 177
- age, 171, 172
- clinical features, 169, 170, 179
- ataxia, 167, 170
- dysarthria, 170
- incoordination, 170
- sensory loss, 170
- control chromosomes, 173
- d(CTA) tract, 175, 176, 180
- d(CGT) tract, 171, 172, 175
- disease penetrance, 167, 168, 171–174, 175–177, 180
- environmental modifiers, 173
- expansion size, 172, 173
- genetic mechanism, 167
- RNA gain-of-function, 167, 170, 179
- genetic modifiers, 173, 179
- imaging, 173
- inheritance pattern
- dominant, 167, 172
- modifiers, 173, 175, 180
- recessive, 172
- sporadic, 172
SCA 10
- allele length, 260
- brain, 258
- d(CTTCT) repeats, 185
- epidemiology, 258
- germ line mosaicism, 189
- inter-tissue variability, 190
- linkage disequilibrium, 189
- neuron-specific instability, 189
- progressive cerebellar dysfunctions, 186
- nuclear foci, 193
- r(CCUG) RNA repeats, 193
- single nucleotide polymorphisms, 189
SCA 12
- allele length, 260
- brain, 258
- epidemiology, 258
- essential tremor, 254
- haplotype, 259
- inclusions, 258
- India, 259
- neuroimages, 258
- neuropathy, 256
- parietal lobe, 258
- parkinsonism, 256
- PP2A, 268
- *PPP2R2B*, 263, 266
- Purkinje cells, 258
- tremor, 256
SCA 17, 202, 204, 205, 213
- transgenic mice, 202
SCN5A, 152
SIX5, 147
Small interfering RNAs (siRNAs), 192
Spinocerebellar ataxia 1–17, see SCA 1–17

TNNT2, 152
TorsinA, 256
Transcription, 63
- elevated mRNA, 65
- start site, 64
Transcription factors, 63, 64, 67, 208
- CBP, 209
- CREB, 209, 214
- Sp1, 209
- TAF4, 209
- TBP, 213

Ubiquitin–Proteasome System (UPS), 206, 207
Unusual DNA structures
- hairpins, 5, 279
- parallel-stranded DNA, 11
- tetraplexes (quadruplexes), 10, 279
- triplexes, 9, 108–110, 279
- unpaired DNA, 190
- unwound DNA, 10

Voltage gated Ca^{2+} channel, 233–235, 241
- α1A subunit, 233–235, 241
- type P/Q, 234
- isoforms, 235

ZNF9, 143

Printing: Krips bv, Meppel
Binding: Stürtz, Würzburg